C0-BQF-340

Lecture Notes in Physics

The Lecture Notes in Physics

The series Lecture Notes in Physics (LNP), founded in 1969, reports new developments in physics research and teaching – quickly and informally, but with a high quality and the explicit aim to summarize and communicate current knowledge in an accessible way. Books published in this series are conceived as bridging material between advanced graduate textbooks and the forefront of research to serve the following purposes:

• to be a compact and modern up-to-date source of reference on a well-defined topic;

• to serve as an accessible introduction to the field to postgraduate students and nonspecialist researchers from related areas;

• to be a source of advanced teaching material for specialized seminars, courses and schools.

Both monographs and multi-author volumes will be considered for publication. Edited volumes should, however, consist of a very limited number of contributions only. Proceedings will not be considered for LNP.

Volumes published in LNP are disseminated both in print and in electronic formats, the electronic archive is available at springerlink.com. The series content is indexed, abstracted and referenced by many abstracting and information services, bibliographic networks, subscription agencies, library networks, and consortia.

Proposals should be sent to a member of the Editorial Board, or directly to the managing editor at Springer:

Dr. Christian Caron
Springer Heidelberg
Physics Editorial Department I
Tiergartenstrasse 17
69121 Heidelberg/Germany
christian.caron@springer-sbm.com

I. Galanakis P.H. Dederichs (Eds.)

Half-Metallic Alloys

Fundamentals and Applications

Editors

Iosif Galanakis
Peter H. Dederichs

Forschungszentrum Jülich GmbH
Institute of Solid State Research
Leo-Brandt-Straße, 52428 Jülich
Germany
E-mail: i.galanakis@fz-juelich.de
p.h.dederichs@fz-juelich.de

I. Galanakis and P.H. Dederichs, *Half-Metallic Alloys*,
Lect. Notes Phys. 676 (Springer, Berlin Heidelberg 2005), DOI 10.1007/b137760

Library of Congress Control Number: 2005930448

ISSN 0075-8450
ISBN-10 3-540-27719-6 Springer Berlin Heidelberg New York
ISBN-13 978-3-540-27719-4 Springer Berlin Heidelberg New York

Springer is a part of Springer Science+Business Media
springeronline.com

Printed in The Netherlands

Typesetting: by the author using a Springer LaTeX macro package

Printed on acid-free paper SPIN: 11506256 54/Techbooks 5 4 3 2 1 0

Preface

In 1903 Heusler, in his attempt to synthesize new materials for the steel industry, discovered Cu_2MnAl. Thereafter, a whole class of new materials with the same lattice have been successfully grown and they were named after Heusler himself. They attracted considerable attention mainly due to the diversity of the magnetic phenomena which they presented like itinerant and localized magnetism, antiferromagnetism, helimagnetism, Pauli paramagnetism or heavy-fermionic behavior. Recently, the rapid development of the field of magnetoelectronics revived the interest on such materials since some of them where found to possess a newly discovered property, the so-called half-metallicity.

Half-metals are particular ferromagnetic materials, which can be considered as hybrids between metals and semiconductors. The existence of these half-metallic materials was predicted using abinitio calculations by de Groot et al. in 1983. They found that in the Heusler alloy NiMnSb the majority spin band structure shows a metallic behavior, while the minority band structure exhibits a band gap at the Fermi level like in a semiconductor. Due to the gap for one spin direction, electrons at the Fermi level show a 100% spinpolarization.

For a long time such half-metals were considered as exotic materials, interesting, but not important. This completely changed due to the appearance of the new field of magneto or spin electronics. While in conventional electronics the charge of the electrons plays the essential role and the spin is only used for information storage, in the future spin electronics both the charge and the spin will play an equally important role. The development of this field was initiated by the discovery of the giant magnetoresistance effect in 1988 by the groups of Fert in Paris and Grünberg at Jülich. It soon became clear that in such a spin-dependent electronics half-metals can play an important role, since the efficiency of any spin-dependent device will be largest, if the current will be 100% spin polarized. Therefore, half-metals can be considered as the ideal materials for spin electronics.

In the last 10–15 years many materials have been found to be half-metallic. Besides the so-called half-Heusler alloys like NiMnSb a large number of full-Heusler alloys like Co_2MnGe are predicted to be half-metals. Ab-initio calculations also yield half-metallicity for some oxides like CrO_2 and Fe_3O_4

(magnetite), for manganites (e.g. $La_{0.7}Sr_{0.3}MnO_3$), double perovskites (e.g. Sr_2FeReO_6), pyrites (e.g. CoS_2), transition metal chalcogenides (e.g. CrSe) and pnictides (e.g. CrAs) in the zinc-blende or wurtzite structures. Also the much debated diluted magnetic semiconductors belong to this class of materials.

This book is fully devoted to the half-metallic Heusler alloys and the structurally very similar transition metal chalcogenides and pnictides. Heusler alloys are particularly interesting for realistic applications due to their very high Curie temperatures and the similarity between their crystal structure and the zinc-blende structure adopted by the III–V and II–VI compound semiconductors widely used in applications. However, these alloys also offer many problems and challenges for research and applications. Typically, the alloys have a lot of point defects and substitutional disorder. These lead to impurity states in the gap which, due to the high defect concentration, strongly affect the half-metallic properties. Another problem occurs at the interfaces to the semiconductors. Here dangling bond states, i.e. interface states in the minority gap, occur which strongly reduce the spin polarization. Basically, due to these problems it is at present not clear if they can be overcome and if Heusler alloys will be useful materials for spin electronics and also whether they will find applications, e.g. in spin injection, in magnetic tunnel junctions or in GMR sensors.

For these reasons this book summarizes the latest advancements in the understanding and applications of Heusler alloys and related compounds. In Chap. 1 a theoretical study of the electronic and magnetic properties of the Heusler alloys is presented, and in Chap. 2 the effect of defects and interfaces on the properties of these compounds is investigated. Chapter 3 discusses the properties of the spinvalves and Chap. 4 analyzes the magnetic properties of films. Chapters 5 and 6 are devoted on the growth mechanisms of half-Heusler alloys (NiMnSb) and full-Heusler alloys Co_2MnGe. Chapter 7 investigates the phenomena occurring during the creation of surfaces and Chap. 8 the properties of the tunnel junctions with a Heusler alloy as the magnetic electrode. Finally, Chaps. 9 and 10 present a theoretical and experimental investigation of the transition-metal chalcogenides and pnictides which crystallize in a similar structure with the Heusler alloys. We should also note that each chapter is self-contained and can be read independently. The reader is referred to introductions in Chaps. 1 and 3 for a survey of the theoretical and experimental studies on Heusler compounds, while Chap. 9 includes a historical survey of the research on the transition-metal pnictides.

We hope that the book encourages further research on these complex materials leading to successful applications.

Jülich,
May 2005

Iosif Galanakis
Peter H. Dederichs

List of Contributors

Iosif Galanakis and Peter H. Dederichs
Institut für Festkörperforschung,
Forschungszentrum Jülich, D-52425
Jülich, Germany
I.Galanakis@fz-juelich.de and
P.H.Dederichs@fz-juelich.de

Silvia Picozzi
CASTI-INFM Regional Lab. and
Dip. Fisica,
Univ. L'Aquila, 67010 Coppito (Aq),
Italy

Alessandra Continenza
CASTI-INFM Regional Lab. and
Dip. Fisica,
Univ. L'Aquila, 67010 Coppito (Aq),
Italy

Arthur J. Freeman
Dept. of Physics and Astronomy,
Northwestern University,
60208 Evanston, Il, USA

K. Westerholt, A. Bergmann, J. Grabis, A. Nefedov, and H. Zabel
Institut für Experimentalphysik IV,
Ruhr-Universität Bochum, D-44780
Bochum, Germany
kurt.westerholt@ruhr-uni-bochum.de

Claudia Felser
Institut für Anorganische Chemie
und Physikalische
Chemie, Johannes Gutenberg-
Universität Mainz, 55099 Mainz,
Germany
felser@uni-mainz.de

Hans-Joachim Elmers
Institut für Physik,
Johannes Gutenberg-Universität
Mainz, 55099 Mainz, Germany
elmers@uni-mainz.de

Gerhard H. Fecher
Institut für Anorganische
Chemie und Physikalische Chemie,
Johannes Gutenberg-Universität
Mainz, 55099 Mainz, Germany
fecher@uni-mainz.de

Willem Van Roy
IMEC, Kapeldreef 75, B-3001
Leuven, Belgium
vanroy@imec.be

Marek Wójcik
Institute of Physics-Polish Academy
of Sciences, Al. Lotnikow 32/46,
02-668 Warszawa, Poland
wojci@ifpan.edu.pl

Thomas Ambrose and Oleg Mryasov
Seagate Research, 1251 Waterfront
Place, Pittsburgh, PA, USA
thomas.f.ambrose@seagate.com
and oleg.mryasov@seagate.com

Hae-Kyung Jeong
Department of Physics, University of Nebraska-Lincoln, USA
hjeong@unlserve.unl.edu and acaruso1@bigred.unl.edu

Anthony Caruso
Department of Physics, University of Nebraska-Lincoln, USA
hjeong@unlserve.unl.edu and acaruso1@bigred.unl.edu

Camelia N. Borca
Institute of Applied Optics, Swiss Federal Institute of Technology, Switzerland
cborca@jilau1.colorado.edu

Andreas Hütten, Sven Kämmerer, Jan Schmalhorst, and Günter Reiss
Department of Physics, University of Bielefeld, P.O.
Box 100131, D-33501 Bielefeld, Germany
huetten@physik.uni-bielefeld.de

Bang-Gui Liu
Institute of Physics, Chinese Academy of Sciences, P.O. Box 603, Beijing 100080, China;
Beijing National Laboratory for Condensed Matter Physics, Beijing 100080, China
bgliu@aphy.iphy.ac.cn
and
Department of Physics, University of California at Berkeley, Berkeley, CA 94720, USA
bgliu@berkeley.edu

Hiro Akinaga
Nanotechnology Research Institute, National Institute
of Advanced Industrial Science and Technology, 1-1-1 Umezono, Tsukuba, Ibaraki 305-8568, Japan
akinaga.hiro@aist.go.jp

Masaki Mizuguchi
Nanotechnology Research Institute, National Institute
of Advanced Industrial Science and Technology, 1-1-1 Umezono, Tsukuba, Ibaraki 305-8568, Japan
akinaga.hiro@aist.go.jp

Kazutaka Nagao
Research Institute of Electrical Communication, Tohoku University, 2-1-1 Katahira, Aoba-ku, Sendai 980-8577, Japan
shirai@riec.tohoku.ac.jp

Yoshio Miura
Research Institute of Electrical Communication, Tohoku University, 2-1-1 Katahira, Aoba-ku, Sendai 980-8577, Japan
shirai@riec.tohoku.ac.jp

Masafumi Shirai
Research Institute of Electrical Communication, Tohoku University, 2-1-1 Katahira, Aoba-ku, Sendai 980-8577, Japan
shirai@riec.tohoku.ac.jp

Contents

Half-Metallicity and Slater-Pauling Behavior in the Ferromagnetic Heusler Alloys

Iosif Galanakis and Peter H. Dederichs

Institut für Festkörperforschung, Forschungszentrum Jülich, D-52425 Jülich, Germany
I.Galanakis@fz-juelich.de and P.H.Dederichs@fz-juelich.de

Abstract. A significant number of the intermetallic Heusler alloys have been predicted to be half-metals. In this contribution we present a study of the basic electronic and magnetic properties of both Heusler families; the so-called half-Heusler alloys like NiMnSb and the the full-Heusler alloys like Co_2MnGe. Based on *ab-initio* results we discuss the origin of the gap which is fundamental for the understanding of their electronic and magnetic properties. We show that the total spin magnetic moment M_t scales linearly with the number of the valence electrons Z_t, such that $M_t = Z_t - 24$ for the full Heuslers and $M_t = Z_t - 18$ for the half Heuslers, thus opening the way to engineer new half-metallic alloys with the desired magnetic properties. Although at the surfaces and interfaces the half-metallic character is in general lost, we show that for compounds with Cr as the metallic element the large enhancement of the Cr surface moment can lead to a high polarization at the surface. Moreover we discuss the role of the spin-orbit coupling, which in principle destroys the half-metallic gap, but in practice only slightly reduces the 100% spin polarization at E_{F}.

1 Introduction

Half-metallic ferromagnets represent a class of materials which attracted a lot of attention due to their possible applications in spintronics (also known as magnetoelectronics) [1]. Adding the spin degree of freedom to the conventional electronic devices has several advantages like non-volatility, increased data processing speed, decreased electric power consumption and increased integration densities [2]. The current advances in new materials and especially in the half-metals are promising for engineering new spintronic devices in the near future [2]. In these materials the two spin bands show a completely different behavior. While the majority spin band (referred also as spin-up band) shows the typical metallic behavior, the minority spin band (spin-down band) exhibits a semiconducting behavior with a gap at the Fermi level. Therefore such half-metals are ferromagnets and can be considered as hybrids between metals and semiconductors. A schematic representation of the density of states of a half-metal as compared to a normal metal and a normal semiconductor is shown in Fig. 1. The spinpolarization at the Fermi level is 100% and therefore these compounds should have a fully spinpolarised

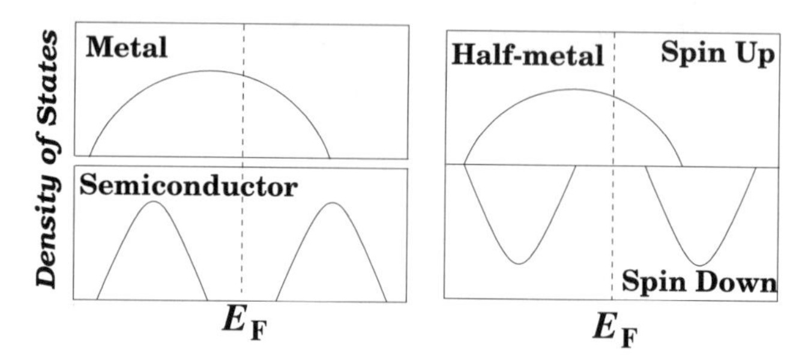

Fig. 1. Schematic representation of the density of states for a half-metal with respect to normal metals and semiconductors

current and might be able to yield a 100% spininjection and thus to maximize the efficiency of magnetoelectronic devices [3].

Heusler alloys [4] have attracted during the last century a great interest due to the possibility to study in the same family of alloys a series of interesting diverse magnetic phenomena like itinerant and localized magnetism, antiferromagnetism, helimagnetism, Pauli paramagnetism or heavy-fermionic behavior [5–8]. The first Heusler alloys studied were crystallizing in the $L2_1$ structure which consists of 4 fcc sublattices. Afterwards, it was discovered that it is possible to leave one of the four sublattices unoccupied ($C1_b$ structure). The latter compounds are often called half- or semi-Heusler alloys, while the $L2_1$ compounds are referred to as full-Heusler alloys. NiMnSb belongs to the half-Heusler alloys [9]. In 1983 de Groot and his collaborators [10] showed by using first-principles electronic structure calculations that this compound is in reality half-metallic, i.e. the minority band is semiconducting with a gap at the Fermi level E_F, leading to 100% spin polarization at E_F as shown in Fig. 1. Other known half-metallic materials except the half- and full-Heusler alloys [11–14] are some oxides (e.g. CrO_2 and Fe_3O_4) [15], the manganites (e.g. $La_{0.7}Sr_{0.3}MnO_3$) [15], the double perovskites (e.g. Sr_2FeReO_6) [16], the pyrites (e.g. CoS_2) [17], the transition metal chalcogenides (e.g. CrSe) and pnictides (e.g. CrAs) in the zinc-blende or wurtzite structures [18–21], the europium chalcogenides (e.g. EuS) [22] and the diluted magnetic semiconductors (e.g. Mn impurities in Si or GaAs) [23, 24]. Although thin films of CrO_2 and $La_{0.7}Sr_{0.3}MnO_3$ have been verified to present practically 100% spin-polarization at the Fermi level at low temperatures [15,25], the Heusler alloys remain attractive for technical applications like spin-injection devices [26], spin-filters [27], tunnel junctions [28], or GMR devices [29] due to their relatively high Curie temperature compared to these compounds [5].

The half-metallic character of NiMnSb in single crystals seems to have been well-established experimentally. Infrared absorption [30] and spin-polarized positron-annihilation [31] gave a spin-polarization of ∼100% at the Fermi level. Recently it has also become possible to grow high quality films of Heusler alloys, and it is mainly NiMnSb that has attracted the attention [32–34]. Unfortunately these films were found not to be half-metallic [15,35–38]; a maximum value of 58% for the spin-polarization of NiMnSb was obtained by Soulen et al. [15]. These polarization values are consistent with a small perpendicular magnetoresistance measured for NiMnSb in a spin-valve structure [39], a superconducting tunnel junction [28] and a tunnel magnetoresistive junction [40]. Ristoiu et al. showed that during the growth of the NiMnSb thin films, Sb and then Mn atoms segregate to the surface, which is far from being perfect, thus decreasing the obtained spin-polarization [41]. But when they removed the excess of Sb by flash annealing, they managed to get a nearly stoichiometric ordered alloy surface being terminated by a MnSb layer, which presented a spin-polarization of about 67 ± 9% at room temperature [41].

Several groups have verified the half-metallic character of bulk NiMnSb using first-principles calculations [42, 43]. Larson et al. have shown that the actual structure of NiMnSb is the most stable with respect to an interchange of the atoms [44] and Orgassa et al. showed that a few percent of disorder induce states within the gap but do not destroy the half-metallicity [45]. Recently, Galanakis has shown by first-principle calculations that NiMnSb surfaces do not present 100% spin-polarization [46] but Wijs and de Groot proposed that at some interfaces it is possible to restore the half-metallic character of NiMnSb [47]. These results were also confirmed by Debernardi et al. who studied the interface between NiMnSb and GaAs [48]. Jenkins and King studied by a pseudopotential technique the MnSb terminated (001) surface of NiMnSb and showed that there are two surface states at the Fermi level, which are well localized at the surface layer [49] and they persist even when the MnSb surface is covered by a Sb overlayer [50]. Finally Kübler calculated the Curie temperature of NiMnSb [51] which was in excellent agreement with the experimental value of 770 K [5]. A similar technique has been also employed to study the Curie temperature of some full-Heusler alloys which are not half-metallic [52].

Webster and Ziebeck [53] and Suits [54] were the first to synthesize full-Heusler alloys containing Co and Rh, respectively. Kübler et al. studied the mechanisms stabilizing the ferro- or the antiferromagnetism in these compounds [55]. Ishida and collaborators have proposed that the compounds of the type Co_2MnZ, where Z stands for Si and Ge, are half-metals [56, 57]. Also the Heusler alloys of the type Fe_2MnZ have been proposed to show half-metallicity [58]. But Brown et al. [59] using polarized neutron diffraction measurements have shown that there is a finite very small spin-down density of states (DOS) at the Fermi level instead of an absolute gap in agreement with the *ab-initio* calculations of Kübler et al. for the Co_2MnAl and Co_2MnSn compounds [55]. Recently, several groups managed to grow Co_2MnGe and Co_2MnSi thin films on various substrates [60–62], and there also exist first-principles calculations for the (001) surface of such an alloy [63]. Geiersbach and collaborators have grown (110) thin films of Co_2MnSi, Co_2MnGe and Co_2MnSn using a metallic seed on top of a MgO(001) substrate [64] and studied also the transport properties of multilayers of these compounds with normal metals [65]. But as Picozzi et al. have shown the interfaces of such structures are not half-metallic [66]. Finally, Kämmerer and collaborators managed to built magnetic tunnel junctions based on Co_2MnSi and found a tunneling magnetoresistance effect much larger than when the $Ni_{0.8}Fe_{0.2}$ or $Co_{0.3}Fe_{0.7}$ are used as magnetic electrodes [67]. Similar experiments have been undertaken by Inomata and collaborators using $Co_2Cr_{0.6}Fe_{0.4}Al$ as the magnetic electrode [68].

In this contribution, we present a study of the basic electronic and magnetic properties of the half-metallic Heusler alloys. Analyzing the *ab-initio* results using the group-theory and simple models we explain the origin of the gap in both the half- and full-Heusler alloys, which is fundamental for understanding their electronic and magnetic properties. For both families of compounds the total spin magnetic moment scales with the number of valence electron, thus opening the way to engineer new half-metallic Heusler alloys with the desired magnetic properties. Although in general the surfaces loose the half-metallic character and show only a small degree of spin-polarization, we show that in the case of compounds containing Cr, the very large Cr moments at the surface reduce the importance of the surface states and the spin-polarization of such surfaces is very high, e.g. 84% for the CrAl-terminated $Co_2CrAl(001)$ surface. Finally we discuss the role of defects and spin-orbit coupling on the half-metallic band gap.

In Sects. 2 and 3 we present the electronic and magnetic properties of the XMnSb (X=Ni,Co,Rh,Pd,Ir or Pt) and Co_2MnZ (Z=Al,Si,Ga,Ge or Sn) compounds, respectively. In Sect. 4 we investigate the effect of compressing or expanding the lattice and in Sect. 5 the properties of the quaternary Heusler alloys. In Sects. 6 and 7 we review the role of defects and of the spin-orbit coupling, respectively and finally in Sect. 8 we review the surface properties of these alloys.

2 Electronic Structure and Magnetism of Half-Heusler Alloys

2.1 Band Structure of Half-Heusler Alloys

In the following we present results for the densities of states of some typical half-Heusler alloys of $C1_b$ structure (see Fig. 2), sometimes also referred to as semi-Heusler alloys. To perform the calculations, we used the Vosko, Wilk and Nusair parameterization [69] for the local density approximation (LDA) to the exchange-correlation potential [70] to solve the Kohn-Sham equations within the full-potential screened Korringa-Kohn-Rostoker (FSKKR) method [71,72]. The prototype example is NiMnSb, the half-metal discovered in 1983

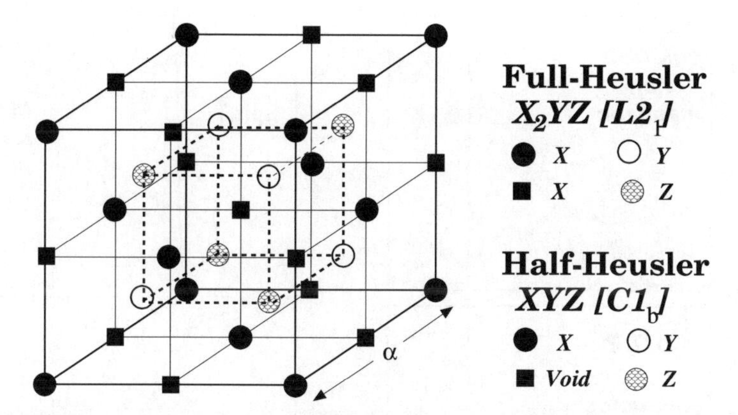

Fig. 2. $C1_b$ and $L2_1$ structures adapted by the half- and full-Heusler alloys. The lattice is consisted of 4 interprenatating f.c.c. lattices. The unit cell is that of a fcc lattice with four atoms as basis, e.g. CoMnSb: Co at (0 0 0), Mn at $(\frac{1}{4}\,\frac{1}{4}\,\frac{1}{4})$, a vacant site at $(\frac{1}{2}\,\frac{1}{2}\,\frac{1}{2})$ and Sb at $(\frac{3}{4}\,\frac{3}{4}\,\frac{3}{4})$ in Wyckoff coordinates. In the case of the full Heusler alloys also the vacant site is occupied by a Co atom. Note also that if all atoms were identical, the lattice would be simply the bcc

by de Groot [10]. Figure 3 shows the density of states (DOS) of NiMnSb in a non-spin-polarized calculation (left upper panel) and in a calculation correctly including the spin-polarization (right panel). Given are the local contributions to the density of states (LDOS) on the Ni-site (dashed), the Mn-site (full line) and the Sb-site (dotted). In the non-magnetic case the DOS of NiMnSb has contributions from 4 different bands: Each Sb atom with the atomic configuration $5s^2 5p^3$ introduces a deep lying s band, which is located at about -12 eV and is not shown in the figure, and three p-bands in the regions between -5.5 and -3 eV. These bands are separated by a deep minimum in the DOS from 5 Ni d bands between -3 and -1 eV, which themselves are separated by a sizeable band gap from the upper 5 d-bands of Mn. Since all atomic orbitals, i.e. the Ni d, the Mn d and the Sb sp orbitals hybridize with each other, all bands are hybrids between these states, being either of bonding or antibonding type. Thus the Ni d-bands contain a bonding Mn d admixture, while the higher Mn d-bands are antibonding hybrids with small Ni d-admixtures. Equally the Sb p-bands exhibit strong Ni d- and somewhat smaller Mn d-contributions.

This configuration for NiMnSb is energetically not stable, since (i) the Fermi energy lies in the middle of an antibonding band and (ii) since the Mn atom can gain considerable exchange energy by forming a magnetic moment. Therefore the spin-polarized results (right figure) show a considerably different picture. In the majority (spin $\uparrow$) band the Mn d states are shifted to lower energies and form a common d band with the Ni d states, while in the minority band (spin $\downarrow$) the Mn states are shifted to higher energies and are unoccupied, so that a band gap at E_F is formed separating the occupied d

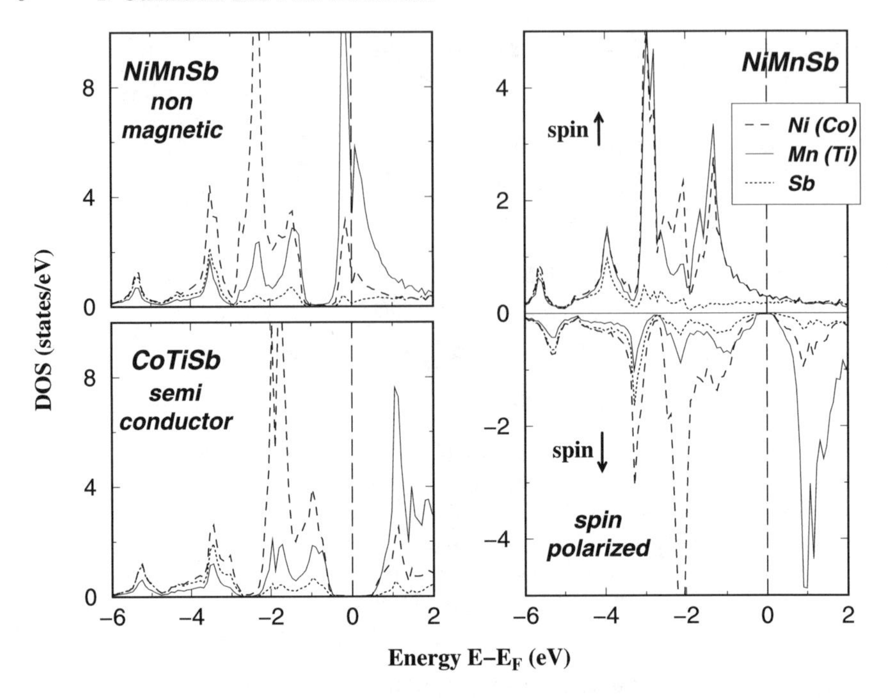

Fig. 3. Atom-resolved density of states (DOS) of NiMnSb for a paramagnetic (*left upper*) and ferromagnetic (*right*) calculation. In the left bottom panel the DOS for the semiconductor CoTiSb. The zero energy value corresponds to the Fermi level E_F

bonding from the unoccupied d-type antibonding states. Thus NiMnSb is a half-metal, with a band gap at E_F in the minority band and a metallic DOS at E_F in the majority band. The total magnetic moment, located mostly at the Mn atom, can be easily estimated to be exactly 4 μ_B. Note that NiMnSb has 22 valence electrons per unit cell, 10 from Ni, 7 from Mn and 5 from Sb. Since, due to the gap at E_F, in the minority band exactly 9 bands are fully occupied (1 Sb-like s band, 3 Sb-like p bands and 5 Ni-like d bands) and accommodate 9 electrons per unit cell, the majority band contains $22 - 9 = 13$ electrons, resulting in a moment of 4 μ_B per unit cell.

The above non-spinpolarized calculation for NiMnSb (Fig. 3 left upper panel) suggests, that if we could shift the Fermi energy as in a rigid band model, a particular stable compound would be obtained if the Fermi level falls for both spin directions into the band gap. Then for both spin directions 9 bands would be occupied, resulting in a semiconductor with 18 valence electrons. Such semiconducting Heusler alloys indeed exist. As a typical example, Fig. 3 shows also the DOS of CoTiSb, which has a gap of 0.8 eV [8]. The gap is indirect corresponding to transitions from the valence band maximum at Γ to the conduction band minimum at X. Other such semiconductors are

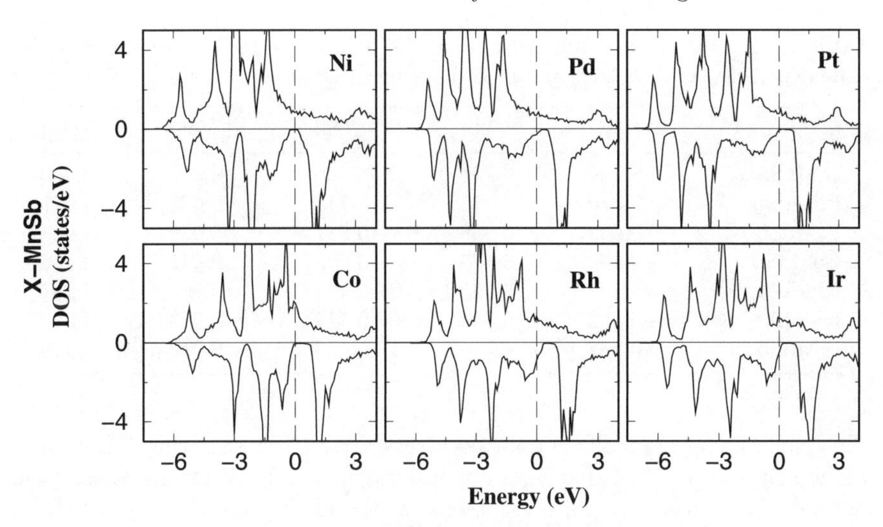

Fig. 4. DOS of XMnSb compounds for X= Ni, Pd, Pt and Co, Rh, Pd

CoZrSb (0.8 eV), FeVSb (0.36 eV) and NiTiSn (0.14 eV) where the values in the bracket denote the size of the gap [8].

2.2 XMnSb Half-Heusler Alloys with X = Ni, Pd Pt and Co, Rh, Ir

Here we present the electronic structure of the half-Heusler alloys of the type XMnSb, with X being an element of the Co or Ni columns in the periodic table. These compounds are known experimentally to be ferromagnets with high Curie temperatures ranging between 500 K and 700 K for the Co, Ni, Pd and Pt compounds, while the Curie temperatures of the Ir and Rh compounds are around room temperature [5]. In Fig. 4 we present the spin-projected total density of states (DOS) for all the six compounds. We remark that all compounds present a gap, which is wider in the compounds containing Co, Rh or Ir than in Ni, Pd or Pt. As above Sb p states occupy the lowest part of the DOS shown in the figure, while the Sb s states are located ~12 eV below the Fermi level. For the Ni compound the Fermi level is at the middle of the gap and for PtMnSb at the left edge of the gap in agreement with previous FPLMTO calculations [42]. In the case of CoMnSb the gap is considerably larger (~1 eV) than in the previous two compounds and the Fermi level is located at the left edge of the spin-down gap. CoMnSb has been studied previously by Kübler, who found similar results by using the augmented spherical waves (ASW) method [73]. For the other three compounds the Fermi level is located below the gap, although in the case of PdMnSb and IrMnSb it is close to the band edge.

The DOS of the different systems are mainly characterized by the large exchange-splitting of the Mn d states which is around 3 eV in all cases.

Table 1. Calculated spin magnetic moments in μ_B for the XMnSb compounds. (The experimental lattice constants [5] have been used)

$m^{spin}(\mu_B)$	X	Mn	Sb	Void	Total
NiMnSb	0.264	3.705	−0.060	0.052	3.960
PdMnSb	0.080	4.010	−0.110	0.037	4.017
PtMnSb	0.092	3.889	−0.081	0.039	3.938
CoMnSb	−0.132	3.176	−0.098	0.011	2.956
RhMnSb	−0.134	3.565	−0.144	<0.001	3.287
IrMnSb	−0.192	3.332	−0.114	−0.003	3.022
FeMnSb	−0.702	2.715	−0.053	0.019	1.979

This leads to large localized spin moments at the Mn site, the existence of which has been verified also experimentally [74]. The localization comes from the fact that although the d electrons of Mn are itinerant, the spin-down electrons are almost excluded from the Mn site. In Table 1 we present the spin magnetic moments at the different sites for all the compounds under study. The moments are calculated by integrating the spin-projected charge density inside every Wigner-Seitz polyhedron. Experimental values for the spin-moment at the Mn site can be deduced from the experiments of Kimura et al. [75] by applying the sum rules to their x-ray magnetic circular dichroism spectra and the extracted moments agree nicely with our results; they found a Mn spin moment of 3.85 μ_B for NiMnSb, 3.95 μ_B for PdMnSb and 4.02 μ_B for PtMnSb. In the case of the Co-, Rh-, and IrMnSb compounds the spin magnetic moment of the X atom is antiparallel to the Mn localized moment and the Mn moment is generally about 0.5 μ_B smaller than in the Ni, Pd and Pt compounds. The Sb atom is here again antiferromagnetically coupled to the Mn atom.

The total magnetic moment in μ_B is just the difference between the number of spin-up occupied states and the spin-down occupied states. As explained above, the number of occupied spin-down states is given by the number of spin down bands, i.e. 9, so that the number of occupied spin-up states is $22 - 9 = 13$ for NiMnSb and the isovalent compounds with Pd and Pt, but $21 - 9 = 12$ for CoMnSb, RhMnSb and IrMnSb and $20 - 9 = 11$ for FeMnSb, provided that the Fermi level stays within the gap. Therefore one expects total moments of $4\mu_B$ for Ni-, Pd- and PtMnSb, 3 μ_B for the compounds with Co, Rh and Ir and 2 μ_B for FeMnSb. In general, for a total number Z_t of valence electrons in the unit cell, the total moment M_t is given by $M_t = Z_t - 18$, since with 9 electron states occupied in the minority band, $Z_t - 18$ is just the number of uncompensated electron spins.

The local moment per unit cell as given in Table 1 is close to 4 μ_B in the case of NiMnSb, PdMnSb and PtMnSb, which is in agreement with the half-metallic character (or nearly half-metallic character in the case of PdMnSb) observed in Fig. 4. Note that due to problems with the ℓ_{max} cutoff the KKR method can only give the correct integer number 4, if Lloyd's formula has been used in the evaluation of the integrated density of states [76], which is not the case in the present calculations. We also find that the local moment of Mn is not far away from the total number of 4 μ_B although there are significant (positive) contributions from the X-atoms and a negative contribution from the Sb atom. The antiferromagnetic coupling between the Sb and Mn moments is due to the different behavior of the p bands created by the Sb atoms discussed in the next section. The minority p bands are more located within the Wigner-Seitz cell of Sb. On the other hand the majority p bands are more expanded in space and contain a larger Mn d-admixture and thus the total Sb spin moment has a negative sign.

In contrast to this we find that for the half-metallic CoMnSb and IrMnSb compounds the total moment is about 3 μ_B. Also the local moment of Mn is reduced, but only by about 0.5 μ_B. The reduction of the total moment to 3 μ_B is therefore accompanied by negative Co and Ir spin moments, *i.e* these atoms couple antiferromagnetically to the Mn moments. The hybridization between Co and Mn is considerably larger than between Ni and Mn being a consequence of the smaller electronegativity difference and the larger extend of the Co orbitals. Therefore the minority valence band of CoMnSb has a larger Mn admixture than the one of NiMnSb whereas the minority conduction band of CoMnSb has a larger Co admixture than the Ni admixture in the NiMnSb conduction band, while the populations of the majority bands are barely changed. As a consequence, the Mn moment is reduced by the increasing hybridization, while the Co moment becomes negative, resulting finally in a reduction of the total moment from 4 to 3 μ_B. The table also shows that further substitution of Fe for Co leads also to a half-metallic alloy with a total spin magnetic moment of 2 μ_B as has been already shown by de Groot et al. in reference [77].

2.3 Origin of the Gap

The inspection of the local DOS shown in Fig. 3 for the ferromagnet NiMnSb as well as for the semiconductor CoTiSb shows that the DOS close to the gap is dominated by d-states: in the valence band by bonding hybrids with large Ni or Co admixture and in the conduction band by the antibonding hybrids with large Mn or Ti admixture. Thus the gap originates from the strong hybridization between the d states of the higher valent and the lower valent transition metal atoms. This is shown schematically in Fig. 5. Therefore the origin of the gap is somewhat similar to the gap in compound semiconductors like GaAs which is enforced by the hybridization of the lower lying As sp-states with the energetically higher Ga sp-states. Note that in the $C1_b$-structure the Ni and Mn sublattices form a zinc-blende structure, which is important for the formation of the gap. The difference with respect to GaAs is only, that 5 d-orbitals, i.e. 3 t_{2g} and 3 e_g orbitals, are involved in the hybridization, instead of 4 sp^3-hybrids in the compound semiconductors.

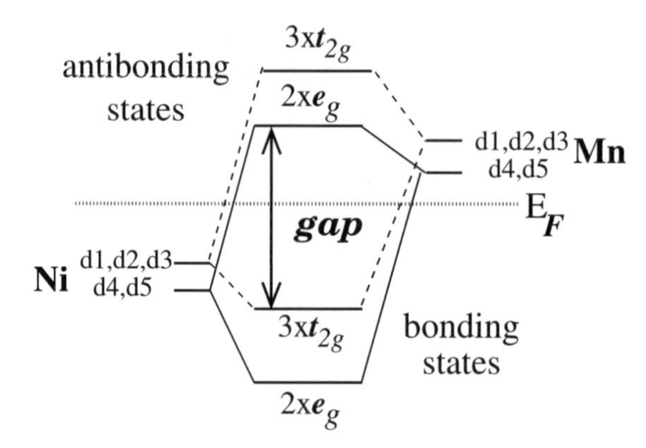

Fig. 5. Schematic illustration of the origin of the gap in the minority band in half-Heusler alloys and in compound semiconductors: The energy levels E_b of the energetically lower lying bonding hybrids are separated from the levels E_{ab} of the antibonding hybrids by a gap, such that only the bonding states are occupied. Due to legibility reasons, we use d1, d2 and d3 to denote the d_{xy}, d_{yx} and d_{zx} orbitals, respectively, and d4, d5 for the d_{r^2}, $d_{x^2-y^2}$ orbitals

Giving these arguments it is tempting to claim, that also a hypothetical zinc-blende compound like NiMn or PtMn should show a half-metallic character with a gap at E_F in the minority band. Figure 6 shows the results of a self-consistent calculation for such zinc-blende NiMn and PtMn, with the same lattice constant as NiMnSb. Indeed a gap is formed in the minority band. In the hypothetical NiMn the Fermi energy is slightly above the gap, however the isoelectronic PtMn compound shows indeed half-metallicity. In this case the occupied minority bands consists of six bands, a low-lying s-band and five bonding d-bands, both of mostly Pt character. Since the total number of valence electrons is 17, the majority bands contain 11 electrons, so that the total moment per unit cell is $11 - 6 = 5\mu_B$, which is indeed obtained in the calculations. This is the largest possible moment for this compound, since in the minority band all 5 Mn d-states are empty while all majority d-states are occupied. The same limit of $5\mu_B$ is also the maximal possible moment of the half-metallic $C1_b$ Heusler alloys.

The gap in the half-metallic $C1_b$ compounds is normally an indirect gap, with the maximum of the valence band at the Γ point and the minimum of the conduction band at the X-point. For NiMnSb we obtain a band gap of about 0.5 eV, which is in good agreement with the experiments of Kirillova and collaborators [30], who, analyzing their infrared spectra, estimated a gap width of ~0.4 eV. As seen already from Fig. 4 the gap of CoMnSb is considerable larger (~1 eV) and the Fermi level is located at the edge of the minority valence band.

As it is well-known, the local density approximation (LDA) and the generalized gradient approximation (GGA) strongly underestimate the values of

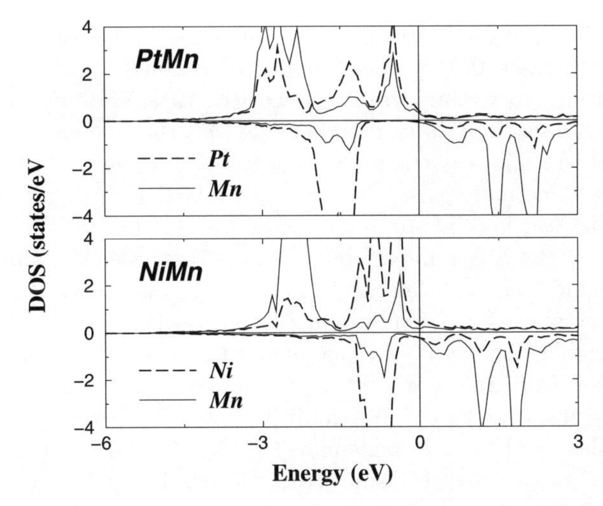

Fig. 6. Atom-resolved DOS for the hypothetical PtMn and NiMn crystallizing in the zinc-blende structure

the gaps in semiconductors, typically by a factor of two. However, very good values for these gaps are obtained in the so-called GW approximation of Hedin and Lundqvist [78], which describes potential in semiconductors very well. On the other hand the minority gap in the half-metallic systems might be better described by the LDA and GGA since in these system the screening is metallic.

2.4 Role of *sp*-Elements

While the *sp*-elements are not responsible for the existence of the minority gap, they are nevertheless very important for the physical properties of the Heusler alloys and the structural stability of the $C1_b$ structure, as we discuss in the following.

While an Sb atom has 5 valence electrons ($5s^2$, $5p^3$), in the NiMnSb compound each Sb atom introduces a deep lying *s*-band, at about -12 eV, and three *p*-bands below the center of the *d*-bands. These bands accommodate a total of 8 electrons per unit cell, so that formally Sb acts as a triple charged Sb^{3-} ion. Analogously, a Te-atom behaves in these compounds as a Te^{2-} ion and a Sn-atom as a Sn^{4-} ion. This does not mean, that locally such a large charge transfer exists. In fact, the *s*- and *p*-states strongly hybridize with the TM *d*-states and the charge in these bands is delocalized and locally Sb even looses about one electron, if one counts the charge in the Wigner-Seitz cells. What counts is that the *s*- and *p*-bands accommodate 8 electrons per unit cell, thus effectively reducing the *d*-charge of the TM atoms.

This is nicely illustrated by the existence of the semiconducting compounds CoTiSb and NiTiSn. Compared to CoTiSb, in NiTiSn the missing *p*-charge of the Sn atom is replace by an increased *d* charge of the Ni atom, so that in both cases all 9 valence bands are occupied.

The *sp*-atom is very important for the structural stability of the Heusler alloys. For instance, it is difficult to imagine that the calculated half-metallic NiMn and PtMn alloys with zinc-blende structure, the LDOS of which are shown in Fig. 6, actually exist, since metallic alloys prefer highly coordinated structures like fcc, bcc, hcp etc. Therefore the *sp*-elements are decisive for the stability of the $C1_b$ compounds. A careful discussion of the bonding in these compounds has been recently published by Nanda and Dasgupta [79] using the crystal orbital Hamiltonian population (COHP) method. For the semiconductor FeVSb they find that while the largest contribution to the bonding arises from the V-*d* – Fe-*d* hybridization, contributions of similar size arise also from the Fe-*d* – Sb-*p* and the V-*d* – Sb-*p* hybridization. Similar results are also valid for the semiconductors like CoTiSb and NiTiSn and in particular for the half-metal NiMnSb. Since the majority *d*-band is completely filled, the major part of the bonding arises from the minority band, so that similar arguments as for the semiconductors apply.

Another property of the *sp*-elements is worthwhile to mention: substituting the Sb atom in NiMnSb by Sn, In or Te destroys the half-metallicity [11]. This is in contrast to the substitution of Ni by Co or Fe, which is documented in Table 1. The total moment of 4 μ_B for NiMnSb is reduced to 3 μ_B in CoMnSb and 2 μ_B in FeMnSb, thus preserving half-metallicity. In NiMnSn the total moment is reduced to 3.3 μ_B (instead of 3) and in NiMnTe the total moment increases only to 4.7 μ_B (instead of 5). Thus by changing the *sp*-element it is rather difficult to preserve the half-metallicity, since the density of states changes more like in a rigid band model [11].

2.5 Slater-Pauling Behavior

As discussed above the total moment of the half-metallic $C1_b$ Heusler alloys follows the simple rule: $M_t = Z_t - 18$, where Z_t is the total number of valence electrons. In short, the total number of electrons Z_t is given by the sum of the number of spin-up and spin-down electrons, while the total moment M_t is given by the difference

$$Z_t = N_\uparrow + N_\downarrow, \quad M_t = N_\uparrow - N_\downarrow \quad \to \quad M_t = Z_t - 2N_\downarrow \tag{1}$$

Since 9 minority bands are fully occupied, we obtain the simple "rule of 18" for half-metallicity in $C1_b$ Heusler alloys

$$M_t = Z_t - 18 \tag{2}$$

the importance of which has been recently pointed out by Jung et al. [80] and Galanakis et al. [11]. It is a direct analogue to the well-known Slater-Pauling behavior of the binary transition metal alloys [81]. The difference with respect

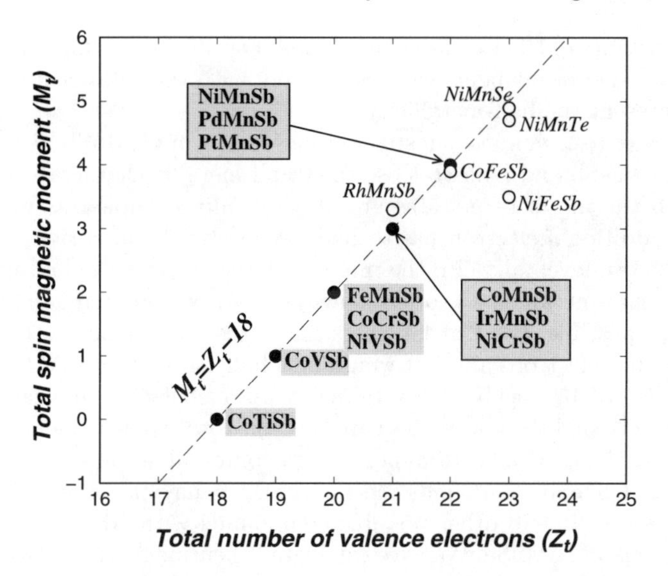

Fig. 7. Calculated total spin moments for all the studied half Heusler alloys. The dashed line represents the Slater-Pauling behavior. With open circles we present the compounds deviating from the SP curve. Some experimental values for bulk systems near the SP curve from reference [5]: NiMnSb 3.85 μ_B, PdMnSb 3.95 μ_B, PtMnSb 4.14 μ_B and finally CoTiSb non-magnetic

to these alloys is, that in the half-Heusler alloys the minority population is fixed to 9, so that the screening is achieved by filling the majority band, while in the transition metal alloys the majority band is filled with 5 d-states and charge neutrality is achieved by filling the minority states. Therefore in the TM alloys the total moment is given by $M_t = 10 - Z_t$. Similar rules with integer total moments are also valid for other half-metals, e.g. for the full-Heusler alloys like Co_2MnGe with $L2_1$ structure. For these alloys we will in Sect. 3 derive the "rule of 24": $M_t = Z_t - 24$, with arises from the fact that the minority band contains 12 electrons. For the half-metallic zinc-blende compounds like CrAs the rule is: $M_t = Z_t - 8$, since the minority As-like valence bands accommodate 4 electrons [18]. In all cases the moments are integer.

In Fig. 7 we have gathered the calculated total spin magnetic moments for the half-Heusler alloys which we have plotted as a function of the total number of valence electrons. The dashed line represents the rule $M_t = Z_t - 18$ obeyed by these compounds. The total moment M_t is an integer quantity, assuming the values 0, 1, 2, 3, 4 and 5 if $Z_t \geq 18$. The value 0 corresponds to the semiconducting phase and the value 5 to the maximal moment when all 10 majority d-states are filled. Firstly we varied the valence of the lower-valent (i.e. magnetic) transition metal atom. Thus we substitute V, Cr and Fe for Mn in the NiMnSb and CoMnSb compounds using the experimental lattice constants of the two Mn compounds. For all these compounds we find that the total spin moment scales accurately with the total charge and that they all present the half-metallicity.

As a next test we have substituted Fe for Mn in CoMnSb and NiMnSb, but both CoFeSb and NiFeSb loose their half-metallic character. In the case of NiFeSb the majority *d*-states are already fully occupied as in NiMnSb, thus the additional electron has to be screened by the minority *d*-states, so that the Fermi level falls into the minority Fe states and the half-metallicity is lost; for half-metallicity a total moment of 5 μ_B would be required which is clearly not possible. For CoFeSb the situation is more delicate. This system has 22 valence electrons and if it would be a half-metal, it should have a total spin-moment of 4 μ_B as NiMnSb. In reality our calculations indicate that the Fermi level is slightly above the gap and the total spin-moment is slightly smaller than 4 μ_B. The Fe atom possesses a comparable spin-moment in both NiFeSb and CoFeSb compounds contrary to the behavior of the V, Cr and Mn atoms. Except NiFeSb other possible compounds with 23 valence electrons are NiMnTe and NiMnSe. We have calculated their magnetic properties using the lattice constant of NiMnSb. As shown in Fig. 7, NiMnSe almost makes the 5 μ_B (its total spin moment is 4.86 μ_B) and is nearly half-metallic, while its isovalent, NiMnTe, has a slightly smaller spin moment. NiMnSe and NiMnTe show big changes in the majority band compared to systems with 22 valence electrons as NiMnSb or NiMnAs, since antibonding *p*-*d* states, which are usually above E_F, are shifted below the Fermi level, thus increasing the total moment to nearly 5 μ_B.

3 Full Heusler Alloys

3.1 Electronic Structure of Co_2MnZ with Z = Al, Si, Ga, Ge and Sn

The second family of Heusler alloys, which we discuss, are the full-Heusler alloys. We consider in particular compounds containing Co and Mn, as these are the full-Heusler alloys that have attracted most of the attention. They are all strong ferromagnets with high Curie temperatures (above 600 K) and except the Co_2MnAl they show very little disorder [5]. They adopt the $L2_1$ structure shown in Fig. 2. Each Mn or *sp* atom has eight Co atoms as first neighbors, sitting in an octahedral symmetry position, while each Co has four Mn and four *sp* atoms as first neighbors and thus the symmetry of the crystal is reduced to the tetrahedral one. The Co atoms occupying the two different sublattices are chemically equivalent as the environment of the one sublattice is the same as the environment of the second one but rotated by 90°. The occupancy of two fcc sublattices by Co (or in general by X) atoms distinguish the full-Heusler alloys with the $L2_1$ structure from the half-Heusler compounds with the $C1_b$ structure, like e.g. CoMnSb, where

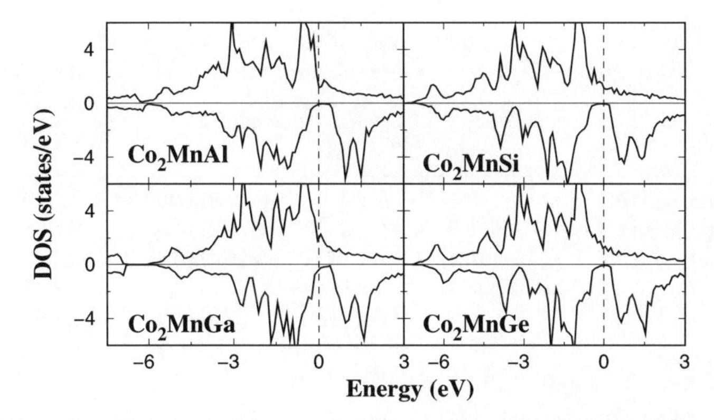

Fig. 8. Atom-resolved DOS for the Co_2MnZ compounds with Z = Al, Si, Ge, Sn compounds

only one sublattice is occupied by Co atoms and the other one is empty. Although in the $L2_1$ structure, the Co atoms are sitting on second neighbor positions, their interaction is important to explain the magnetic properties of these compounds as we will show in the next section.

In Fig. 8 we have gathered the spin-resolved density of states (DOS) for the Co_2MnAl, Co_2MnGa, Co_2MnSi and Co_2MnGe compounds calculated using the FSKKR. Firstly as shown by photoemission experiments by Brown et al. in the case of Co_2MnSn [82] and verified by our calculations, the valence band extends around 6 eV below the Fermi level and the spin-up DOS shows a large peak just below the Fermi level for these compounds. Although Ishida et al. [56] predicted them to be half-metals with small spin-down gaps ranging from 0.1 to 0.3 eV depending on the material, our previous calculations showed a very small DOS at the Fermi level, in agreement with the ASW results of Kübler et al. [81] for Co_2MnAl and Co_2MnSn. However a recalculation of our KKR results with a higher ℓ-cut-off of $\ell_{\max} = 4$ restores the gap and we obtain good agreement with the recent results of Picozzi et al. using the FLAPW method. The gap is an indirect gap, with the maximum of the valence band at Γ and the minimum of the conduction band at the X-point.

In the case of the half-Heusler alloys like NiMnSb the Mn spin magnetic moment is very localized due to the exclusion of the spin-down electrons at the Mn site and amounts to about 3.7 μ_B in the case of NiMnSb. In the case of CoMnSb the increased hybridization between the Co and Mn spin-down electrons decreased the Mn spin moment to about 3.2 μ_B (in Table 2 we have gathered the atomic-resolved and total moments of the Co_2MnZ compounds). In the case of the full-Heusler alloys each Mn atom has eight Co atoms as first neighbors instead of four as in CoMnSb and the above hybridization is very important decreasing even further the Mn spin moment to less than 3 μ_B except in the case of Co_2MnSn where it is comparable to the CoMnSb compound. The Co atoms are ferromagnetically coupled to

Table 2. Calculated spin magnetic moments in μ_B using the experimental lattice constants (see reference [5]) for the Co_2MnZ compounds, where Z stands for the *sp* atom

$m^{spin}(\mu_B)$	Co	Mn	Z	Total
Co_2MnAl	0.768	2.530	−0.096	3.970
Co_2MnGa	0.688	2.775	−0.093	4.058
Co_2MnSi	1.021	2.971	−0.074	4.940
Co_2MnGe	0.981	3.040	−0.061	4.941
Co_2MnSn	0.929	3.203	−0.078	4.984

the Mn spin moments and they posses a spin moment that varies from ∼0.7 to 1.0 μ_B. Note that in the half-metallic $C1_b$ Heusler alloys, the X-atom has a very small moment only, in the case of CoMnSb the Co moment is even negative. However in the full Heusler alloys the Co moment is large and positive and arises basically from two unoccupied Co bands in the minority conduction band, as explained below. Therefore both Co atoms together can have a moment of about 2 μ_B, if all majority Co states are occupied. This is basically the case for Co_2MnSi,Co_2MnGe and Co_2MnSn (see Table 2). In contrast to this the *sp* atom has a very small negative moment which is one order of magnitude smaller than the Co moment. The negative sign of the induced *sp* moment characterizes most of the studied full and half Heusler alloys with very few exceptions. The compounds containing Al and Ga have 28 valence electrons and the ones containing Si, Ge and Sn 29 valence electrons. The first compounds have a total spin moment of $4\mu_B$ and the second ones of 5 μ_B which agree with the experimental deduced moments of these compounds [83]. So it seems that the total spin moment, M_t, is related to the total number of valence electrons, Z_t, by the simple relation: $M_t = Z_t - 24$, while in the half-Heusler alloys the total magnetic moment is given by the relation $M_t = Z_t - 18$. In the following section we will analyze the origin of this rule.

3.2 Origin of the gap in Full-Heusler Alloys

Since, similar to the half-Heusler alloys, the four *sp*-bands are located far below the Fermi level and thus are not relevant for the gap, we consider only the hybridization of the 15 d states of the Mn atom and the two Co atoms. For simplicity we consider only the d-states at the Γ point, which show the full structural symmetry. We will give here a qualitative picture, since a thorough group theoretical analysis has been given in reference [12]. Note that the Co atoms form a simple cubic lattice and that the Mn atoms (and the Ge atoms) occupy the body centered sites and have 8 Co atoms as nearest neighbors. Although the distance between the Co atoms is a second neighbor distance, the hybridization between these atoms is qualitatively very

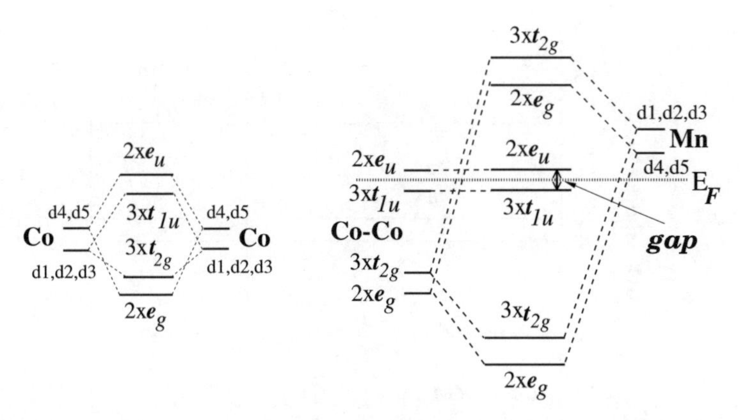

Fig. 9. Schematic illustration of the origin of the gap in the minority band in full-Heusler alloys. Due to legibility reasons, we use d1, d2 and d3 to denote the d_{xy}, d_{yx} and d_{zx} orbitals, respectively, and d4, d5 for the d_{r^2}, $d_{x^2-y^2}$ orbitals

important. Therefore we start with the hybridization between these Co atoms which is qualitatively sketched in Fig. 9. The 5 *d*-orbitals are divided into the twofold degenerate d_{r^2}, $d_{x^2-y^2}$ and the threefold degenerate d_{xy}, d_{yx}, d_{zx} states. The e_g orbitals (t_{2g} orbitals) can only couple with the e_g orbitals (t_{2g} orbitals) of the other Co atom forming bonding hybrids, denoted by e_g (or t_{2g}) and antibonding orbitals, denoted by e_u (or t_{1u}). The coefficients in front of the orbitals give the degeneracy.

In a second step we consider the hybridization of these Co-Co orbitals with the Mn *d*-orbitals. As we show in the right-hand part of Fig. 9, the double degenerated e_g orbitals hybridize with the d_{r^2} and $d_{x^2-y^2}$ of the Mn that transform also with the same representation. They create a doubly degenerate bonding e_g state that is very low in energy and an antibonding one that is unoccupied and above the Fermi level. The $3 \times t_{2g}$ Co orbitals couple to the $d_{xy,yx,zx}$ of the Mn and create 6 new orbitals, 3 of which are bonding and are occupied and the other three are antibonding and high in energy. Finally the $2 \times e_u$ and $3 \times t_{1u}$ Co orbitals *can not* couple with any of the Mn *d*-orbitals since none of these is transforming with the *u* representations and they are orthogonal to the Co e_u and t_{1u} states. With respect to the Mn and the Ge atoms these states are therefore non-bonding. The t_{1u} states are below the Fermi level and they are occupied while the e_u are just above the Fermi level. Thus in total 8 minority *d*-bands are filled and 7 are empty.

Therefore all 5 Co-Mn bonding bands are occupied and all 5 Co-Mn antibonding bands are empty, and the Fermi level falls in between the 5 non-bonding Co bands, such that the three t_{1u} bands are occupied and the two e_u bands are empty. The maximal moment of the full Heusler alloys is therefore 7 μ_B per unit cell, which is achieved, if all majority *d*-states are occupied.

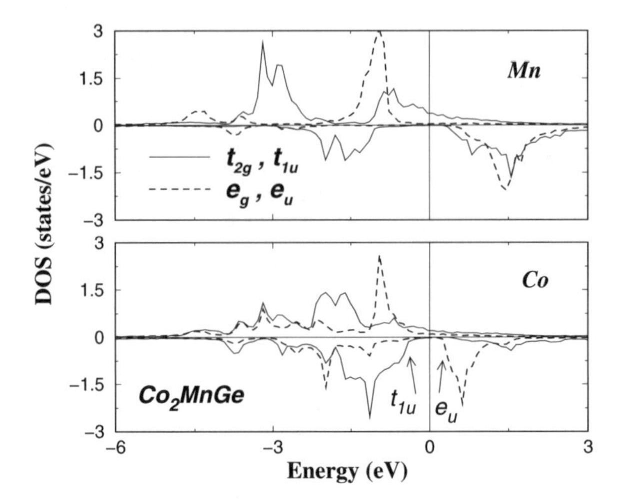

Fig. 10. Atom- and angular momentum – resolved DOS for the Co_2MnGe compound

In order to demonstrate the existence of the t_{1u} and e_u states at the Fermi level, we show in Fig. 10 the LDOS of Co_2MnGe at the Co and Mn sites, which are splitted up into the local d_{xy}, d_{yx} and d_{zx} orbitals (normally referred to as t_{2g}; full lines) and the local d_{r^2} and $d_{x^2-y^2}$ orbitals (normally e_g; dashed). In the nomenclature used above, the d_{xy}, d_{yx} and d_{zx} contributions contain both the t_{2g} and the t_{1u} contributions, while the d_{r^2} and $d_{x^2-y^2}$ orbitals contain the e_g and e_u contributions. The Mn DOS clearly shows a much bigger effective gap at E_F, considerably larger than in CoMnSb (Fig. 4), as one would expect from the stronger hybridization in Co_2MnGe. However the real gap is determined by the Co-Co interaction only, in fact by the $t_{1u} - e_u$ splitting, and is smaller than in CoMnSb. Thus the origin of the gap in the full-Heusler alloys is rather subtle.

3.3 Slater-Pauling Behavior of the Full-Heusler Alloys

Following the above discussion we will investigate the Slater-Pauling behavior and in Fig. 11 we have plotted the total spin magnetic moments for all the compounds under study as a function of the total number of valence electrons. The dashed line represents the half metallicity rule: $M_t = Z_t - 24$ of the full Heusler alloys. This rule arises from the fact that the minority band contains 12 electrons per unit cell: 4 are occupying the low lying s and p bands of the sp element and 8 the Co-like minority d bands ($2 \times e_g$, $3 \times t_{2g}$ and $3 \times t_{1u}$), as explained above (see Fig. 9). Since 7 minority bands are unoccupied, the largest possible moment is 7 μ_B and occurs when all majority d-states are occupied.

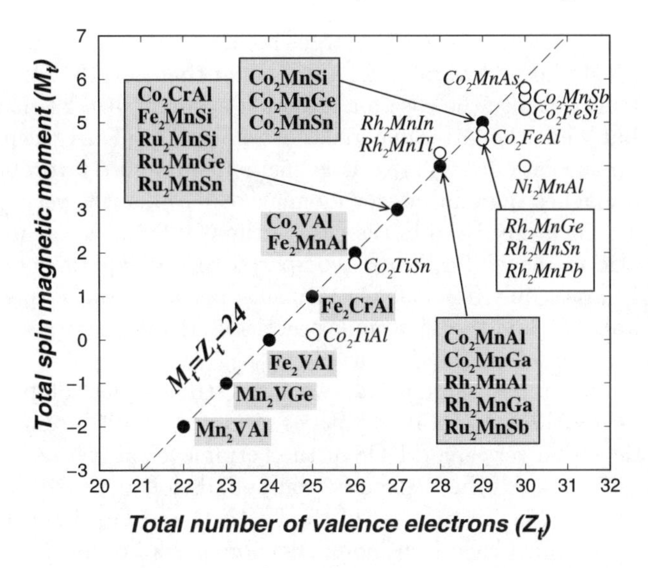

Fig. 11. Calculated total spin moments for all the studied full Heusler alloys. The dashed line represents the Slater-Pauling behavior. With open circles we present the compounds deviating from the SP curve. Some experimental values for bulk systems near the SP curve from reference [5]: Co_2MnAl 4.01 μ_B, Co_2MnSi 5.07 μ_B, Co_2MnGa 4.05 μ_B, Co_2MnGe 5.11 μ_B, Co_2MnSn 5.08 μ_B, Co_2FeSi 5.9 μ_B, Mn_2VAl -1.82 μ_B and finally Fe_2VAl non-magnetic

Overall we see that many of our results coincide with the Slater-Pauling curve. Some of the Rh compounds show small deviations which are more serious for the Co_2TiAl compound. We see that there is no compound with a total spin moment of 7 μ_B or even 6 μ_B. Moreover we found also examples of half-metallic materials with less than 24 electrons, Mn_2VGe with 23 valence electrons and Mn_2VAl with 22 valence electrons. Firstly, we have calculated the spin moments of the compounds Co_2YAl where Y = Ti, V, Cr, Mn and Fe. The compounds containing V, Cr and Mn show a similar behavior. As we substitute Cr for Mn, which has one valence electron less than Mn, we depopulate one Mn spin-up state and thus the spin moment of Cr is around 1 μ_B smaller than the Mn one while the Co moments are practically the same for both compounds. Substituting V for Cr has a larger effect since also the Co spin-up DOS changes slightly and the Co magnetic moment is increased by about 0.1 μ_B compared to the other two compounds and V possesses a small moment of 0.2 μ_B. This change in the behavior is due to the smaller hybridization between the Co atoms and the V ones as compared to the Cr and Mn atoms. Although all three Co_2VAl, Co_2CrAl and Co_2MnAl compounds are on the SP curve as can be seen in Fig. 11, this is not the case for the compounds containing Fe and Ti. If the substitution of Fe for Mn followed the same logic as the one of Cr for Mn then the Fe moment should be around 3.5 μ_B which is a very large moment for the Fe site. Therefore it is energetically more favorable for the system that also the Co moment is increased, as it was also the case for the other systems with 29 electrons like Co_2MnSi, but while the latter one makes it to 5 μ_B, Co_2FeAl reaches a value of 4.9 μ_B. In the case of Co_2TiAl, it is energetically more favorable to have a weak ferromagnet than an integer moment of 1 μ_B as it is very difficult to magnetize the Ti atom. Even in the case of the Co_2TiSn the calculated total spin magnetic moment of 1.78 μ_B (compared to the experimental value of 1.96 μ_B [84]) arises only from the Co atoms as was also shown experimentally by Pendl et al. [85], and the Ti atom is practically nonmagnetic and the latter compound fails to follow the SP curve.

As a second family of materials we have studied the compounds containing Fe. Fe_2VAl has in total 24 valence electrons and is a semi-metal, i.e. nonmagnetic with a very small DOS at the Fermi level, as it is already known experimentally [86]. All the studied Fe compounds follow the SP behavior as can be seen in Fig. 11. In the case of the Fe_2CrAl and Fe_2MnAl compounds the Cr and Mn atoms have spin moments comparable to the Co compounds and similar DOS. In order to follow the SP curve the Fe in Fe_2CrAl is practically nonmagnetic while in Fe_2MnAl it has a small negative moment. When we substitute Si for Al in Fe_2MnAl, the extra electron exclusively populates Fe spin-up states and the spin moment of each Fe atom is increased by 0.5 μ_B contrary to the corresponding Co compounds where also the Mn spin moment was considerably increased. Finally we calculated as a test Mn_2VAl and Mn_2VGe that have 22 and 23 valence electrons, respectively, to see if we can reproduce the SP behavior not only for compounds with more than 24, but also for compounds with less than 24 electrons. As we have already shown Fe_2VAl is nonmagnetic and Co_2VAl, which has two electrons more, has a spin moment of 2 μ_B. Mn_2VAl has two valence electrons less than Fe_2VAl and its total spin moment is -2 μ_B and thus it follows the SP behavior; negative total spin moment means that the "minority" band with the gap has more occupied states than the "majority" one.

As we have already mentioned the maximal moment of a full-Heusler alloy is seven μ_B, and should occur, when all 15 majority d states are occupied. Analogously for a half-Heusler alloy the maximal moment is 5 μ_B. However this limit is difficult to achieve, since due to the hybridization of the d states with empty *sp*-states of the transition metal atoms (sites X and Y in Fig. 2), d-intensity is transferred into states high above E_F, which are very difficult to occupy. Although in the case of half-Heusler alloys, we could identify systems with a moment of nearly 5 μ_B, the hybridization is much stronger in the full-Heusler alloys so that a total moment of 7 μ_B seems to be impossible. Therefore we restrict our search to possible systems with 6 μ_B, i.e. systems with 30 valence electrons, but as shown also in Fig. 11, none of them makes exactly the 6 μ_B. Co_2MnAs shows the largest spin moment: 5.8 μ_B. The basic reason, why moments of 6 μ_B are so difficult to achieve, is that as a result of

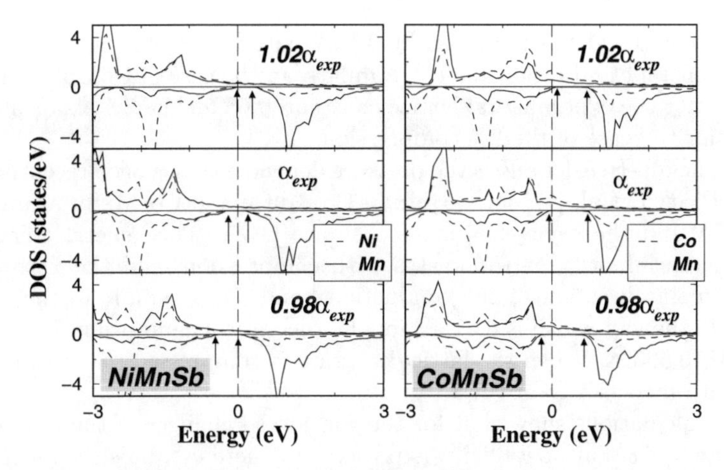

Fig. 12. Atom-resolved DOS for the experimental lattice parameter for NiMnSb and CoMnSb, compared with the once compressed or expanded by 2%. With the small arrow we denote the edges of the minority gap

the strong hybridization with the two Co atoms the Mn atom cannot have a much larger moment than 3 μ_B. While due to the empty e_u-states the two Co atoms have no problem to contribute a total of 2 μ_B, the Mn moment is hybridization limited.

4 Effect of the Lattice Parameter

In this section we will study the influence of the lattice parameter on the electronic and magnetic properties of the $C1_b$ and $L2_1$ Heusler alloys. To the best of our knowledge no relevant experimental study exists. For this reason we plot in Fig. 12 the DOS of NiMnSb and CoMnSb for the experimental lattice parameter and the ones compressed and expanded by 2%. First one sees, that upon compression the Fermi level moves in the direction of the conduction band, upon expansion towards the valence band. In both cases, however, the half-metallic character is conserved. To explain this behavior, we first note, that the Fermi level is determined by the metallic DOS in the majority band. As we believe, the shift of E_F is determined from the behavior of the Sb p-states, in particular by the large extension of these states as compared to the d states. Upon compression the p-states are squeezed and hybridize stronger, thus pushing the d-states and the Fermi level to higher energies, i.e. towards the minority conduction band. In addition the Mn d and Ni or Co d states hybridize stronger, which increases the size of the gap. Upon expansion the opposite effects are observed. In the case of NiMnSb and for the experimental lattice constant the gap-width is ~0.4 eV. When the lattice is expanded by 2% the gap shrinks to 0.25 eV and when compressed by 2% the gap-width is increased by 0.1 eV with respect to the experimental lattice constant and is now 0.5 eV. Similarly in the case of CoMnSb, the gap is 0.8 eV for the experimental lattice constant, 0.65 for the 2% expansion and 0.9 eV for the case of the 2% compression.

For the full-Heusler alloys the pressure dependence has been recently studied by Picozzi et al. [87] for Co_2MnSi, Co_2MnGe and Co_2MnSn, using both the LDA and the somewhat more accurate GGA. The general trends are similar: the minority gap increases with compression, and the Fermi level moves in the direction of the conduction band. For example in the case of Co_2MnSi the gap-width is 0.81 eV for the theoretical equilibrium lattice constant of 10.565 Å. When the lattice constant is compressed to ~10.15 Å, the gap-width increases to about 1 eV.

The calculations show that for the considered changes of the lattice constants of ± 2%, half-metallicity is preserved. There can be sizeable changes of the local moments, but the total moment remains constant, since E_F stays in the gap.

5 Quaternary Heusler Alloys

We proceed our study by examining the behavior of the so-called quaternary Heusler alloys [13, 90]. In the latter compounds, one of the four sites is occupied by two different kinds of neighboring elements like $Co_2[Cr_{1-x}Mn_x]Al$ where the Y site is occupied by Cr or Mn atoms To perform this study we used the KKR method within the coherent potential approximation (CPA) as implemented by H. Akai [24], which has been already used with success to study the magnetic semiconductors [24]. For all calculations we assumed that the lattice constant varies linearly with the concentration x which has been verified for several quaternary alloys [5,6]. To our knowledge from the systems under study only $Co_2Cr_{0.6}Fe_{0.4}Al$ has been studied experimentally [88, 89].

We calculated the total spin moment for several quaternary alloys taking into account several possible combinations of chemical elements and assuming in all cases a concentration increment of 0.1. We resume our results in Fig. 13. The first possible case is when we have two different low-valent transition metal atoms at the Y site like $Co_2[Cr_{1-x}Mn_x]Al$. The total spin moment varies linearly between the 3 μ_B of Co_2CrAl and the 4 μ_B of Co_2MnAl. In the case of the $Co_2[Cr_{1-x}Fe_x]Al$ and $Co_2[Mn_{1-x}Fe_x]Al$ compounds and up to around $x = 0.6$ the total spin moment shows the SP behavior but for larger concentrations it slightly deviates to account for the non-integer moment value of Co_2FeAl (see Fig. 12). This behavior is clearly seen in Fig. 13 when we compare the lines for the $Co_2[Mn_{1-x}Fe_x]Al$ and $Co_2Mn[Al_{1-x}Sn_x]$ compounds; the latter family follows the SP behavior. The second case is when one mixes the *sp* elements, but as we just mentioned these compounds also obey the rule for the total spin moments. The third and final case is to mix the higher valent transition metal atoms like in $[Fe_{1-x}Co_x]_2MnAl$

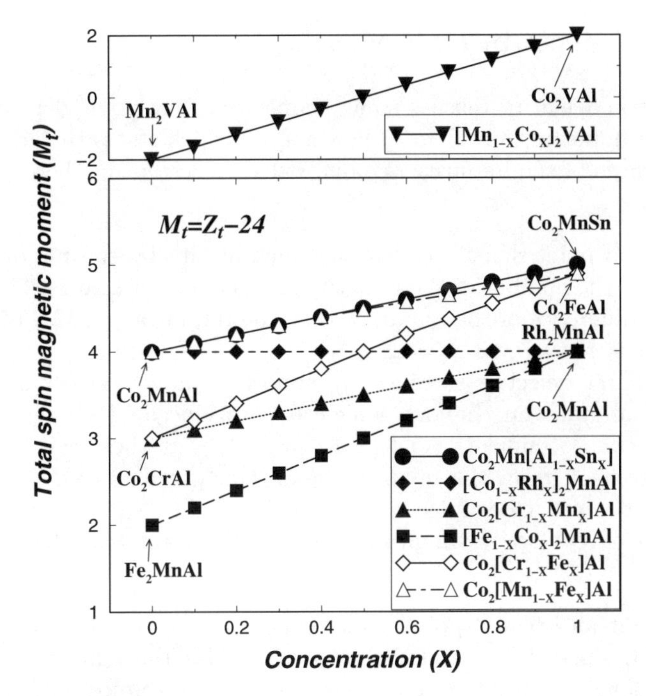

Fig. 13. Calculated total spin moment M_t in μ_B for a variety of compounds as a function of the concentration x ($x = 0, 0.1, 0.2, ..., 0.9, 1$). We assumed that the lattice constant varies linearly with the concentration x. With solid lines the cases obeying the rule $M_t = Z_t - 24$ are shown where Z_t and M_t are the average total number of valence electrons and the average total moment

and $[Rh_{1-x}Co_x]_2MnAl$ alloys. In the first case the total spin moment varies linearly between the 2 and 4 μ_B of Fe_2MnAl and Co_2MnAl compounds, respectively. Rh is isoelectronic to Co and for the second family of compounds we find a constant integer value of 4 μ_B for all the concentrations. A special case is Mn_2VAl which has less than 24 electrons and the total spin moment is -2 μ_B. If now we mix Mn and Co, we get a family of compounds where the total spin moment varies linearly between the -2 μ_B and the 2 μ_B and for $x = 0.5$ we get the case of a paramagnetic compound consisting of magnetic elements. Thus all the compounds obey the rule $M_t = Z_t - 24$, showing the Slater-Pauling behavior regardless of the origin of the extra charge.

As a rule of thumb we expect, that for two half-metallic alloys like XYZ and $X'YZ$ (or $XY'Z, XYZ'$), which both lay on the Slater-Pauling curve, also the mixtures like $X_{1-x}X'_xYZ$ lay on the Slater Pauling curve, with an average moment of $< M_t >= (1-x)M_t^{XYZ} + xM_t^{X'YZ}$. However, if these intermediate structures are stable, is not guaranteed in particular if the parent compounds are not neighbors on the Slater-Pauling curve.

6 Point Defects in Half-Metals

First we would like to discuss some simple rules for point defects in half-metals. An important problem is, how and by which states the additional or missing charge of the impurity or point defect is screened. There are several mechanisms:

1. Either the point defects is screened metallically by the majority states such that the number of the minority states does not change. If ΔZ is the valence difference of the impurity, then the total change ΔM_t of the alloy moment is given by $\Delta M_t = \Delta Z_t$.
2. or the point defect is screened by minority states. These can either be additional states in the gap, which are introduced by the impurity and which, when occupied, lead to $\Delta M_t = -\Delta Z_t$, or these can be localized states split-off from the minority band which lead to $\Delta M_t = +\Delta Z_t$, if these states are unoccupied.
3. or both effects occur simultaneously, which is not expected for simple defects.

Note that an isolated point defect cannot change the band gap nor the Fermi level, since these are bulk properties. Also the number of minority states cannot be changed, except when the defect introduces additional resonances in the minority band, or takes out weight from this band by splitting-off states into the gap. As a result, in the dilute limit the band gap and half-metallicity is preserved, but localized states in the gap, either occupied or empty ones, can occur. An exception occurs if a multifold degenerate gap-state is partially occupied and thus fixed at the Fermi level. Then a symmetry lowering Jahn-Teller splitting of the level is expected to occur.

For finite concentrations, impurity states in the gap overlap and fast broaden to form impurity bands. If the impurities are randomly distributed, one can show by applying the coherent potential approximation (CPA) [24], that the band width scales as $\sqrt{c}$, where c is the impurity concentration. For instance, this means, that the bandwidth is for a concentration of 1% only a factor 3 smaller than for 10%. Therefore the impurity bands broaden very fast with concentration and can soon fill up the band gap, in particular, if the band gap is small and the impurity states are rather extended. Therefore the control of defects and disorder is an important problem for the application of Heusler alloys in spin electronics.

Unfortunately there are very few theoretical investigations for defects in half-metallic Heusler alloys. Picozzi and coworkers have recently studies antisite defects in Co_2MnGe and Co_2MnSi [91]. For details we refer to their review in this volume. Here we address only two screening aspects. A Mn-antisite atom on the Co position represents an impurity with $\Delta Z_t = -2$. It can either be screened by minority states by pulling two states out-off the minority valence band in the energy region above E_F. However, this is very difficult, since the minority states are basically Co states with small Mn admixture. Therefore the Mn antisite is screened by shifting 2 Mn-states out-off the occupied majority band above the Fermi level, thus decreasing the total moment, such that $\Delta M_t = -2$. A Co antisite atom on a Mn position is another interesting case with $\Delta Z_t = +2$. Since the majority band is nearly filled and a further filling would lead to an increase of the moment, which a Co atom cannot sustain, the additional charge is provided by the minority states by pulling a doubly degenerate e_u-state from the empty Co e_u-band in the energy region slightly below E_F, thus decreasing the moment by $-2\mu_B$.

Orgassa and coworkers [45] have investigated the effect of disorder on the electronic structure of NiMnSb. In particular they discuss the effect of impurity gap states on the minority DOS and how the impurity bands broaden and fill the gap for higher concentrations of antisites. While the Mn antisite on Ni position is screened by majority states, not leading to a state in the gap, the Ni antisite on the Mn position introduces a (three-fold degenerate) d-level below the Fermi level, which broadens into an impurity band with increasing concentration. Already for 1% of antisite pairs this band comes close to E_F and at 5% half-metallicity disappears and the spin polarization at E_F decreases from 100% to 52%. Thus the general impression is that already a disorder concentration of 1% is dangerous for the gap and that point defects represent a serious problem for half-metallicity. More theoretical studies are needed.

7 Effect of Spin-orbit Coupling

As discussed above in a real crystal defects will destroy the half-metallic band gap since they destroy the perfect crystal periodicity and thus the covalent hybridization leading to the gap. Moreover at finite temperature thermally activated spin-flip scattering, e.g. spin waves, will also induce states within the gap [92, 93]. But even in an ideally prepared single crystal at zero temperature, the spin-orbit coupling will introduce states in the half-metallic gap of the minority states (for the spin-down electrons), which are produced by spin-flip scattering of the majority states (with spin-up direction). The KKR calculations presented up to this point were obtained using the scalar-relativistic approximation, thus all-relativistic effects have been taken into account with the exception of the spin-orbit coupling. To study the latter effect the KKR method has been extended to a fully relativistic treatment by solving the Dirac equation for the cell-centered potentials [72]. Thus the spin-orbit coupling, which is a relativistic effect, is automatically taken into account. For more details we refer to reference [94].

Although in our method the Dirac equation is solved, it is easier to understand the spin-orbit effect within perturbation theory using the Schrödinger equation. In this framework, we remind that the spin-orbit coupling of the two spin channels is related to the unperturbed potential $V(r)$ around each atom via the angular momentum operator $\boldsymbol{L}$ and the Pauli spin matrix $\boldsymbol{\sigma}$:

$$V_{\mathrm{so}}(r) = \frac{1}{2m^2c^2}\frac{\hbar}{2}\frac{1}{r}\frac{dV}{dr}\,\boldsymbol{L}\cdot\boldsymbol{\sigma} = \begin{pmatrix} V_{\mathrm{so}}^{\uparrow\uparrow} & V_{\mathrm{so}}^{\uparrow\downarrow} \\ V_{\mathrm{so}}^{\downarrow\uparrow} & V_{\mathrm{so}}^{\downarrow\downarrow} \end{pmatrix} \tag{3}$$

The 2×2 matrix form is understood in spinor basis. With $\uparrow$ and $\downarrow$, the two spin directions are denoted. The unperturbed crystal Hamiltonian eigenvalues for the two spin directions are $E_{n\boldsymbol{k}}^{0\uparrow}$ and $E_{n\boldsymbol{k}}^{0\downarrow}$, and the unperturbed Bloch eigenfunctions as $\Psi_{n\boldsymbol{k}}^{0\uparrow}$ and $\Psi_{n\boldsymbol{k}}^{0\downarrow}$. Then, noting that within the energy range of the spin-down gap there exist no unperturbed solutions $\Psi_{n\boldsymbol{k}}^{0\downarrow}$ and $E_{n\boldsymbol{k}}^{0\downarrow}$, the first order solution of the Schrödinger equation for the perturbed wavefunction $\Psi_{n\boldsymbol{k}}^{\downarrow}$ reads for states in the gap:

$$\Psi_{n\boldsymbol{k}}^{(1)\downarrow}(\boldsymbol{r}) = \sum_{n'} \frac{\langle \Psi_{n'\boldsymbol{k}}^{0\downarrow} | V_{\mathrm{so}}^{\downarrow\uparrow} | \Psi_{n\boldsymbol{k}}^{0\uparrow} \rangle}{E_{n\boldsymbol{k}}^{0\uparrow} - E_{n'\boldsymbol{k}}^{0\downarrow}} \Psi_{n'\boldsymbol{k}}^{0\downarrow}(\boldsymbol{r}). \tag{4}$$

Here, the summation runs only over the band index n' and not over the Bloch vectors $\boldsymbol{k}'$, because Bloch functions with $\boldsymbol{k}' \neq \boldsymbol{k}$ are mutually orthogonal. Close to the crossing point $E_{n\boldsymbol{k}}^{0\uparrow} = E_{n'\boldsymbol{k}}^{0\downarrow}$ the denominator becomes small and the bands strongly couple. Then one should also consider higher orders in the perturbation expansion. Since at the gap edges there exist spin-down bands of the unperturbed Hamiltonian, this effect can become important near the gap edges. Apart from that, the important result is that in the gap region the spin-down spectral intensity is a weak image of the spin-up one, determined by the majority states $E_{n\boldsymbol{k}}^{0\uparrow}$ in the energy region of the gap. Since the spin-down DOS is related to $|\Psi_{n\boldsymbol{k}}^{(1)\downarrow}|^2$, it is expected that within the gap the DOS has a quadratic dependence on the spin-orbit coupling strength: $n_{\downarrow}(E) \sim (V_{\mathrm{so}}^{\downarrow\uparrow})^2$.

In Table 3 we present the results for several Heusler alloys. In addition to NiMnSb, the cases of FeMnSb, CoMnSb, PdMnSb, and PtMnSb have been studied. Although the last two are not half-metallic (a spin-down gap exists, but E_F enters slightly into the valence band), it is instructive to examine the DOS in the gap region and see how the spinpolarization decreases as one changes to heavier elements (Ni→Pd→Pt). Here the spinpolarization is $P(E)$ is defined by the ratio

Table 3. Calculated differential spinpolarization at the Fermi level [$P(E_F)$] and in the middle of the spin-down gap [$P(E_M)$], for various Heusler alloys. The alloys PdMnSb and PtMnSb present a spin-down gap, but are not half-metallic, as E_F is slightly below the gap

Compound	FeMnSb	CoMnSb	NiMnSb	PdMnSb	PtMnSb
$P(E_F)$	99.3%	99.0%	99.3%	40.0%	66.5%
$P(E_M)$	99.4%	99.5%	99.3%	98.5%	94.5%

$$P(E) = \frac{n^{\uparrow}(E) - n^{\downarrow}(E)}{n^{\uparrow}(E) + n^{\downarrow}(E)} \tag{5}$$

For the Heusler alloys CoMnSb, FeMnSb, NiMnSb, PdMnSb, and PtMnSb, two quantities are shown: the differential spinpolarization at E_F, $P(E_F)$, and in the middle of the gap, $P(E_M)$. The last quantity reflects the strength of the spin-orbit induced spin flip scattering, while the first is relevant to our considerations only when E_F is well within the gap (which is not the case for PdMnSb and PtMnSb). Clearly, the compounds including 3d transition elements (NiMnSb, CoMnSb, and FeMnSb) show high spin polarization with small variation as we move along the 3d row of the periodic table from Ni to Fe. In contrast, when we substitute Ni with its isoelectronic Pd the number of the induced minority states at the middle of the gap increases and $P(E_M)$ drops drastically, and much more when we change Pd for the isoelectronic Pt. This trend is expected, since it is known that heavier elements are characterized by stronger spin-orbit coupling. But in all cases the alloys retain a region of very high spin-polarization, instead of a real gap present in the scalar relativistic calculations, and thus this phenomenon will not be important for realistic applications. Defects, thermally activated spin-flip scattering, surface or interfaces states will have a more important effect on the spin-polarization.

Since we can include the spin-orbit coupling to the calculations, we are also able to determine the orbital moments of the alloys (see reference [95] for more details). The minority valence bands are completely occupied while for the majority ones the density of states at the Fermi level is usually very small since most of the *d* states are occupied. Thus the orbital moments are almost completely quenched and their absolute values are negligible with respect to the spin magnetic moments. Also Picozzi and collaborators have studied the orbital magnetism of Co_2Mn-Si, -Ge and -Sn compounds and have reached to similar conclusions [87]. This also means that the magnetic anisotropy of the half-metallic Heusler alloys is expected to be rather small.

8 Surface Properties

Surfaces can change the bulk properties severely, since the coordination of the surface atoms is strongly reduced. Based on the experiences from ferromagnets and semiconductors, two effects should be particularly relevant for the surfaces of half-metals: (i) for ferromagnets the moments of the surface atoms are strongly enhanced due to the missing hybridization with the cut-off neighbors, and (ii) for semiconductors surface states appear in the gap, such that the surface often becomes metallic. Also this is a consequence of the reduced hybridization, leading to dangling bond states in the gap.

In this section we review some recent calculations for the (001) surfaces of half- and full-Heusler alloys. We will not touch the more complex problem of the interfaces with semiconductors, since this is treated in the review chapter of Picozzi et al. in this book. As a model we firstly consider the (001)-surface of NiMnSb. Jenkins and King have investigated the MnSb-terminated

NiMnSb(001) surface in detail and have shown that the surface relaxations are very small; Sb atoms relax outwards while Mn atoms relax inwards with a total buckling of 0.06Å [49]. Therefore we neglect the relaxations and assume an "ideal" epitaxy. The second possible termination, with a Ni-atom and an unoccupied site at the surface, is not considered since the configuration is likely to be unstable or to show large relaxations of the Ni-surface atoms.

As mentioned above, in the case of the MnSb-terminated (001) NiMnSb surface, Jenkins and King have shown that there are two surface states [49]. The lower lying state (at 0.20 eV above the $\bar{\Gamma}$ point) is due to the interaction between e_g-like dangling bond states located at the Mn atoms. The second surface state, which is also higher in energy (0.44 eV above the $\bar{\Gamma}$ point) rises from the hybridization between t_{2g}-like orbitals of Mn with p-type orbitals of Sb. The first surface state disperses downwards along the [110] direction while the second surface state disperses upwards along the same direction. Their behavior is inversed along the $[1\bar{1}0]$ direction. The two surface states cross along the $[1\bar{1}0]$ direction bridging the minority gap between the valence and the conduction band. Along the other directions anticrossing occurs leading to band-gaps. Of interest are also the saddle-like structures around the zone center which manifest as van Hove singularities in the DOS.

In Fig. 14, we present the atom- and spin-projected density of states (DOS) for the Mn and Sb atoms in the surface layer and the Ni and vacant site in the subsurface layer for the MnSb terminated NiMnSb(001) surface. We compare the surface DOS with the bulk calculations (dashed line). With the exception of the gap region, the surface DOS is very similar to the bulk case. The Ni atom in the subsurface layer presents practically a half-metallic behavior with an almost zero spin-down DOS, while for the bulk there is an absolute gap. The spin-down band of the vacant site also presents a very small DOS around the Fermi level. The Mn and Sb atoms in the surface layer show more pronounced differences with respect to the bulk, and within the gap there is a very small Mn-d and Sb-p DOS. These intensities are due to the two surface states found by Jenkins and King [49]. These two surface states are strongly localized at the surface layer as at the subsurface layer there is practically no states inside the gap. Our results are in agreement with the experiments of Ristoiu et al. [41] who in the case of a MnSb well ordered (001) surface measured a high spin-polarization. Finally we should mention that the spin moment of the Mn atom at the surface is increased by 0.3 μ_B with respect to the bulk NiMnSb and reaches the $\sim$4 μ_B due to the missing hybridization with the cut-off Ni neighbors.

It is also interesting to examine the spin-polarization at the Fermi level. In Table 4 we have gathered the number of spin-up and spin-down states at the Fermi level for each atom at the surface and the subsurface layer for the MnSb-terminated surfaces for different compounds. We calculated the spin-polarization as the ratio between the number of spin-up states minus the number of spin-down states over the total DOS at the Fermi level. P_1 corresponds to the spin-polarization when we take into account only the surface

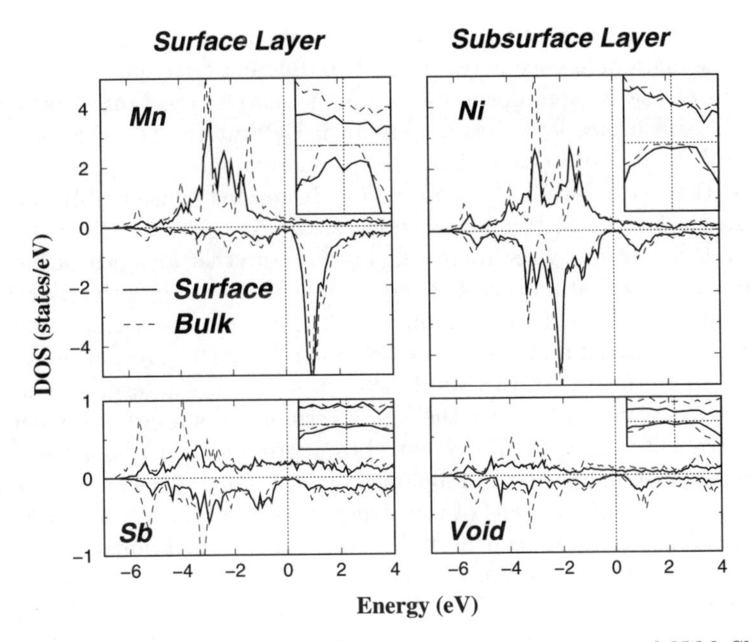

Fig. 14. Spin- and atom-projected DOS for the MnSb-terminated NiMnSb(001) surface. In the insets we have blown up the region around the gap (between −0.5 and 0.5 eV). The dashed lines give the local DOS of the atoms in the bulk

Table 4. Atomic-resolved spin-up and spin-down local DOS at the Fermi level in states/eV units. Polarization ratios at the Fermi level are calculated taking into account only the surface layer P_1, and the sum of the surface and subsurface layers P_2

MnSb (MnGe or CrAl) -termination						
	Surface Layer		Subsurface Layer			
	Mn (↑ / ↓)	Sb (↑ / ↓)	Ni[Co,Pt] (↑ / ↓)	Void (↑ / ↓)	P_1 ($\frac{\uparrow-\downarrow}{\uparrow+\downarrow}$)	P_2 ($\frac{\uparrow-\downarrow}{\uparrow+\downarrow}$)
NiMnSb	0.16/0.19	0.17/0.03	0.28/0.05	0.05/0.02	26%	38%
CoMnSb	0.23/0.27	0.16/0.07	0.91/0.15	0.07/0.02	6%	46%
PtMnSb	0.21/0.24	0.31/0.06	0.38/0.04	0.08/0.02	26%	46%

layer and P_2 if we sum the DOS of the surface and subsurface layers. P_2 simulates reasonably well the experimental situation as the spin-polarization in the case of films is usually measured by inverse photoemission which probes only the surface of the sample [96]. In all cases the inclusion of the subsurface layer increased the spin-polarization since naturally the second layer is expected to be more bulk-like. In the case of the Ni terminated surface, the spin-up DOS at the Fermi level is equal to the spin-down DOS and the net local polarization P_2 is zero. In the case of the MnSb terminated surface the spin-polarization increases and now P_2 reaches a value of 38%, which means that the spin-up DOS at the Fermi level is about two times the spin-down DOS. The main difference between the two different terminations is the contribution of the Ni spin-down states. In the case of the MnSb surface the Ni in the subsurface layer has a spin-down DOS at the Fermi level of 0.05 states/eV, while in the case of the Ni-terminated surface the Ni spin-down DOS at the Fermi level is 0.40 states/eV decreasing considerably the spin-polarization for the Ni terminated surface; the Ni spin-up DOS is the same for both terminations. It is interesting also to note that for both surfaces the net Mn spin-polarization is close to zero while Sb atoms in both cases show a large spin-polarization and the number of the Sb spin-up states is similar to the number of Mn spin-up states, so that Sb and not Mn is responsible for the large spin-polarization of the MnSb layer in both surface terminations. The calculated P_2 value of 38% for the MnSb terminated surface is smaller than the experimental value of 67% obtained by Ristoiu and collaborators [41] for a thin-film terminated in a MnSb stoichiometric alloy surface layer. But experimentally no exact details of the structure of the film are known and the comparison between experiment and theory is not straightforward.

In the case of the full-Heusler alloys containing Mn the results are similar to the ones obtained for the NiMnSb compound. For the MnGe-terminated (001) surface the induced Mn minority surface states locally completely kill the spin polarization. This is clearly seen in the bottom panel of Fig. 15 where

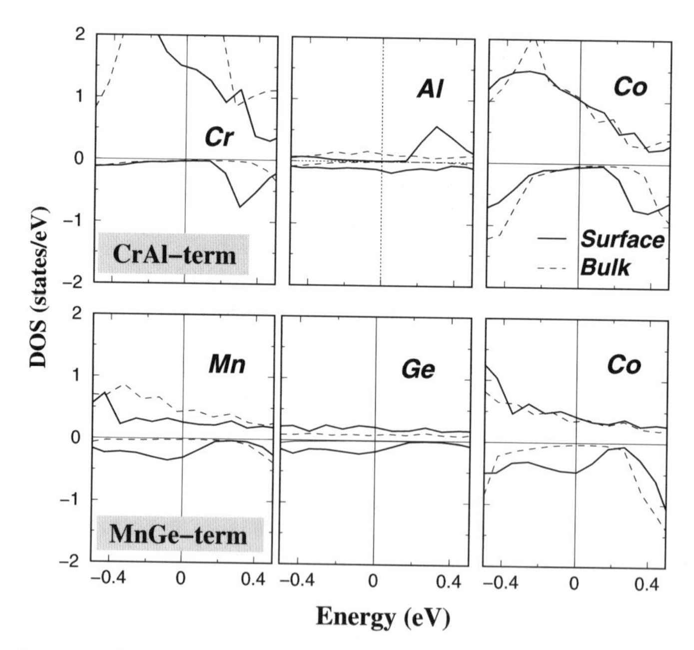

Fig. 15. Atom- and spin-projected DOS for the MnGe and CrAl terminated (001) surfaces of Co_2MnGe and Co_2CrAl, respectively. With dashed line: the bulk results

we have plotted the DOS around the Fermi level for the Mn and Ge atoms at the surface layer and the Co atoms at the subsurface layer. Note that also for Co at the subsurface the local net spin polarization is zero. But in the case of Co_2CrAl results differ considerably. In line with the reduction of the total valence electrons by 2, the Cr moment is rather small (1.54 μ_B) yielding a total moment of only 3 μ_B instead of 5 μ_B for Co_2MnGe. The Co terminated $Co_2CrAl(001)$ surface shows a similar behavior as the corresponding surface of Co_2MnGe, being in both cases dominated by a strong Co peak in the gap region of the minority band. However the CrAl terminated Co_2CrAl surface behaves very differently, being driven by the large surface enhancement of the Cr moment from 1.54 μ_B to 3.12 μ_B. As a consequence the splitting of the Cr peaks in the majority and minority bands is even enlarged and in particular in the minority band the pseudogap is preserved. Thus this surface is a rare case, since for all the other surfaces studied in this paper, the half-metallicity is destroyed by surface states. If we look closer at the gap region (see Fig. 15) for the CrAl surface we find that the Al DOS still has some weight in the gap region. Thus compared to NiMnSb there is still one surface state left, which has only an Al *p*-component, but no *d*-admixture from the Cr atom. However in total the surface keeps a high degree of differential spinpolarization: i.e. P_2 is 84%.

The increase of the spin magnetic moment of the surface atoms due to their lower coordination with respect to the bulk can be better understood in the case of the transition-metal pnictides and chalcogenides which crystallize in the zinc-blende structure. Using CrAs and CrSe as examples, Galanakis and Mavropoulos have shown that the Cr-terminated (001) surfaces of these alloys retain the half-metallicity of the bulk compounds [18]. Contrary to the CrAl-terminated $Co_2CrAl(001)$ there is no *sp* atom at the surface layer now to induce a surface state. Thus the situation is simpler to understand as compared to the Heusler alloys. In the bulk case, Cr has four As or Se atoms as first neighbors. Thus the bonding can be described in terms of four directional bonds around each Cr or As(Se) atom, similar to the case of binary semiconductors where sp^3 hybrids are being formed. Each Cr atom provides 0.75 e^- per Cr-As bond and 0.5 e^- per Cr-Se bond. The Cr atom at the surface looses two out of its four As or Se first neighbors and thus regains 1.5 electrons in the case of CrAs(001) and 1 electron in the case of CrSe(001). The extra electrons now fill up only spin up states and thus the total spin moment at the surface (adding the spin moments of the Cr surface atom and of the As or Se at the subsurface layer) is enhanced by exactly 1.5 μ_B and 1 μ_B with respect to the bulk for CrAs and CrSe, respectively [18]. This rule for the total spin moment can be also generalized to the case of the interfaces with semiconductors with the condition that half-metallicity is preserved [18].

9 Summary and Outlook

In this review we have given an introduction into the electronic structure and the resulting magnetic properties of half-metallic Heusler alloys, which represent interesting hybrids between metallic ferromagnets and semiconductors. Many unusual features arise from the half-metallicity induced by the gap in the minority band, and therefore the understanding of the gap is of central importance.

For the half-Heusler alloys like NiMnSb, crystallizing in the $C1_b$ structure, the gap arises from the hybridization between the *d*-wavefunctions of the lower-valent transition metal atom (e.g. Mn) with the *d*-wavefunctions of the higher-valent transition metal atom (e.g. Ni). Thus the *d-d* hybridization leads to 5 occupied bonding bands, which have a larger Ni and smaller Mn admixture. These states form the valence band, being separated by a band gap from the conduction band which is formed by five antibonding hybrids with a large Mn *d*- and a small Ni *d*-admixture. The role of the *sp* atoms like Sb is very different. Firstly they are important for the bonding, in particular for the stabilization of the $C1_b$ structure. Secondly the *sp* atom creates for each spin direction one *s* and three *p* bands in the energy region below the *d* states which by hybridization can accommodate also transition metal electrons, such that e.g. Sb formally acts like a Sb^{3-} and Sn as a Sn^{4-} anion. In this way the effective number of valence *d*-electrons can be changed by the valence of the *sp* elements.

Since the minority valence band consist of 9 bands, compounds with 18 valence electrons like CoTiSb have the same density of states for both spin directions and are semiconductors. More general, compounds with a total number of Z_t valence electrons per unit cell are ferromagnets and have an integer total spin moment of $M_t = Z_t - 18$, since $Z_t - 18$ is the number of uncompensated spins. For instance, NiMnSb has 22 valence electrons and therefore a total moment of exactly 4 μ_B. This relation is similar to the well known Slater-Pauling behavior observed for binary transition-metal alloys and allows to classify the half-metallic $C1_b$ Heusler alloys into classes with integer moments between 0 and 5 μ_B. The maximum moment of 5 μ_B is difficult to achieve, since it requires that all majority *d*-sates are occupied.

In the case of the full-Heusler alloys like Co_2MnGe, there are, in addition to the Co-Mn bonding and antibonding *d*-hybrids, also Co states which cannot hybridize with both the Mn and the Ge atoms and are exclusively localized at the two Co sublattices. Thus in addition to the 5 Co-Mn bonding and 5 Co-Mn antibonding bands, there exist 5 such "non-bonding" bands which are only splitted-up by the weaker Co-Co hybridization into 3 occupied *d* states of t_{1u} symmetry and 2 unoccupied e_u states, which are located just below and just above the Fermi level such that the indirect gap in these materials is smaller than in the half-Heuslers. Due to the additional 3 occupied t_{1u} cobalt bands, the full-Heusler alloys have 12 occupied minority bands instead of 9 in the case of the half-Heusler compounds and their relation for the total spin magnetic moment becomes $M_t = Z_t - 24$. Thus systems like Fe_2VAl with 24 valence electrons are semiconductors, Co_2VAl (26 valence electrons) has a total spin moment of 2 μ_B, Co_2CrAl 3 μ_B, Co_2MnAl 4 μ_B and finally Co_2MnSi which has 29 valence electrons has a total spin moment of 5 μ_B. The maximal total spin moment for these alloys is 7 μ_B, but as has been shown even the 6 μ_B are unlikely to be achieved.

Having understood the basic elements of the electronic structure, there is still a long way to go for understanding the half-metallic behavior of the real materials. Since the existence of the minority gap is central for any application of half-metals in spintronics, it is of great importance to understand and control all mechanisms that can destroy the gap. Firstly we have discussed that spin-orbit interaction couples the two spin-bands and induces states in the gap; however this effect is weak and the spinpolarization remains in most cases as high as ∼99%. While this effect exists already in the ground state, there are, secondly, excitation effects leading to states in the gap. In the simplest approach one can consider in the adiabatic approximation "static spin waves", which are superpositions of spin-up and spin-down states. At higher temperatures spinwave excitations will smear out the gap [93]. These excitations drive the system to the paramagnetic state above the Curie temperature. At low temperatures the interaction of the electrons with magnons leads to non-quasiparticle excitations in the minority gap above the Fermi level [97]. Note that spin wave excitations lead to new states in the gap above and below the Fermi level, whereas at low temperatures the non-quasiparticle states introduce only additional states at and above E_F.

Thirdly and most importantly all kind of defects are expected to lead to states in the gap. We have discussed this qualitatively for the case of point defects arising from substitutional disorder and refer to the contribution of Picozzi et al. in this volume for realistic calculations. We feel that many more calculations and experiments are needed; the aim is to find systems, which either do not lead to states in the gap (which is presumably not possible) or systems with particularly high defect formation energies or sufficiently low annealing temperatures. Equally important is the control of surface and interface states in the gap, the latter are in particular important for interfaces to semiconductors. Here it should be possible to find junctions which do not have interface states at the Fermi level. The case of Co_2CrAl discussed in this paper shows, that the occurrence of the transition-metal induced surface state at the Fermi level can be suppressed by the increase of the Cr spin moment at the surface. Finally we add that from point of view of transport a single interface state does not affect the magnetoconductance since the wavefunction is orthogonal to all bulk states incident to the interface. It is the interaction with other defect states in bulk systems and/or with surface defects which make these states conducting.

References

1. I. Žutić, J. Fabian, and S. Das Sarma: Rev. Mod. Phys. **76**, 323 (2004)
2. S.A. Wolf, D.D. Awschalom, R.A. Buhrman, J.M. Daughton, S. von Molnár, M.L. Roukes, A.Y. Chtchelkanova, and D.M. Treger: Science **294**, 1488 (2001); G.A. Prinz: Science **282**, 1660 (1998); G.A. Prinz: J. Magn. Magn. Mater. **200**, 57 (1999)
3. J. de Boeck, W. van Roy, J. Das, V. Motsnyi, Z. Liu, L. Lagae, H. Boeve, K. Dessein, and G. Borghs: Semicond. Sci. Tech. **17**, 342 (2002); J. de Boeck, W. van Roy, V. Motsnyi, Z. Liu, K. Dessein, and G. Borghs: Thin Solid Films **412**, 3 (2002)
4. F. Heusler: Verh. Dtsch. Phys. Ges. **5**, 219 (1903)
5. P.J. Webster and K.R.A. Ziebeck. In: *Alloys and Compounds of d-Elements with Main Group Elements. Part 2.*, Landolt-Börnstein, New Series, Group III, vol 19c, ed by H.R.J. Wijn, (Springer, Berlin 1988) pp 75–184
6. K.R.A. Ziebeck and K.-U. Neumann. In: *Magnetic Properties of Metals*, Landolt-Börnstein, New Series, Group III, vol 32/c, ed by H.R.J. Wijn, (Springer, Berlin 2001) pp 64–414
7. J. Pierre, R.V. Skolozdra, J. Tobola, S. Kaprzyk, C. Hordequin, M.A. Kouacou, I. Karla, R. Currat, and E. Lelièvre-Berna: J. Alloys Comp. **262-263**, 101 (1997)
8. J. Tobola, J. Pierre, S. Kaprzyk, R.V. Skolozdra, and M.A. Kouacou: J. Phys.: Condens. Matter **10**, 1013 (1998); J. Tobola and J. Pierre: J. Alloys Comp. **296**, 243 (2000); J. Tobola, S. Kaprzyk, and P. Pecheur: Phys. St. Sol. (b) **236**, 531 (2003)
9. K. Watanabe: Trans. Jpn. Inst. Met. **17**, 220 (1976)
10. R.A. de Groot. F.M. Mueller, P.G. van Engen, and K.H.J. Buschow: Phys. Rev. Lett. **50**, 2024 (1983)
11. I. Galanakis, P.H. Dederichs, and N. Papanikolaou: Phys. Rev. B **66**, 134428 (2002)
12. I. Galanakis, P.H. Dederichs, and N. Papanikolaou: Phys. Rev. B **66**, 174429 (2002)
13. I. Galanakis: J. Phys.: Condens. Matter **16**, 3089 (2004)
14. M. Zhang, X. Dai, H. Hu, G. Liu, Y. Cui, Z. Liu, J. Chen, J. Wang, and G. Wu: J. Phys: Condens. Matter **15**, 7891 (2003); M. Zhang, Z. Liu, H. Hu, G. Liu, Y. Cui, G. Wu, E. Brück, F.R. de Boer, and Y. Li : J. Appl. Phys. **95**, 7219 (2004)
15. R.J. Soulen Jr., J.M. Byers, M.S. Osofsky, B. Nadgorny, T. Ambrose, S.F. Cheng, P.R. Broussard, C.T. Tanaka, J. Nowak, J.S. Moodera, A. Barry, and J.M.D. Coey: Science **282**, 85 (1998)
16. H. Kato, T. Okuda, Y. Okimoto, Y. Tomioka, K. Oikawa, T. Kamiyama, and Y. Tokura: Phys. Rev. B **69**, 184412 (2004)
17. T. Shishidou, A.J. Freeman, and R. Asahi: Phys. Rev. B **64**, 180401 (2001)
18. I. Galanakis: Phys. Rev. B **66**, 012406 (2002); I. Galanakis and Ph. Mavropoulos: Phys. Rev. B **67**, 104417 (2003); Ph. Mavropoulos and I. Galanakis: J. Phys.: Condens. Matter **16**, 4261 (2004)
19. S. Sanvito and N.A. Hill: Phys. Rev. B **62**, 15553 (2000); A. Continenza, S. Picozzi, W.T. Geng, and A.J. Freeman: Phys. Rev. B **64**, 085204 (2001); B.G. Liu: Phys. Rev. B **67**, 172411 (2003); B. Sanyal, L. Bergqvist, and O. Eriksson:

Phys. Rev. B **68**, 054417 (2003); W.-H. Xie, B.-G. Liu, and D.G. Pettifor: Phys. Rev. B **68**, 134407 (2003); W.-H. Xie, B.-G. Liu, and D.G. Pettifor: Phys. Rev. Lett. **91**, 037204 (2003); Y.Q. Xu, B.-G. Liu, and D.G. Pettifor: Phys. Rev. B **68**, 184435 (2003); M. Zhang et al.: J. Phys: Condens. Matter **15**, 5017 (2003); C.Y. Fong, M.C. Qian, J.E. Pask, L.H. Yang, and S. Dag: Appl. Phys. Lett. **84**, 239 (2004); J.E. Pask, L.H. Yang, C.Y. Fong, W.E. Pickett, and S. Dag: Phys. Rev. B **67**, 224420 (2003); J.-C. Zheng and J.W. Davenport: Phys. Rev. B **69**, 144415 (2004)
20. H. Akinaga, T. Manago, and M. Shirai: Jpn. J. Appl. Phys. **39**, L1118 (2000); M. Mizuguchi, H. Akinaga, T. Manago, K. Ono, M. Oshima, and M. Shirai: J. Magn. Magn. Mater. **239**, 269 (2002); M. Mizuguchi, H. Akinaga, T. Manago, K. Ono, M. Oshima, M. Shirai, M. Yuri, H.J. Lin, H.H. Hsieh, and C.T. Chen: J. Appl. Phys. **91**, 7917 (2002); M. Mizuguchi, M. K. Ono, M. Oshima, J. Okabayashi, H. Akinaga, T. Manago, and M. Shirai: Surf. Rev. Lett. **9**, 331 (2002); M. Nagao, M. Shirai, and Y. Miura: J. Appl. Phys. **95**, 6518 (2004); K. Ono, J. Okabayashi, M. Mizuguchi, M. Oshima, A. Fujimori, and H. Akinaga: J. Appl. Phys. **91**, 8088 (2002); M. Shirai: Physica E **10**, 143 (2001); M. Shirai: J. Appl. Phys. **93**, 6844 (2003)
21. J.H. Zhao, F. Matsukura, K. Takamura, E. Abe, D. Chiba, and H. Ohno: Appl. Phys. Lett. **79**, 2776 (2001); J.H. Zhao, F. Matsukura, K. Takamura, E. Abe, D. Chiba, Y. Ohno, K. Ohtani, and H. Ohno: Mat. Sci. Semicond. Proc. **6**, 507 (2003)
22. M. Horne, P. Strange, W.M. Temmerman, Z. Szotek, A. Svane, and H. Winter: J. Phys.: Condens. Matter **16**, 5061 (2004)
23. A. Stroppa, S. Picozzi, A. Continenza, and A.J. Freeman: Phys. Rev. B **68**, 155203 (2003)
24. H. Akai: Phys. Rev. Lett. **81**, 3002 (1998)
25. J.-H. Park, E. Vescovo, H.-J. Kim, C. Kwon, R. Ramesh, and T. Venkatesan: Nature **392**, 794 (1998)
26. S. Datta and B. Das: Appl. Phys. Lett. **56**, 665 (1990)
27. K.A. Kilian and R.H. Victora: J. Appl. Phys. **87**, 7064 (2000)
28. C.T. Tanaka, J. Nowak, and J.S. Moodera: J. Appl. Phys. **86**, 6239 (1999)
29. J.A. Caballero, Y.D. Park, J.R. Childress, J. Bass, W.-C. Chiang, A.C. Reilly, W.P. Pratt Jr., and F. Petroff: J. Vac. Sci. Technol. A **16**, 1801 (1998); C. Hordequin, J.P. Nozières, and J. Pierre: J. Magn. Magn. Mater. **183**, 225 (1998)
30. M.M. Kirillova, A.A. Makhnev, E.I. Shreder, V.P. Dyakina, and N.B. Gorina: Phys. Stat. Sol. (b) **187**, 231 (1995)
31. K.E.H.M. Hanssen and P.E. Mijnarends: Phys. Rev. B **34**, 5009 (1986); K.E.H.M. Hanssen, P.E. Mijnarends, L.P.L.M. Rabou, and K.H.J. Buschow: Phys. Rev. B **42**, 1533 (1990)
32. W. van Roy, M. Wojcik, E. Jdryka, S. Nadolski, D. Jalabert, B. Brijs, G. Borghs, and J. De Boeck: Appl. Phys. Lett. **83**, 4214 (2003); W. van Roy, J. de Boeck, B. Brijs, and G. Borghs: Appl. Phys. Lett. **77**, 4190 (2000); J.-P. Schlomka, M. Tolan, and W. Press: Appl. Phys. Lett. **76**, 2005 (2000)
33. P. Bach, C. Rüster, C. Gould, C.R. Becker, G. Schmidt, and L.W. Molenkamp: J. Cryst. Growth **251**, 323 (2003); P. Bach, A.S. Bader, C. Rüster, C. Gould, C.R. Becker, G. Schmidt, L.W. Molenkamp, W. Weigand, C. Kumpf, E. Umbach, R. Urban, G. Woltersdorf, and B. Heinrich: Appl. Phys. Lett. **83**, 521 (2003)

34. J. Giapintzakis, C. Grigorescu, A. Klini, A. Manousaki, V. Zorba, J. Androulakis, Z. Viskadourakis, and C. Fotakis: Appl. Surf. Sci. **197**, 421 (2002); J. Giapintzakis, C. Grigorescu, A. Klini, A. Manousaki, V. Zorba, J. Androulakis, Z. Viskadourakis, and C. Fotakis: Appl. Phys. Lett. **80**, 2716 (2002); C.E.A. Grigorescu, S.A. Manea, M. Mitrea, O. Monnereau, R. Rotonier, L. Tortet, R. Keschawarz, J. Giapintzakis, A. Klini, V. Zorba, J. Androulakis, and C. Fotakis: Appl. Surf. Sci. **212**, 78 (2003); S. Gardelis, J. Androulaki, P. Migiakis, J. Giapintzakis, S.K. Clowes, Y. Bugoslavsky, W.R. Branford, Y. Miyoshi, and L.F. Cohen: J. Appl. Phys. **95**, 8063 (2004)
35. F.B. Mancoff, B.M. Clemens, E.J. Singley, and D.N. Basov: Phys. Rev. B **60**, *R*12 565 (1999)
36. W. Zhu, B. Sinkovic, E. Vescovo, C. Tanaka, and J.S. Moodera: Phys. Rev. B **64**, *R*060403 (2001)
37. G.L. Bona, F. Meier, M. Taborelli, E. Bucher, and P.H. Schmidt: Sol. St. Commun. **56**, 391 (1985)
38. S.K. Clowes, Y. Mioyoshi, Y. Bugoslavsky, W.R. Branford, C. Grigorescu, S.A. Manea, O. Monnereau, and L.F. Cohen: Phys. Rev. B **69**, 214425 (2004)
39. J.A. Caballero, A.C. Reilly, Y. Hao, J. Bass, W.P. Pratt, F. Petroff, and J.R. Childress: J. Magn. Magn. Mat. **198–199**, 55 (1999); R. Kabani, M. Terada, A. Roshko, and J.S. Moodera: J. Appl. Phys. **67**, 4898 (1990)
40. C.T. Tanaka, J. Nowak, and J.S. Moodera: J. Appl. Phys. **81**, 5515 (1997)
41. A.N. Caruso, C.N. Borca, D. Ristoiu, J.P. Nozieres and P.A. Dowben: Surf. Sci. **525**, L109 (2003); D. Ristoiu, J.P. Nozières, C.N. Borca, T. Komesu, H.-K. Jeong, and P.A. Dowben: Europhys. Lett. **49**, 624 (2000); D. Ristoiu, J.P. Nozières, C.N. Borca, B. Borca, and P.A. Dowben: Appl. Phys. Lett. **76**, 2349 (2000); T. Komesu, C.N. Borca, H.-K. Jeong, P.A. Dowben, D. Ristoiu, J.P. Nozières, Sh. Stadler, and Y.U. Idzerda: Phys. Lett. A **273**, 245 (2000)
42. I. Galanakis, S. Ostanin, M. Alouani, H. Dreyssé, and J.M. Wills: Phys. Rev. B **61**, 4093 (2000)
43. E. Kulatov and I.I. Mazin: J. Phys.: Condens. Matter **2**, 343 (1990); S.V. Halilov and E.T. Kulatov: J. Phys.: Condens. Matter **3**, 6363 (1991); X. Wang, V.P. Antropov, and B.N. Harmon: IEEE Trans. Magn. **30**, 4458 (1994). S.J. Youn and B.I. Min: Phys. Rev. B **51**, 10 436 (1995); V.N. Antonov, P.M. Oppeneer, A.N. Yaresko, A.Ya. Perlov, and T. Kraft: Phys. Rev. B **56**, 13012 (1997)
44. P. Larson, S.D. Mahanti, and M.G. Kanatzidis: Phys. Rev. B **62**, 12574 (2000)
45. D. Orgassa, H. Fujiwara, T.C. Schulthess, and W.H. Butler: Phys. Rev. B **60**, 13237 (1999)
46. I. Galanakis: J. Phys.:Condens. Matter: **14**, 6329 (2002)
47. G.A. Wijs and R.A. de Groot: Phys. Rev. B **64**, *R*020402 (2001)
48. A. Debernardi, M. Peressi, and A. Baldereschi: Mat. Sci. Eng. C-Bio S **23**, 743 (2003)
49. S.J. Jenkins and D.A. King: Surf. Sci. **494**, L793 (2001)
50. S.J. Jenkins and D.A. King: Surf. Sci. **501**, L185 (2002)
51. J. Kübler: Phys. Rev. B **67**, 220403 (2003)
52. E. Sasioglou, L. Sandratskii, and P.Bruno: Phys. Rev. B **70**, 024427 (2004)
53. P.J. Webster: J. Phys. Chem. Solids **32**, 1221 (1971); K. R.A. Ziebeck and P.J. Webster: J. Phys. Chem. Solids **35**, 1 (1974)
54. J.C. Suits: Phys. Rev. B **14**, 4131 (1976)
55. J. Kübler, A.R. Williams, and C.B. Sommers: Phys. Rev. B **28**, 1745 (1983)

56. S. Ishida, S. Akazawa, Y. Kubo, and J. Ishida: J. Phys. F: Met. Phys. **12**, 1111 (1982); S. Ishida, S. Fujii, S. Kashiwagi, and S. Asano: J. Phys. Soc. Jpn. **64**, 2152 (1995)
57. S. Fujii, S. Sugimura, S. Ishida, and S. Asano: J. Phys.: Condens. Matter **2**, 8583 (1990)
58. S. Fujii, S. Asano, and S. Ishida: J. Phys. Soc. Jpn. **64**, 185 (1995)
59. P.J. Brown, K.U. Neumann, P.J. Webster, and K.R.A. Ziebeck: J. Phys.: Condens. Matter **12**, 1827 (2000)
60. F.Y. Yang, C.H. Shang, C.L. Chien, T. Ambrose, J.J. Krebs, G.A. Prinz, V.I. Nikitenko, V.S. Gornakov, A.J. Shapiro, and R.D. Shull: Phys. Rev. B **65**, 174410 (2002); T. Ambrose, J.J. Krebs, and G.A. Prinz: Appl. Phys. Lett. **76**, 3280 (2000); T. Ambrose, J.J. Krebs, and G.A. Prinz: J. Appl. Phys. **87**, 5463 (2000); T. Ambrose, J.J. Krebs, and G.A. Prinz: J. Appl. Phys. **89**, 7522 (2001)
61. M.P. Raphael, B. Ravel, M.A. Willard, S.F. Cheng, B.N. Das, R.M. Stroud, K.M. Bussmann, J.H. Claassen, and V.G. Harris: Appl. Phys. Lett. **70**, 4396 (2001); M.P. Raphael, B. Ravel, Q. Huang, M.A. Willard, S.F. Cheng, B.N. Das, R.M. Stroud, K.M. Bussmann, J.H. Claassen, and V.G. Harris: Phys. Rev. B **66**, 104429 (2002); B. Ravel, M.P. Raphael, V.G. Harris, and Q. Huang: Phys. Rev. B **65**, 184431 (2002)
62. L. Ritchie, G. Xiao, Y. Ji, T.Y. Chen, C.L. Chien, M. Chang, C. Chen, Z. Liu, G.Wu, and X.X. Zhang: Phys. Rev. B **68**, 104430 (2003); Y.J. Chen, D. Basiaga, J.R. O'Brien, and D. Heiman: Appl. Phys. Lett. **84**, 4301 (2004)
63. S. Ishida, T. Masaki, S. Fujii, and S. Asano: Physica B **245**, 1 (1998)
64. U. Geiersbach, A. Bergmann, and K. Westerholt: J. Magn. Magn. Mater. **240**, 546 (2002); U. Geiersbach, A. Bergmann, and K. Westerholt: Thin Solid Films **425**, 225 (2003)
65. K. Westerholt, U. Geirsbach, and A. Bergmann: J. Magn. Magn. Mater. **257**, 239 (2003)
66. S. Picozzi, A. Continenza, and A.J. Freeman: J. Appl. Phys. **94**, 4723 (2003); S. Picozzi, A. Continenza, and A.J. Freeman: J. Phys. Chem. Solids **64**, 1697 (2003)
67. S. Kämmerer, A. Thomas, A. Hütten, and G. Reiss: Appl. Phys. Lett. **85**, 79 (2004); J. Schmalhorst, S. Kämmerer, M. Sacher, G. Reiss, A. Hütten, and A. Scholl: Phys. Rev. B **70**, 024426 (2004)
68. K. Inomata, S. Okamura, R. Goto, and N. Tezuka: Jpn. J. Appl. Phys. **42**, L419 (2003); K. Inomata, N. Tezuka, S. Okamura, H. Kurebayashi, and H. Hirohata: J. Appl. Phys. **95**, 7234 (2004)
69. S.H. Vosko, L. Wilk, and N. Nusair: Can. J. Phys. **58**, 1200 (1980)
70. P, Hohenberg and W. Kohn: Phys. Rev. **136**, B864 (1964); W. Kohn and L.J. Sham: Phys. Rev. **140**, A1133 (1965)
71. R. Zeller, P.H. Dederichs, B. Újfalussy, L. Szunyogh, and P. Weinberger: Phys. Rev. B **52**, 8807 (1995)
72. N. Papanikolaou, R. Zeller, and P.H. Dederichs: J. Phys.: Condens. Matter **14**, 2799 (2002)
73. A.R. Williams, J. Kübler, and C.D. Gelatt: Phys. Rev. B **19**, 6094 (1979)
74. M.V. Yablonskikh, V.I. Grebennikov, Yu.M. Yarmoshenko, E.Z. Kurmaev, S.M. Butorin, L.-C. Duda, C. Såthe, T. Kämbre, M. Magnuson, J. Nordgren, S. Plogmann, and M. Neumann: Solid State Commun. **117**, 79 (2001); M.V. Yablonskikh, Yu.M. Yarmoshenko, V.I. Grebennikov, E.Z. Kurmaev, S.M. Butorin,

L.-C. Duda, J. Nordgren, S. Plogmann, and M. Neumann: Phys. Rev. B **63**, 235117 (2001)
75. A. Kimura, S. Suga, T. Shishidou, S. Imada, T. Muro, S.Y. Park, T. Miyahara, T. Kaneko, T. Kanomata: Phys. Rev. B **56**, 6021 (1997)
76. R. Zeller, J. Phys.: Condens. Matter **16**, 6453 (2004)
77. R.A. de Groot, A.M. van der Kraan, and K.H.J. Buschow: J. Magn. Magn. Mater. **61**, 330 (1986)
78. L. Hedin and S. Lundqvist, In: Solid State Physics, vol 23, ed by F. Seitz, D. Turnbull, and H. Ehrenreich, (Academic Press, New York and London 1969) pp 1–181
79. B. R. K. Nanda and I. Dasgupta: J. Phys.: Condens. Matter **15**, 7307 (2003)
80. D. Jung, H.-J. Koo, and M.-H. Whangbo: J. Mol. Struct. (Theochem) **527**, 113 (2000)
81. J. Kübler: Physica B **127**, 257 (1984)
82. D. Brown, M.D. Crapper, K.H. Bedwell, M.T. Butterfield, S.J. Guilfoyle, A.E.R. Malins, and M. Petty: Phys. Rev. B **57**, 1563 (1998)
83. R.A. Dunlap and D.F. Jones: Phys. Rev. B **26**, 6013 (1982); S. Plogmann, T. Schlathölter, J. Braun, M. Neumann, Yu.M. Yarmoshenko, M.V. Yablonskikh, E.I. Shreder, E.Z. Kurmaev, A. Wrona, and A. Ślebarski: Phys. Rev. B **60**, 6428 (1999)
84. P.G. van Engen, K.H.J. Buschow, and M. Erman: J. Magn. Magn. Mater. **30**, 374 (1983)
85. W. Pendl Jr., R.N. Saxena, A.W. Carbonari, J. Mestnik Filho, and J. Schaff: J. Phys.: Condens. Matter **8**, 11317 (1996)
86. Ye Feng, J.Y. Rhee, T.A. Wiener, D.W. Lynch, B.E. Hubbard, A.J. Sievers, D.L. Schlagel, T.A. Lograsson, and L.L. Miller: Phys. Rev. B **63**, 165109 (2001); C.S. Lue, J.H. Ross Jr., K.D.D. Rathnayaka, D.G. Naugle, S.Y. Wu and W.-H. Li: J. Phys.: Condens. Matter **13**, 1585 (2001); Y. Nishino, H. Kato, M. Kato, U. Mizutani: Phys. Rev. B **63**, 233303 (2001); A. Matsushita, T. Naka, Y. Takanao, T. Takeuchi, T. Shishido, and Y. Yamada: Phys. Rev. B **65**, 075204 (2002)
87. S. Picozzi, A. Continenza, and A.J. Freeman: Phys. Rev. B **66**, 094421 (2002)
88. C. Felser, B. Heitkamp, F. Kronast, D. Schmitz, S. Cramm, H.A. Dürr, H.-J. Elmers, G.H. Fecher, S. Wurmehl, T. Block, D. Valdaitsev, S.A. Nepijko, A. Gloskovskii, G. Jakob, G. Schonhense, and W. Eberhardt: J. Phys.:Condens. Matter **15**, 7019 (2003); N. Auth, G. Jakob, T. Block T, and C. Felser: Phys. Rev. B **68**, 024403 (2003); H.J. Elmers, G.H. Fecher, D. Valdaitsev, S.A. Nepijko, A. Gloskovskii, G. Jakob, G. Schonhense, S. Wurmehl, T. Block, C. Felser, P.-C. Hsu, W.-L. Tsai, and S. Cramm: Phys. Rev. B **67**, 104412 (2003); T. Block, C. Felser, G. Jakob, J. Ensling, B. Muhling, P. Gutlich, and R.J. Cava: J. Sol. St. Chem. **176**, 646 (2003)
89. R. Kelekar and B.M. Klemens: J. Appl. Phys. **96**, 540 (2004)
90. Y. Miura, K. Nagao, and M. Shirai: Phys. Rev. B **69**, 144113 (2004); Y. Miura, M. Shirai, and K. Nagao: J. Appl. Phys. **95**, 7225 (2004)
91. S. Picozzi, A. Continenza, and A.J. Freeman: Phys. Rev. B **69**, 094423 (2004)
92. C.N. Borca, T. Komesu, H.-K. Jeong, P.A. Dowben, D. Ristoiu, Ch. Hordequin, J.P. Nozières, J. Pierre, Sh. Stadler, and Y.U. Idzerda: Phys. Rev. B **64**, 052409 (2001)
93. P.A. Dowben and R. Skomski: J. Appl. Phys. **93**, 7948 (2003); P.A. Dowben and R. Skomski: J. Appl. Phys. **95**, 7453 (2004)

94. Ph. Mavropoulos, K. Sato, R. Zeller, P. H. Dederichs, V. Popescu, and H. Ebert: Phys. Rev. B **69**, 054424 (2004); Ph. Mavropoulos, I. Galanakis, V. Popescu, and P. H. Dederichs: J. Phys.: Condens. Matter **16**, S5759 (2004)
95. I. Galanakis: Phys. Rev. B **71** 012413 (2005)
96. C.N. Borca, T. Komesu, and P.A. Dowben: J. Electron. Spectrocs. **122**, 259 (2002)
97. L. Chioncel, M.I. Katsnelson, R.A. de Groot, and A.I. Lichtenstein: Phys. Rev. B **68**, 144425 (2003)

Role of Structural Defects on the Half-Metallic Character of Heusler Alloys and Their Junctions with Ge and GaAs

Silvia Picozzi[1], Alessandra Continenza[1], and Arthur J. Freeman[2]

[1] CASTI-INFM Regional Lab. and Dip. Fisica, Univ. L'Aquila, 67010 Coppito (Aq), Italy
[2] Dept. of Physics and Astronomy, Northwestern University, 60208 Evanston, Il, USA

Abstract. Heusler–alloys, such as Co_2MnGe and Co_2MnSi, predicted from first-principles to be half-metallic, have recently attracted great attention for spin-injection purposes. However, spin polarizations of only 50%–60% were experimentally obtained for Heusler thin films – a decrease attributed to defects in the Mn and Co sublattices. We performed *ab-initio* FLAPW calculations in order to determine the effects of several types of defects (Co and Mn antisites, atomic swaps, etc.) on the electronic and magnetic properties of the bulk Heusler compounds. Our findings, in general agreement with experiments, show that Mn antisites have the lowest formation energy and retain the half-metallic character. On the other hand, Co antisites have a slightly higher formation energy and a dramatic effect on the electronic properties, due to the defect states that locally destroy half-metallicity. Moreover, technologically relevant properties, such as potential discontinuity and half–metallic behavior, are determined for GaAs and Ge interfaces. Our results show that the character of the contact is dramatically affected by the semiconductor side (changing from ohmic-for-holes in Ge to Schottky-like in GaAs) and that interface states appear in both sides of the junctions, so that half-metallicity is locally lost.

1 Introduction

One of the key properties in magneto–electronics is the so–called half–metallicity, i.e. conduction electrons that are 100% spin–polarized – due to a gap at the Fermi level, E_F, in the minority spin channel – and a finite density of states at E_F for the majority spin channel. Co_2MnGe and Co_2MnSi Heusler alloys were predicted from first–principles [1, 2] to be half–metallic, along with a remarkably high Curie temperature (∼900 K). Therefore, in principle, they appear as ideal materials for spintronic applications, such as tunneling or giant magnetoresistance elements. Within this framework, the efficiency of spintronics devices is often crucially determined by a successful though challenging spin-injection, i.e. the transport of spin–polarized electrons and holes from one material to another (the Heusler compound and the semiconductor, respectively). *Ab–initio* calculations were first carried out for NiMnSb/semiconductor interfaces [3, 4], mainly showing the loss of half–metallicity at the interfaces, except in the case of NiMnSb/CdS. Very recently, the Co_2CrAl/InP interfaces [5] were investigated by means of first principles calculations, suggesting mechanisms to achieve a high spin–polarization in proximity to the junction. An important ingredient that largely affects the expected efficiency of the injection process [6] is the potential line–up between the "source" and the "sink" of spin–polarized carriers (i.e. magnetic and non–magnetic materials, respectively). Hence, a first–principles study of the Schottky barrier height (SBH), in terms of dependence on the interface geometry and semiconducting (SC) material, is of great interest. We therefore performed extensive *ab-initio* calculations for [001] ordered GaAs/Co_2MnGe and Ge/Co_2MnGe junctions [7], focusing in particular on the interface–related effects on the technologically relevant electronic and magnetic properties, such as potential discontinuity and half–metallic behavior.

The experimental interest in Co_2MnGe and Co_2MnSi Heusler alloys has been renewed recently. Despite the predicted half–metallicity, spin polarizations of the order of only 50–60% [8–10] have so far been measured, thus hindering their practical use. Ambrose et al. [11] deposited Co_2MnGe films on a GaAs substrate by molecular beam epitaxy; their measurements revealed the perfect crystallinity of the Heusler films up to a 350 Å thickness and a large magnetization along with a small magnetic anisotropy. The magnetic, structural, and transport properties of Co_2MnSi were reported for sputtered thin films and single crystals [10]. Neutron diffraction showed the disorder to be zero for the Mn-Si antisite, but extensive for Co–Mn disorder; in particular, as much as 14% of the Mn sites are occupied by Co and 5–7% of the Co sites are occupied by Mn atoms. Similar resuls were obtained from EXAFS and neutron diffraction for Co_2MnSi large grain or powder samples [12]. Atomic disorder in Co_2MnGe was measured by anomalous X–ray diffraction: about 12.7% of the Mn sublattice atoms were occupied by Co, consistent with the observed deviation from stoichiometry.

From the theoretical point of view, pioneering Korringa-Kohn–Rostoker (KKR) calculations performed for Co_2MnGe and Co_2MnSi bulk compounds predicted half–metallicity for the first time in these compounds [1]. More recently, calculations performed by the same authors for Co_2MnSi and Co_2MnGe surfaces [13], showed that the half–metallic character was strongly dependent on the type (i.e. Co_2MnSi films are generally half–metallic, Co_2MnGe films are not), surface termination (Mn and Si–terminated [001] surfaces are generally half–metallic, Co–terminated ones are not), growth direction and thickness of the film. Within this same framework, density–functional layer KKR simulations within the coherent potential approximation by Orgassa et al. [14] for the half–Heusler compound NiMnSb, showed that atomic disorder (i.e. interchange of Ni and Mn or vacancies) experimentally observed in thin-films leads to minority spin states at E_F, resulting in the loss of the half–metallic character. Since atomic disorder was suggested as a mechanism to reduce spin–polarization, we systematically investigate the effects of several kind of defects (such as swaps and antisites) in both Co_2MnSi and Co_2MnGe hosts, in terms of formation energy and defect–induced electronic and magnetic properties [15]. In particular, according to experiment, the most likely defects are: *i)* Mn antisites where a Co atom is replaced by Mn, *ii)* Co antisites where a Mn atom is replaced by Co, *iii)* Co-Mn swaps, where a Mn-Co nearest–neighbor pair shows exchanged positions compared [15] to the ideal bulk and *iv)* Mn–Si swaps in Co_2MnSi , where a Mn–Si second–nearest–neighbor pair shows exchanged positions compared to the bulk. Now, Mn–Ge swaps in Co_2MnGe were not considered, since experiments did not reveal the occurrence of this kind of disorder and, in addition, our calculations performed for Co_2MnSi suggested that this kind of swap is extremely unfavorable (see below). Recall that for antisites (swaps) the bulk stoichiometry is not preserved (preserved), since, for example, Co antisites in Co_2MnSi lead to a $Co_2Co_xMn_{1-x}Si$ compound.

This chapter is therefore organized as follows: in Sect. 2 we briefly recall the structural and magnetic properties of the bulk Co_2MnX (X = Si, Ge, Sn) compounds (for a thorough discussion of their bulk properties we refer to the introductory chapter by Galanakis and Dederichs). In Sect. 3, we focus on the electronic and magnetic properties of the Co_2MnGe/GaAs and Co_2MnGe/Ge interfaces, in terms of potential line–up and density of states. In Sect. 4, we discuss the formation energy and electronic and magnetic properties of structural defects (such as antisites and atomic swaps). Finally, we draw our conclusions and suggest future perspective works.

As for the computational details, all the simulations were performed within density–functional theory using the all-electron full–potential linearized augmented plane wave (FLAPW) method [16]. Both the local spin density approximation (LSDA) [17] and the generalized gradient approximation (GGA) [18] were used for the exchange–correlation potential. In particular, in the case of interfaces, we performed LSDA calculations; this choice was motivated by the fact that experimental semiconductor lattice constants are generally closer to LSDA values, whereas GGA tends to overestimate them. On the other hand, GGA works better when a high percentage of magnetic ions is present in the unit cell. We therefore chose to perform GGA calculations in the case of defects. Muffin tin radii were chosen to be $R_{MT} = 2.1$ a.u. for all the atoms; angular momenta (wave vector) up to $l_{max} = 8$ ($K_{max} =$ 3.6 a.u.) were used for the expansions of both wave functions and charge density. The Brillouin zone was sampled using special **k** points according to the Monkhorst–Pack scheme [19].

2 Brief Review of Bulk Properties of Co_2MnX (X = Si, Ge, Sn) Compounds

The equilibrium lattice constant and bulk modulus were calculated using both LSDA and GGA for the Heusler compounds in the cubic $L2_1$ structure (space group $Fm3m$); the results are compared with available experimental data [29, 30] in Table 1. A comparison between the equilibrium lattice constants predicted using LSDA and GGA for the exchange and correlation functional shows that GGA is essential to accurately reproduce the equilibrium structural properties of these Heusler alloys. In fact, LSDA underestimates the equilibrium volume by about 7% with respect to experiment, whereas the GGA error is at most 1%, following the general tendency found for the 3*d* elements: due to its functional form, GGA gives larger lattice constants, leading to a better agreement with experiment compared to LSDA [24]. As expected, the lattice constant increases as we increase the X atomic number (by 1.8% when substituting Ge for Si and by 3.7% upon substituting Sn for Ge). As for the lattice matching with respect to standard semiconductors to be used in spintronic devices, Co_2MnSi seems the most promising material. In fact, its lattice mismatch with GaAs and AlGaAs (used, for example, as quantum wells and side layers, respectively, in the first spin–light emitting diode [25]) is less than 0.4%; however, this is not the only parameter that determines a successful spin-injection, which is also crucially controlled by band line-ups and transport properties across the Heusler/semiconductor interface, as we will discuss.

As for the band structures of the different alloys (not shown here [26]) the majority spin band structure is strongly metallic, while the minority spin bands shows a semiconducting gap around the Fermi level, E_F, except in Co_2MnSn, where the minority "valence band maximum" (VBM) is just slightly above E_F, leading to a "nearly half-metallic" behavior [27]. On the other hand, half–metallicity is marked in Co_2MnSi and Co_2MnGe: the GGA calculated indirect $\Gamma - X$ band gap for minority carriers is $E_{gap}(Co_2MnSi) = 0.81$ eV and $E_{gap}(Co_2MnGe) = 0.54$ eV, respectively, while the Fermi level lies 0.33 eV and 0.03 eV above the minority spin VBM, respectively. This last and very important quantity is often referred to as the "spin–gap", *i.e.* the minimum energy required to flip a minority spin electron from the VBM to the majority spin Fermi level [28]. These results are in good agreement with those obtained from similar calculations by Galanakis and Dederichs and reported in the introductory chapter of this volume.

The calculated LSDA, GGA and experimental magnetic moments for the compounds and for the single constituents are also reported in Table 1 and compared with available experiments [10, 11, 29, 30, 32]. The Mn atom carries the largest moment (around 3 μ_B), while Co has a positive moment of about 1 μ_B. The X atom is slightly antiferromagnetically spin–polarized with respect to Mn and Co. The total moment in all three compounds is 5 μ_B, which is in good agreement with experimental values [10, 11]; the integer value of the total magnetic moment is an indication of the half-metallic character (complete, as in the Si and Ge compounds, or "nearly" complete, as in the Sn compound). The Mn magnetic moment appears to be slightly underestimated with respect to experiment for Co_2MnSn [33].

Table 1. Equilibrium structural properties, and total spin moment and predicted spin magnetic moments calculated within the muffin tin spheres at their respective equilibrium, within LSDA, GGA and experiment (where available). The lattice constant a, the bulk modulus B and the magnetic moments are in a.u., in GPa, and Bohr magnetons respectively

	Co_2MnSi			Co_2MnGe			Co_2MnSn		
	LSDA	GGA	exp	LSDA	GGA	exp	LSDA	GGA	exp
a	10.42	10.65	10.68^a	10.61	10.84	10.85^a	11.04	11.27	11.34^a, 11.30^b
μ_{tot}	5.0	5.0	$5.01 \leq \mu \leq 5.15^c$	5.0	5.0	5.11^d	5.0	5.0	5.08 ± 0.05^a
μ_{Mn}	2.81	2.92	2.7^e	2.88	2.98	–	2.98	3.09	3.6^f, 4^b
μ_{Co}	1.07	1.06	1.3^e	1.02	1.02	–	0.98	0.99	0.75^f
μ_X	−0.02	−0.04	–	−0.02	−0.03	–	−0.03	−0.05	–

a. reference [29]
b. reference [30]
c. reference [10]
d. reference [11]
e. reference [31]
f. reference [32]

As for the trends with the X atomic number, we note that the Mn magnetic moment increases slightly along the Si→Ge→Sn series; this can be ascribed to the increasingly larger lattice constant of the Mn sublattice, which results [24] in a smaller $d-d$ overlap (and smaller band width) and a consequently larger exchange interaction. On the other hand, the Co magnetic moment is reduced as the anion atomic number increases, keeping the global magnetic moment very close to 5 μ_B.

3 Electronic and Magnetic Properties of Co_2MnGe/GaAs Interfaces

In order to simulate the Heusler/semiconductor interfaces, we considered supercells containing up to 13 Co_2MnGe + 11 semiconductor (SC) layers, with the Heusler compound tetragonally strained to match the semiconducting substrate; all the atomic positions were fully relaxed. We focus mainly on the Mn–Ge termination of the Heusler side of the interfaces, and on the As–termination in GaAs. In addition, in order to address the dependence of the results on the interface geometry, calculations were also performed for other terminations (i.e. Co/As, Mn-Ge/Ga, Co/Ga in Co_2MnGe/GaAs).

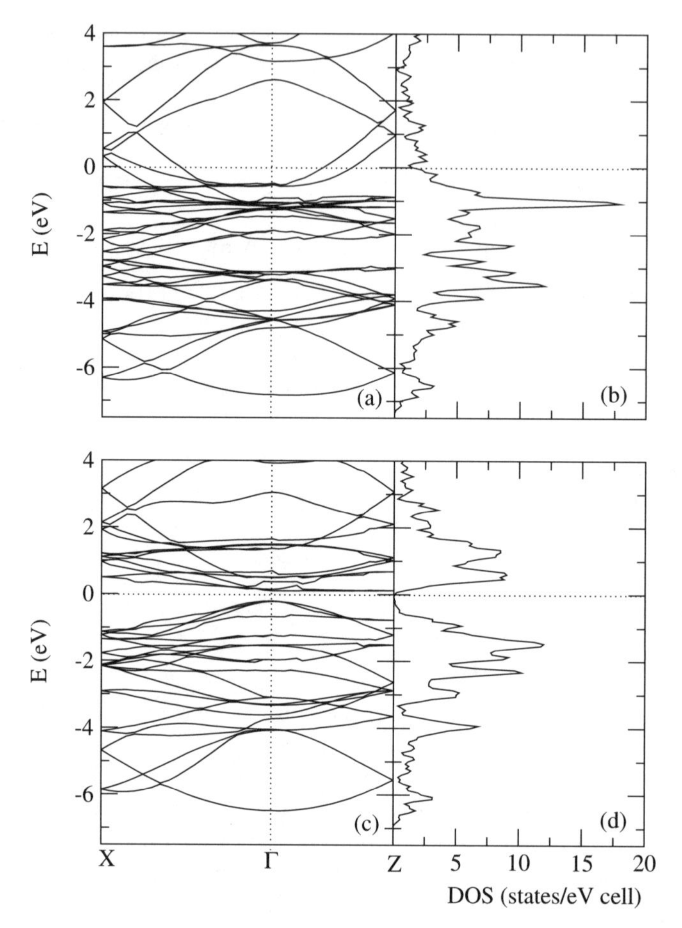

Fig. 1. (**a**) Band structure and (**b**) density of states for the majority spin carriers in Co_2MnGe strained on a GaAs substrate; (**c**) and (**d**) show the band structure and density of states for the minority spin carriers

3.1 Co_2MnGe Strained on a GaAs Substrate

As discussed in reference [2], the Co_2MnGe compound shows half–metallicity in its equilibrium bulk phase. However, it is not obvious that the half–metallicity is kept under strain conditions and within the LSDA approximation. Therefore, first of all we show in Fig. 1 the LSDA calculated spin-polarized band structure (for the (a) up and (c) down spin states) and the total density of states (for the up (b) and down (d) spin states) of the Heusler compound in the same strain conditions as in the supercells considered. The lower binding energy part of the band structure (below −7.5 eV, not shown) is mainly due to Ge s states, which are almost unaffected by the exchange coupling of Mn and Co. On the other hand, all the upper and dispersed bands have mainly a d character, due to hybridization between Mn $3d$, Co $3d$ and Ge $4p$ states. The region around −1 eV of the majority band structure, showing very dense and undispersed bands (corresponding to the high peak in the total density of states – see Fig. 1(b)) is mainly due to Mn and Co d states. The spin–up band structure has a metallic character, whereas the spin-down band structure shows a gap in proximity of the E_F. It is therefore clear that the half–metallic character is kept under strain, resulting in a spin–gap (i.e. the difference between the Fermi level and minority valence band maximum) of about 0.2 eV. The half–metallicity is consistent with the calculated integer total magnetic moment of 5 μ_B/formula-unit.

3.2 Density of States and Interface States at the Junction

In Fig. 2, we show the density of states projected (PDOS) on the interface atoms of the Mn–Ge/As terminated junction. The PDOS for the other terminations do not differ qualitatively from this representative Mn–Ge/As case and are therefore not shown. A comparison with the bulk PDOS (dashed line) shows that the interface represents a strong "perturbation" for these atoms, resulting in appreciable changes of their density of states. The most spectacular of these effects is the loss of half–metallicity in the Heusler side, clearly shown in Fig. 2: as in the case of other semiconducting/Heusler junctions [3], interface states appear in the minority semiconducting gap, in particular allowing both spin states at the E_F. The presence of interface states suggests that in the device design, one must take into account not only the properties of the bulk material (in this case, the Heusler compound exhibiting a 100% spin polarization), but also the complications introduced by the presence of "perturbations", such as surfaces, interfaces and defects. An analysis of the angular character of the gap–states on both sides of the junction gives further insights about hybridization mechanisms at the interface, that finally lead to the loss of half–metallicity. As expected, there is strong hybridization between the d states of the transition metal atom (i.e. Co or Mn, depending on the termination) and the p states of the interface semiconducting atom. Moreover, recall that each plane in the Co_2MnGe side has two atoms; in particular, at the interface, both the "bridge" site (that occupies the "ideal" zincblende position) and the "anti–bridge" site (that occupies the "hollow" zincblende position) are occupied. Our calculations show that in the case of the Co–terminated junctions the $p - d$ hybridization is stronger for the "bridge" site, even though an appreciable density of gap states is present on the "antibridge" site, too.

Furthermore, to show how the interface states decay within both the Heusler compound and the SC we plot the density of states (DOS) at E_F as a function of the distance from the interface. The DOS at E_F ($D(E_F, z)|_{z=z_{at}}$) on the different atomic planes, z_{at}, perpendicular to the [001] growth direction for both spin components is shown in Fig. 3. In proximity to the junction ($z = z_{junc}$), the majority and minority spin contributions to the DOS at

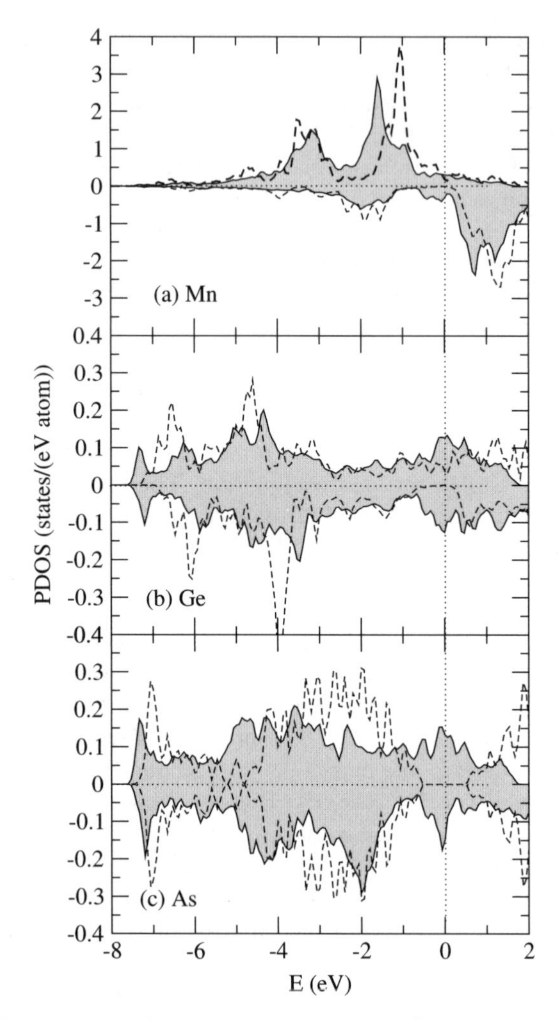

Fig. 2. PDOS on the interface atoms in the Mn-Ge/As terminated junction: (**a**) Mn, (**b**) Ge and (**c**) As. The solid line (and dashed area) shows the interface PDOS, whereas the dashed line shows the PDOS in their respective bulk phase

E_F are basically equivalent, so that the net spin–polarization is close to zero – which may degrade the spin-injection efficiency. The decreasing trend of the minority spin DOS, $D(E_F, z)^{\downarrow}|_{z=z_{at}}$, away from the interface shows that interface states decay into the Co_2MnGe side within a few atomic layers, completely restoring half–metallicity far from the junction. Equivalently, in the GaAs side, gap–states are even more efficiently screened. This shows that *i*) the size of the supercell is sufficiently large to recover bulk conditions on both sides of the interface and *ii*) it is possible to gain further insights about the gap states by looking at their exponential decay [35,36] in both the SC and

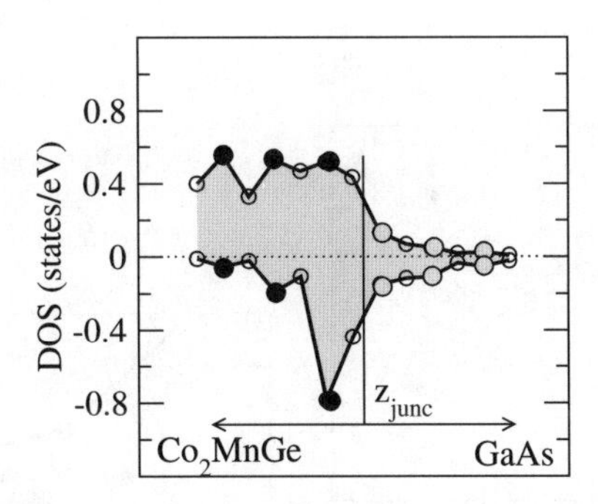

Fig. 3. DOS at E_F calculated on the atomic planes as a function of the distance from the interface, for the up–spin ($D(E_F,z)^{\uparrow}|_{z=z_{at}}$, positive y-axis) and down–spin ($D(E_F,z)^{\downarrow}|_{z=z_{at}}$, negative y-axis) configuration. The interface plane is evidenced by the vertical solid line and denoted as z_{junc}. Circles denote the positions of the atomic planes along the growth direction: on the Co_2MnGe side – left side of the junction – filled black (empty) circles denote Co (Mn and Ge) atoms; on the GaAs side – right side – filled gray (empty) circles denote As (Ga) atoms

Co_2MnGe regions (minority component only) as a function of the distance from the interface (i.e. $D(E_f,z) \propto \exp(-|z - z_{junc}|/\lambda)$, λ being the decay length). Our estimated interface–states decay lengths are $\lambda^{\uparrow}$(GaAs) ~2.7 Å ($\lambda^{\downarrow}$(GaAs) ~3.3 Å) in the SC side for the majority (minority) component and $\lambda^{\downarrow}$(Co_2MnGe) ~2.5 Å in the Heusler side of the junction. Therefore, the gap states extend for a couple of interface layers, almost disappearing in deeper layers, with a similar behavior in GaAs and Co_2MnGe.

We point out that our calculated values are consistent with the decay length reported for unpolarized GaAs interfaced with Al ($\lambda \sim$ 3.0 Å) [35] and confirm that λ is a "bulk" property and therefore not dependent on the deposited material [35, 36] In the Co_2MnGe/Ge case, an estimate of the decay length in the Co_2MnGe side leads to $\lambda^{\downarrow}$(Co_2MnGe) ~2.8 Å (and therefore consistent with the Co_2MnGe/GaAs case) while in the Ge side leads to $\lambda^{\uparrow}$(Ge) ~$\lambda^{\downarrow}$(Ge) ~4.5 Å. Therefore, the penetration depth for Ge is much larger than that obtained for GaAs ($\lambda \sim$ 3 Å), in agreement with the qualitative idea that λ is smaller in materials with larger energy band–gap and smaller dielectric constants [36].

3.3 Magnetic Moments

In Fig. 4 we show the magnetic moments of the different atomic species as a function of the distance from the interface. First of all, note that all the atom, sufficiently far from the junction recover (within at most 0.02 μ_B) their bulk magnetic moment, both in the strained Heusler compound ($\mu_{Mn} = 2.88$ μ_B, $\mu_{Co} = 0.99$ μ_B and $\mu_{Ge} = -0.006\,\mu_B$) and in the semiconductor ($\mu_{Ga} = \mu_{As} = 0$ μ_B), again showing that the cell dimension is sufficiently large for

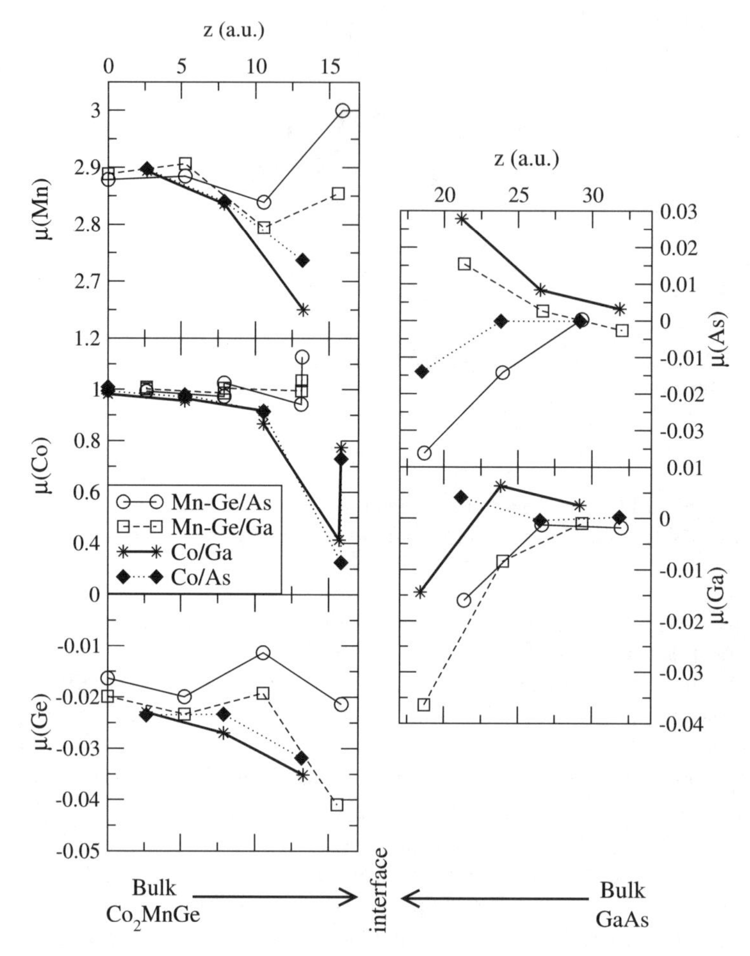

Fig. 4. Magnetic moments (in μ_B) in the (**a**) Mn, (**b**) Co, (**c**) Ge, (**d**) As and (**e**) Ga muffin tin spheres in the four junctions considered as a function of the distance (in a.u.) from the interface

the present purposes. As for Mn, an enhancement with respect to the bulk value is present in the interface atom of the Mn-Ge/As case; for the other terminations (and more markedly for Co/Ga), Mn generally shows a magnetic moment decreasing with distance from the interface. As expected, for Co we find that the changes from the bulk value in the two Co–terminated interfaces are much larger than in the Mn–Ge case, even reaching deviations of about 0.7 μ_B. Moreover, it is remarkable how the different coordination and geometry of the two different sites for the interface Co atoms affect their magnetic properties: Co in the ideal zincblende "bridge" site has a much smaller magnetic moment (by as much as 0.4 μ_B) compared to Co in the zincblende hollow "anti–bridge" site, resulting in a strong reduction of the magnetic polarization of the latter. As expected, this tendency is more marked for the Co- rather than for the Mn-Ge terminations. The small induced negative magnetic moment in Ge oscillates around the bulk value in all the junctions, generally showing a very slight reduction for the interface atom. As for the semiconducting atoms, the interface atoms are always antiferromagnetically polarized; the As subinterface atoms in the Ga–terminated junctions are polarized slightly positively, whereas their Ga counterpart in the As–terminated junctions do not show a definite tendency.

It is instructive to compare our results with those reported in a recent first–principles study of the Fe/GaAs(001) interface [37]. It was found that *i)* local moments for surface Fe atoms were significantly enhanced (to 3.0 μ_B) with respect to their bulk value (i.e. 2.33 μ_B); *ii)* an adlayer of Ga or As on top of the Fe film suppresses the Fe magnetic moments, the effect being particularly pronounced in the As–capped case, due to strong covalent bonding between the As and Fe atoms and *iii)* substrate GaAs atoms up to 3–4 layers deep have small induced moments, always with negative spin moments for the interface atoms. In our case: *i)* the spin moment enhancement of the interface atoms is not a general feature and, when present (i.e. Mn–Ge termination) is not as marked as in the Fe/GaAs case. This is consistent with much smaller changes of the Mn–Co *d* bands with respect to the Fe *d* bands, due to the interaction with the GaAs substrate. As for *ii)*, a similar mechanism leading to magnetism quenching can be outlined in our case: changes in coordination and symmetry, resulting in changes of the Fe *d* bands in the presence of an As adlayer have a similar effect, as already pointed out, on the Co "bridge" atoms, with respect to their "antibridge" counterpart. Finally, as for *iii)*, our GaAs calculated induced moments show exactly the same features of the Fe/GaAs interface, both in sign and magnitude.

3.4 Spin Resolved Band Line-Up

Let us now discuss the potential discontinuity at the Co_2MnGe/SC junction, evaluated as usual in all–electron calculations by taking core levels as reference energies and considering both *bulk* and *interface* contributions [38]. Our results for the Co_2MnGe/SC junctions are shown in Fig. 5, where we plot *i)* the position of E_F with respect to the SC VBM, giving rise to a *p*-type SBH for the majority spin carriers, $\Phi^{\uparrow}_{B_p}$ and *ii)* the position of the VBM for the minority–spin states with respect to the SC VBM, giving rise to a valence band offset for the minority spin carriers, VBO$^{\downarrow}$. The position of E_F with respect to the SC conduction band minimum (CBM), denoted as $\Phi^{\uparrow}_{B_n}$, has been obtained considering the experimental energy gap (due to the band–gap problem within DFT) for GaAs and Ge (E_{gap} = 1.5 eV and E_{gap} = 0.7 eV, respectively) [32] so that $\Phi^{\uparrow}_{B_n} = E_{gap} - \Phi^{\uparrow}_{B_p}$. The potential line–up is remarkably different in the two cases: for GaAs, E_F lies more or less in the center of the band gap, whereas for Ge, E_F is close to the VBM. Our results show that *i)* the charge rearrangement at the interface is largely dependent on the SC material, *ii*) in both SCs, the E_F level is *pinned* at an energy position similar to that obtained with other non–magnetic metals [*eg.* the SBH is 0.07 (0.18) eV for Ge/Au (Ge/Al), 0.52 (0.62) eV for GaAs/Au (GaAs/Al)] [34]. This suggests that spin-polarization effects do not play an important role in the potential line–up, which is therefore predictable on the basis of non–spin–polarized SBH. On the other hand, first–principles calculations are essential to predict spin-polarized–metal/SC junctions where interface gap states do not reduce the spin polarization at the junction [3]. Furthermore, consistent with the pinning mechanism previously outlined for GaAs, different atomic terminations in Co_2MnGe/GaAs result in changes of at most 0.2 eV in the potential line–up, not affecting the rectifying character of the case shown. Therefore, in the case of Heusler/GaAs junctions, the presence of a Schottky barrier may be useful for spin–injection purposes: the rectifying region was one of the mechanisms proposed to reduce the conductivity mismatch [22] between the spin–polarized source and the semiconductor, which is believed to hinder a high efficiency for the injection of highly polarized currents. The contact character shown in Fig. 5(b) may be useful for spin–injection purposes, Co_2MnGe being in this case a source of spin–polarized holes to be injected in the Ge side. However, it has to be pointed out that electrons have longer lifetimes and higher mobilities than holes so that, so far, greater interest has been devoted to materials that could inject spin–polarized electrons rather than holes.

4 Structural Defects

4.1 Technical and Structural Details

The calculations were performed within the GGA–FLAPW method, using a 32 atom unit–cell, obtained by considering the $L2_1$ bulk unit cell (with four atoms) and doubling its Bravais vectors (i.e. the defective cell is $2\times2\times2$ times the $L2_1$ unit–cell). In order to check the convergence of the results as a function of the unit cell dimensions, we performed some test calculations using 64 atoms; we obtained only small quantitative changes in the relevant properties, but, qualitatively, the physical results are the same as those obtained for the smaller 32–atoms cell. The lattice constant was kept fixed to the calculated bulk equilibrium value [2] – within the generalized gradient approximation (GGA) [18] –, i.e. $a(Co_2MnGe) = 10.84$ a.u. and $a(Co_2MnSi) = 10.65$ a.u. Internal degrees of freedom were fully relaxed according to ab–initio atomic forces.

The formation energy is estimated as [20]

$$\Delta E = E^{def} - E^{id} + n_{Mn}\mu^0_{Mn} + n_{Co}\mu^0_{Co} + n_X\mu^0_X \tag{1}$$

where E^{def} and E^{id} are the total energies of the unit cell with and without defect, respectively; the n_i take into account, in forming the defect, that n_i

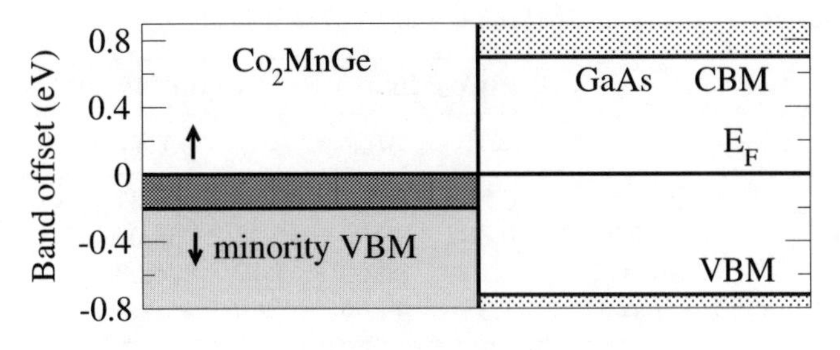

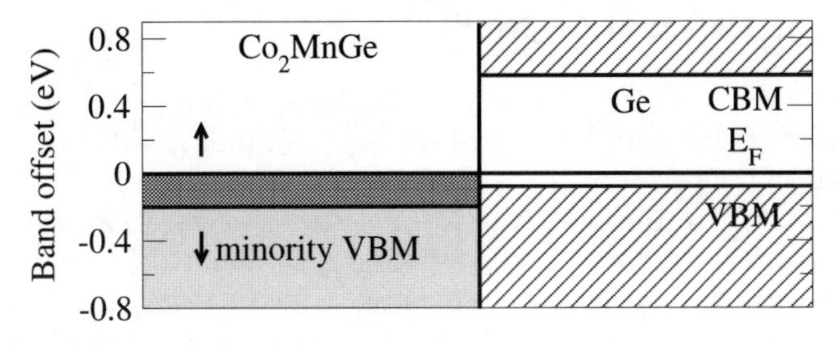

Fig. 5. Potential line–up at the interface for the Mn-Ge terminated (**a**) Co_2MnGe/GaAs and (**b**) Co_2MnGe/Ge junction. The Fermi level is set to zero of the ordinate scale

atoms are transferred to or from a chemical reservoir that has a characteristic energy, μ_i^0. In our case, we choose as the stable phase for element i [i = Mn, Co, X (= Si, Ge)] the [001]–ordered fcc antiferromagnetic Mn [24], ferromagnetic hcp Co [24] and the Si (or Ge) diamond–like structure, respectively. Charged defects or competing crystalline phases that might be formed during defective growth are not taken into account.

Moreover, the equilibrium concentration of the defect at temperature T, considering a Boltzmann–like distribution, can be estimated as [21]

$$D_{def} = N_{sites} \exp\left(-\frac{\Delta E}{k_B T}\right) \tag{2}$$

where N_{sites} is the number of available sites.

4.2 Results and Discussion

In Table 2, we report the calculated formation energies for all the defects considered in the two different matrices; we also show the total magnetic moment in the unit cell. Recall that both Co_2MnGe and Co_2MnSi in their bulk phase show a total magnetic moment of 5 μ_B (or, equivalently, a magnetic moment of 40 μ_B in the 32–atoms unit cell used for the defect calculations) [2].

Table 2. Formation energy (in eV) and total magnetic moments (in Bohr magnetons) for different defects in Co_2MnGe and Co_2MnSi hosts

	Co_2MnSi		Co_2MnGe	
	ΔE	M_{tot}	ΔE	M_{tot}
Co antisite	0.80	38.01	0.84	38.37
Mn antisite	0.33	38.00	0.33	38.00
Co–Mn swap	1.13	36.00	1.17	36.00
Mn–Si swap	1.38	40.00	–	–

Mn Antisite Defect in Co_2MnSi

The considerably smaller value obtained ($\Delta E = 0.33$ eV) suggests that this kind of defect is likely to be formed during Co_2MnSi growth. In particular, considering a temperature T = 1523 K, suitable for a triarc Czochralski growth of bulk Co_2MnSi [10], the Mn antisite concentration can reach the high value of $0.36 \cdot 10^{22}$ cm^{-3}, corresponding to about 8%. This is in good agreement with the reported experimental value of 5–7% [10]. However, we point out that ours is only a rough quantitative estimate, since equation 2 is strictly valid in the limit of very low concentrations, that are highly exceeded when considering more than a few percent.

We now discuss the electronic and magnetic properties of the system, focussing in particular on the defect–induced changes on half–metallicity. In Fig. 6 we show the total density of states (DOS) of both the defective and the ideal systems. The comparison shows that the defect induces only minor modifications – such as features at -1.5 eV and -0.8 eV for the majority spin states; on the other hand, due to the Mn antisite, a rigid shift of about 0.05 eV towards higher binding energies occurs in the minority spin channel. This results in a very small increase of the spin–gap (0.34 eV and 0.39 eV in the ideal and defective systems, respectively). However, it is clear that half–metallicity is kept even in the presence of the antisite, the total magnetic moment being 38 μ_B and 40 μ_B in the defective and ideal unit cells, respectively. We remark that, interestingly, the half–metallic character is kept thanks to a charge balance granted by the majority component only. In fact, in this case, the total charge differs from the ideal case by $\Delta Z = -2$ electrons; since the number of occupied states in the minority channel does not change (see Fig. 6, $\Delta N_- = 0$), then the variation of occupied states in the majority component must account for the total ΔZ (i.e. $\Delta N_+ = -2$ electrons). In particular, the two-electron states appear as a new peak at $\sim$0.3 eV above E_F, almost entirely due to the Mn–antisite d states, as shown by the corresponding feature in the DOS projected (PDOS) on the Mn defective site (see shaded area in Fig. 6). Our results are particularly relevant in the spin–injection framework: even if the Mn antisite defects – due their low formation energies – are most likely to occur during Co_2MnSi growth, the

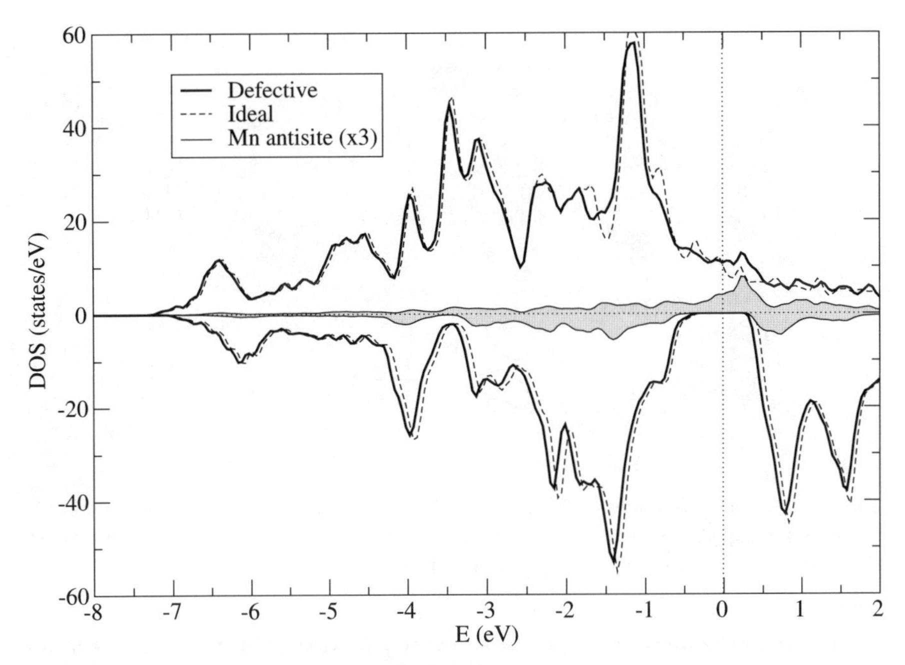

Fig. 6. Total DOS for defective (*solid bold line*) and ideal (*dashed line*) Co_2MnSi with Mn antisite in Co_2MnSi; majority spin shown positive and minority spin negative. The Mn–antisite PDOS (multiplied by a factor of 3) is also shown (*grey shaded area*)

most important property of this Heusler alloy, i.e. its half–metallicity, is kept. Therefore, experimental results reporting a polarization lower than the expected 100% need to be explained invoking surface effects [13] or other types of bulk defects (such as Co–antisites [8, 9] – see below –, vacancies, interstitials, etc). Further insights can be gained from the magnetic moments of the different atomic species. In Fig. 7, we report a schematic view of the region close to the defect (i.e. the Mn antisite and its first nearest–neighbors along the [110] direction) as compared with the same region in the ideal Heusler alloy, including the relative magnetic moments in the MT spheres. Here and in the following similar figures, we report only these few atoms since *i*) other symmetry directions (i.e. $[1\bar{1}0]$) show an equivalent behavior and *ii*) farther atomic shells do not show variations of their magnetic moments larger than 1% with respect to their bulk values and/or are below our numerical precision ($\sim$0.01 μ_B). The central Mn antisite accounts for all the defect–induced reduction of 2 μ_B in the total magnetic moment, since it has a magnetic moment of -0.87 μ_B and substitutes for a Co having a magnetic moment of 1.06 μ_B. The large change in the atomic magnetization (which also reverses its sign) on the defect site shows that a large rearrangement of the charge and spin densities occurs when the defect is introduced, as shown by

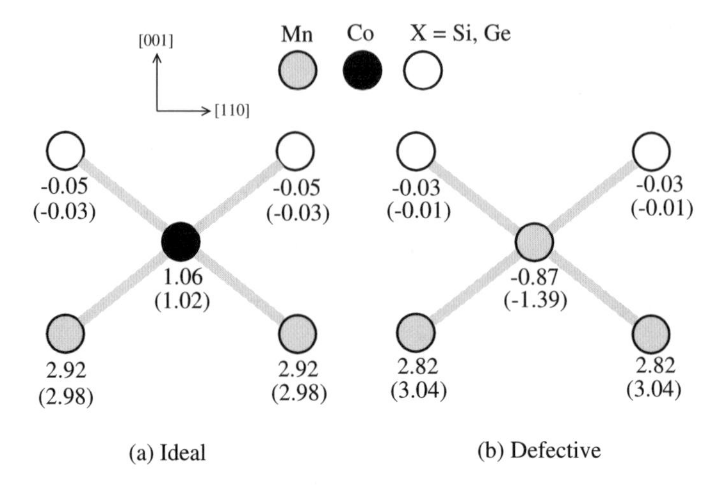

Fig. 7. Magnetic moments within MT spheres for the (**a**) ideal and (**b**) defective systems for the Mn antisite in Co_2MnSi around the defect. Values in parenthesis denote magnetic moments for equivalent atoms in Co_2MnGe

the completely different projected DOS (not reported) on the Mn defect site compared to Mn in bulk Co_2MnSi. Only small changes are observed on its nearest neighbors, which shows that the defect–induced effects are efficiently screened out in the Co_2MnSi matrix; incidentally, these findings support our computational choice of a relatively small unit cell.

Co Antisite Defect in Co_2MnSi

A slightly higher formation energy is calculated for the Co antisite. This increase, however results, according to (2), in a concentration which is almost two order of magnitude smaller than that obtained for the Mn antisite. Therefore, although present, these defects are expected to have a relatively small density. This is actually at variance with experiments, which found Co and Mn antisites occurring with more or less the same concentrations. We think that, as pointed out, some errors may arise from the quantitative estimate of the concentrations for a high defect density: (2) shows that small uncertainties in ΔE (due to the finite size of the unit cell, **k**-point sampling or other computational details) result in large errors in the defect concentration.

An analysis of the total DOS (see Fig. 8) shows a defect–induced dramatic change in the conducting character: half–metallicity is lost, due to a very sharp peak located just in proximity to E_F. As shown in Fig. 8, there is an almost exact superposition between the ideal and defective total DOS, except for the peak at E_F. The projected density of states (not shown) reveals that this is almost entirely due to the antisite Co d states. In particular, these states accommodate two electrons (see below) and are double degenerate at the Brillouin-zone center with e_g symmetry. At variance with the Mn antisite

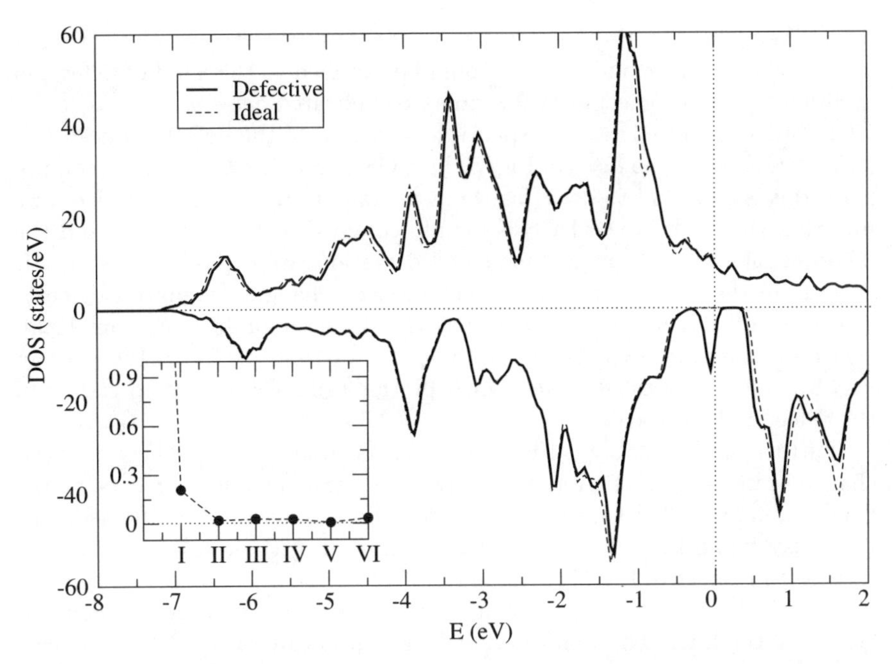

Fig. 8. Total DOS for defective (solid bold line) and ideal (dashed line) Co_2MnSi with Co antisite in Co_2MnSi. The inset shows the minority DOS at E_F projected on the different neighbors (denoted as Roman numbers) as one moves away from the Co–antisite defect

defect, the total charge difference is $\Delta Z = +2$ electrons; being the majority DOS component almost unaffected with respect to the ideal case (see Fig. 8), the total charge variation is taken over by the peak in the DOS just below E_F in the minority component only ($\Delta N_- = +2$ electrons). In this case, the spin polarization at the Fermi level, defined as:

$$P = \frac{N^{\uparrow}(E_F) - N^{\downarrow}(E_F)}{N^{\uparrow}(E_F) + N^{\downarrow}(E_F)} \tag{3}$$

where $N^{\uparrow}(E_F)$ and $N^{\downarrow}(E_F)$ respectively denote the up- and down-spin component of the total DOS at E_F, is as low as 6%. Moreover, since the tunnelling current in experiments is dominated by the s electrons, we also calculated the s-component of the spin polarization at E_F, defined as:

$$P_s = \frac{N_s^{\uparrow}(E_F) - N_s^{\downarrow}(E_F)}{N_s^{\uparrow}(E_F) + N_s^{\downarrow}(E_F)} \tag{4}$$

Our result is $P_s = 55\%$, in (surprising) excellent quantitative agreement with experiment [8]. Therefore, we suggest that, due to the possibility of these defects being formed, the decreased value of the measured spin polarizations – with respect to their bulk value – could be ascribed to this kind of defect, in addition to surface effects or other more complicated defects.

In order to investigate the spatial localization of the defect–induced gap states, we show in the inset of Fig. 8 the DOS at E_F for the minority component; this is expected to decay as we move away from the defect. Indeed, we find that the peak in the DOS is mainly confined to the region around the defective site. Even though some small deviations from the bulk also occur away from the defect, the main defect–induced changes are quite efficiently screened out. Therefore, we infer that the defect–induced states are localized both spatially – as shown by the decay of DOS at E_F in the inset of Fig. 8 – and energetically – as shown by the < 0.1 eV energy–spreading of the defect–induced peak.

Further insights about the spatial localization of the defect–induced changes can be gained from an analysis of the magnetic moments of the Co defect and of their Co first-nearest-neighbors (shown in Fig. 9). As in the previous Mn–antisite case, most of the variation of the total magnetic moment with respect to the ideal value is due to the defect atomic site, whereas already the first–nearest neighbor shell largely recovers the magnetic moment typical of the bulk. This confirms the localized nature of the defect–induced changes.

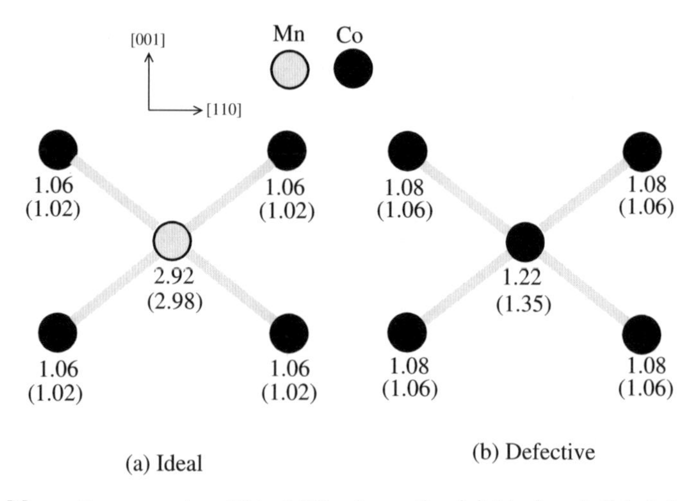

Fig. 9. Magnetic moments within MT spheres for (**a**) ideal and (**b**) defective systems for the Co antisite in Co_2MnSi around the defect. Values in parenthesis denote magnetic moments for equivalent atoms in Co_2MnGe

Co-Mn Swaps in Co_2MnSi

We now focus on atomic interchanges and consider Mn-Co swaps; this defect can also be viewed as the sum of two different Mn and Co atomic antisites that tend to aggregate. This defect shows a pretty high formation energy; however, this is of the same order of magnitude as the sum of the separated defects (Mn and Co antisites) – which might indicate that point defects have more or less the same probability to cluster, leading to this kind of disorder, or to remain isolated.

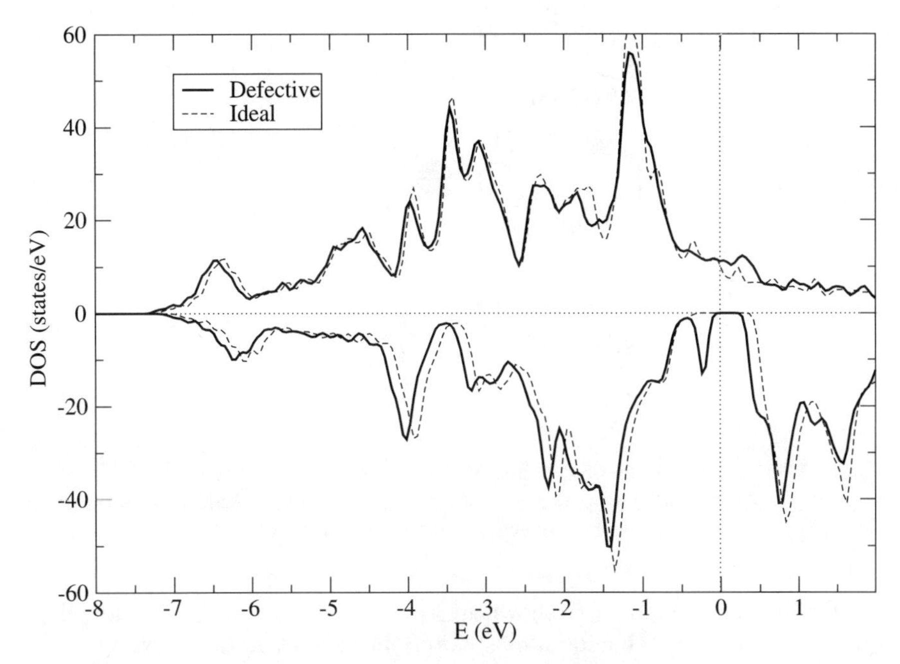

Fig. 10. Total DOS for defective (*solid bold line*) and ideal (*dashed line*) Co_2MnSi with Co–Mn swap in Co_2MnSi

The calculated integer total magnetic moment suggests half–metallic behavior. Indeed, this is shown by the total DOS (see Fig. 10). A comparison with the ideal situation shows that the majority DOS is basically unaffected; on the other hand, the occupied minority DOS shows a small shift (about 0.1−0.2 eV) towards higher binding energies, along with a defect–induced peak located at −0.2 eV below E_F. Therefore, due to its energy position, this peak is not as crucial as in the Co–antisite case and half–metallicity is kept by Co-Mn swaps. Moreover, our results are in overall agreement with those reported by Orgassa et al. for NiMnSb [14]: in the case of $Ni_{1-x}Mn_xSb$ disorder – similar to the Mn-Co swap considered here – a peak just below E_F arises in the minority spin component of the total DOS. This peak broadens in energy as the percentage of defects is increased, finally reaching E_F and resulting in the loss of half–metallicity (see Fig. 1 of [14]). In our case, we have a very similar behavior, except for the more "energetically–localized" nature of the peak; we think that this quantitative disagreement might be due to the computational method – layer Korringa-Kohn-Rostoker in conjunction with the coherent potential approximation – used in [14].

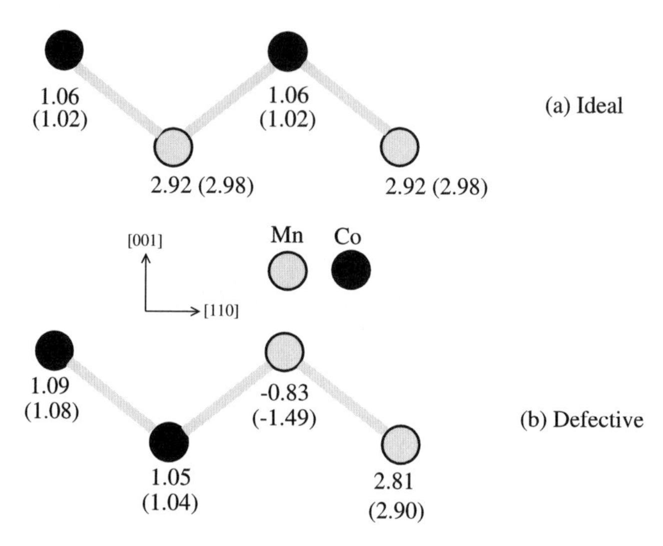

Fig. 11. Magnetic moments within MT spheres for (**a**) ideal and (**b**) defective systems for the Co–Mn swap in Co_2MnSi around the defect. Values in parenthesis denote magnetic moments for equivalent atoms in Co_2MnGe

The magnetic moments (shown in Fig. 11) in the defective region show quite an interesting behavior: almost two Bohr magnetons are given up by the Co substituting for Mn and another two Bohr magnetons are given up by the Mn substituting for Co. As a result, the total magnetic moment is reduced by 4 μ_B (*i.e.*, the difference between the bulk and defective total magnetic moments, 40 μ_B and 36 μ_B, respectively). It is also interesting to note that the swaps induce a behavior similar to that of the single point defects; in fact, in this case, the magnetic moment for Mn substituting for Co is -0.83 μ_B, compared with -0.87 μ_B for the Mn antisite. Similarly, the magnetic moment for Co substituting for Mn is 1.05 μ_B, to be compared with 1.22 μ_B for the Co antisite. This shows that the first coordination shell (similar in the case of isolated point defects and for the swaps [23]) mainly dictates the behavior of the magnetic moments.

Mn-Si Swap in Co_2MnSi

As shown by Table 2, the highest formation energy among the cases studied is shown by the Mn-Si swap; therefore, the occurrence of an appreciable concentration of this kind of defect can definitely be ruled out. This is consistent with experiments which show that the Si site is fully occupied by Si, indicating that any disorder in Co_2MnSi does not involve Si. On the other hand, our findings are at variance with other first–principles predictions for the NiMnSb half–Heusler alloy: according to Orgassa et al. [14], the Mn–Sb (analogous to the Mn-Si here considered) disorder seems likely.

As far as the electronic and magnetic properties are concerned, a comparison between the total DOS (see Fig. 12) in the ideal and defective cells shows only minor defect–induced changes. In particular, in the energy region around E_F, the DOS is very similar for both minority and majority spin components, resulting in the same band–gap and half–metallic character as for the pure bulk. As a result, the Mn-Si swap system shows a total magnetic moment equal to that of the ideal Heusler alloy. This is consistent with the atomic magnetic moments shown in Fig. 13: within 0.04 μ_B, the defect only results in a swap between the Mn and Si magnetic moments. Again, this shows that the first coordination shell – which in this case is exactly the same as the one in the bulk for both the exchanged atoms – is the most relevant

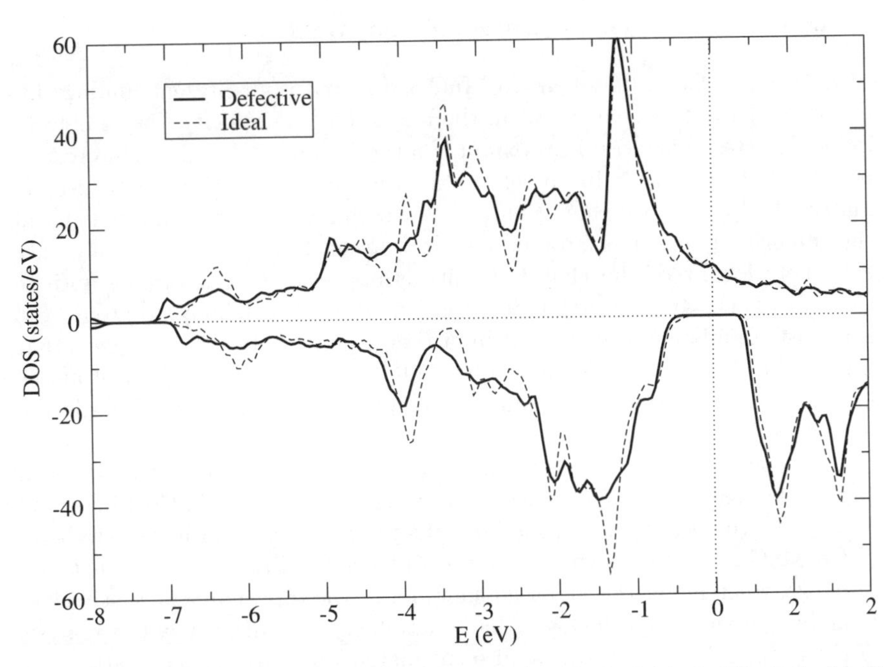

Fig. 12. Total DOS for defective (*solid bold line*) and ideal (*dashed line*) Co_2MnSi with Mn–Si swap in Co_2MnSi

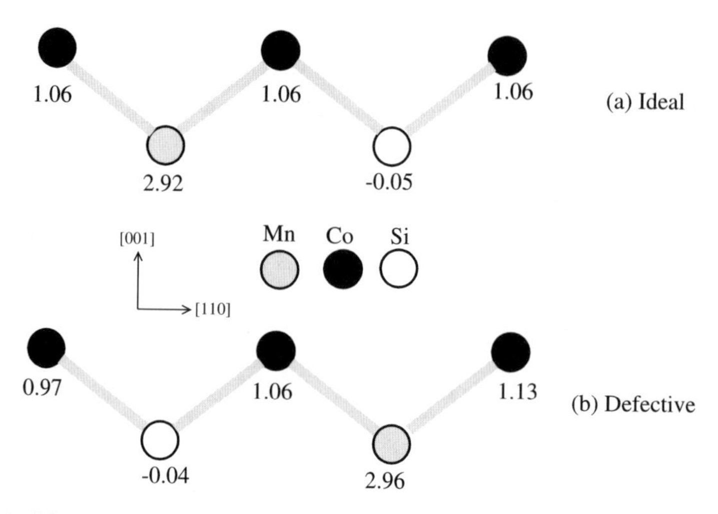

Fig. 13. Magnetic moments within MT spheres for (**a**) ideal and (**b**) defective systems for the Mn–Si swap in Co_2MnSi around the defect

in the formation of the bonds, local charge and spin density rearrangement and resulting magnetic moments.

Comparison between Co_2MnSi and Co_2MnGe

In the case of Co antisites in Co_2MnGe, similar to our present findings for Co_2MnSi, gap–states can result in the loss of half–metallicity due to defect–induced states that arise in proximity to the Fermi level. It is of interest to investigate how these induced defect states are screened within the Co_2MnGe matrix. In Fig. 14, we show the spin–down charge density contour plots of the minority gap states around E_F. It is clear that the defect states are extremely localized: the charge density is appreciable only on the antisite Co, while it is significatively reduced on the nearest–neighbor Co atoms and is almost completely screened out in farther atomic shells. This suggests that this kind of defect may locally destroy half–metallicity and, if present at the Co_2MnGe/SC interface, may hinder the spin–polarization and degrade the spin–injection efficiency.

We remark that both the spatial– and energy–behavior of defect–induced gap states suggests that Co antisites show a similar behavior in Co_2MnGe and Co_2MnSi. We remark that in the Co–antisite case, the loss of half–metallicity in Co_2MnGe is also shown by the non–integer total magnetic moment (see Table 2), whereas for the Si–based compounds the total moment is still close to an integer value and the loss of half–metallicity is evidenced by the analysis of the total DOS. Moreover, we also calculated the defect concentrations for Co_2MnGe considering two different growth temperatures: $T_1 = 825$ K, as reported in reference [9] for films via pulsed laser deposition, and $T_2 = 165$ K,

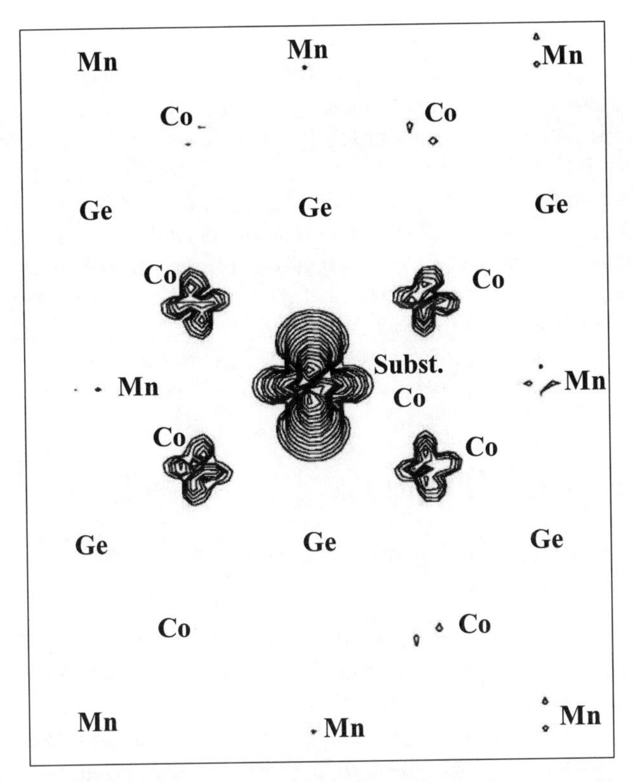

Fig. 14. Spin–down charge density contour plots around E_F in the [110] plane for Co_2MnGe in the presence of a Co antisite. Contour plots start at $1 \times 10^{-3} e/a.u.^3$ and increase successively by a factor of $2^{1/2}$

as reported for films grown via molecular beam epitaxy (MBE). As expected, the two different temperatures result in very different concentrations: in the case of the MBE growth, all the defects have an almost negligible probability of being formed, whereas in the higher temperature growth the concentration of Mn atoms occupying the Co site is of the order of $< 1\%$ and other defects have much smaller densities.

In order to compare the behavior of the defects considered in Co_2MnSi and Co_2MnGe, we report in parentheses in Figs. 7, 9 and 11 the value of the atomic magnetic moments in the region around the defect site. The two different hosts globally show a very similar behavior in terms of magnetic moments, suggesting that the larger lattice constant or the smaller minority band–gap of the Ge–based compound are not really relevant in the final determination of the magnetic properties. Quantitatively, the largest differences are shown by the Co–antisite case: there is about a 0.4 μ_B difference between the total magnetic moments in Co_2MnSi and Co_2MnGe (see Table 2). Figure 9 shows that this difference is almost entirely due to differences in the central

Co antisite, which changes from about -0.9 μ_B in Co_2MnSi to about -1.4 μ_B in Co_2MnGe, due to a very small difference in the energy–position (<0.05 eV) of the defect–induced peak with respect to E_F.

Finally, it is interesting to note that the position of the defect–induced peak in the DOS with respect to E_F – and the related energy position of the conduction band minimum in the minority spin component – is very similar: for both Co_2MnSi and Co_2MnGe, the peak is basically coincident with E_F in the case of the Co–antisite, whereas it lies at about -0.2 eV in the case of the Co–Mn swap. Except for very small differences that can be traced back to differences in the ideal bulk hosts (see ([2]) for details), the calculated DOS for antisite and swap defects in Co_2MnGe are not shown, due to behavior very similar to Co_2MnSi.

5 Conclusions and Outlook

Our main results for the Co_2MnGe/GaAs and Co_2MnGe/Ge junctions can be summarized as follows:

- the character of the contact is strongly dependent on the SC: for GaAs, E_F is pinned well within the band–gap (irrespective of the interface geometry), whereas for Ge, E_F lies very close to the Ge VBM. Thus, in the first case, the junction might be of use in tunneling contacts, whereas the second interface could be used as a source of spin–polarized holes;
- half–metallicity, peculiar of bulk Co_2MnGe, is lost in proximity to the junction. Interface states appear in both sides of the junction; however, the Heusler compound (semiconductor) recover its half–metallic character within a couple of layers away from the interface;
- the study of the magnetic moments as a function of the distance from the interface shows that significative deviations are present only for the Co–terminated case, where due to symmetry and coordination the Co on the bridge site, shows a strong depression of the magnetic moment.

As for the different defects in Co_2MnSi and Co_2MnGe, we were able to show that both Co and Mn antisites, due to low formation energies, are likely to be formed in a concentration as high as 8%; on the other hand, atomic swaps (such as Co-Mn and Mn–Si swaps) have lower defect–densities. Moreover, half–metallicity, typical of the ideal Heusler alloy, is preserved in all cases, except for the Co antisite, where a defect–induced peak arises at the Fermi level. Our findings suggest that the loss of half–metallicity found in experiments could be due to the presence of Co antisites, since the s-component of the spin-polarization at E_F in Co_2MnSi is in excellent agreement with tunneling measurements.

As a general conclusion, we point out that interfaces of Heusler alloys with semiconductors as well as with insulators are currently far from being understood, both from the electronic structure as well as from the transport properties point of view. Ideally, the theoretical analysis should be accompanied by complementary experimental investigations, still in their infancy in this field.

Finally, since Heusler alloys at the present stage are considered highly defective materials, much work remains to be done in the field of point defects, where a comprehensive investigation focused on the whole class of Heusler alloys is needed to elucidate the mechanisms that result in losing or keeping their most important property, i.e. half–metallicity. Their comprehension (and resulting guide towards optimization of desired properties) will determine the success or failure of the Heusler alloys as basic spintronic materials in the near future.

References

1. S. Fujii, S. Sugimura, Ishida, and S. Asano: J. Phys.: Condens. Matter **2** 8583 (1990)
2. S. Picozzi, A. Continenza, and A.J. Freeman: Phys. Rev. B **66**, 094421 (2002)
3. G.A.de Wijis and R.A. de Groot: Phys. Rev. B **64**, 020402 (2001)
4. A. Debernardi, M. Peressi, and A. Baldereschi: Mat. Sci. Eng. C **23**, 743 (2003)
5. I. Galanakis: J. Phys.: Condens. Matter **16**, 8007 (2004)
6. J.E. Sullivan and S.C. Erwin: unpublisehd; preprint arXiv::cond-mat/0110474
7. S. Picozzi, A. Continenza, and A.J. Freeman: J. Appl. Phys. **94**, 4723 (2003); J. Phys. Chem. Solids **64**, 1697 (2003)
8. M. Raphael, B. Ravel, M. Willard, S. Cheng, B. Das, R. Stroud, K. Bussmann, J. Claassen, and V. Harris: Appl. Phys. Lett. **79**, 4396 (2001)
9. B. Ravel et al.: Appl. Phys. Lett. **81**, 2812 (2002)
10. M.P. Raphael, S.F. Cheng, B.N. Das, B. Ravel, B. Nadgorny, G. Trotter, E.E. Carpenter and V.G. Harris: MRS Proceedings, Spring meeting 2001
11. T. Ambrose, J.J. Krebs, and G.A. Prinz: J. Appl. Phys. **87**, 5463 (2000)
12. B. Ravel, M.P.Raphael, V.G.Harris, and Q. Huang: Phys. Rev. B **65**, 184431 (2002)
13. S. Ishida, T. Masaki, S. Fujii, and S. Asano: Physica B **245**, 1 (1998)
14. D. Orgassa, H. Fujiwara, T.C. Schulthess, and W.H. Butler: Phys. Rev. B **60**, 13237 (1999); D. Orgassa, H. Fujiwara, T.C. Schulthess,and W. H. Butler: J. Appl. Phys. **87**, 5870 (2000)
15. S. Picozzi, A. Continenza, and A.J. Freeman: Phys. Rev. B **69**, 094423 (2004)
16. H.J.F. Jansen, and A.J. Freeman: Phys. Rev. B **30**, 561 (1984); E. Wimmer, H. Krakauer, M. Weinert, and A.J. Freeman: Phys. Rev. B **24**, 864 (1981) and references therein
17. U. von Barth and L. Hedin: J. Phys. C **5**, 1629 (1972)
18. J.P Perdew and Y. Wang: Phys. Rev. B **45**, 13244 (1992); J.P. Perdew, K. Burke, and M. Ernzernhof: Phys. Rev. Lett. **77**, 3865 (1996)
19. H.J. Monkhorst and J.D. Pack: Phys. Rev. B **13**, 5188 (1976)
20. S.B. Zhang et al.: Phys. Rev B **57**, 9642 (1998)
21. C.G. Van de Walle et al.: Phys. Rev. B **47**, 9425 (1993)

22. G. Schmidt, D. Ferrand, L.W. Molenkamp, A.T. Filip, and B.J. van Wees: Phys. Rev. B **62**, R4790 (2000)
23. In the isolated point case: *i*) when Co substitutes for Mn, the first coordination shell of the defect is made of 8 Co atoms; *ii*) in the Mn antisite case, the first coordination shell of the defect is made of 4 Mn and 4 Si atoms. On the other hand, in the Co–Mn swap, the exchanged Mn has the first coordination shell made of 3 Mn, 1 Co and 4 Si atoms, whereas the exchanged Co has the first coordination shell made of 7 Co and 1 Mn atoms.
24. T. Asada and K. Terakura: Phys. Rev. B **47**, 15992 (1993); T. Asada and K. Terakura: *ibid.* **46**, 13599 (1992)
25. R. Fiederling, M. Klein, W. Ossau, G. Schmidt, A. Waag, and L.W. Molenkamp, Nature **402**, 787 (1999)
26. For further details and for the band structure plots, we refer to reference [2].
27. The accuracy in the position of E_F with respect to the minority VBM has been checked by increasing the number of **k** points and the k_{max} value, in order to achieve a higher level of convergence. We found that the relative alignment is stable to within 0.02 eV.
28. It was recently argued (see K. Capelle and G. Vignale: Phys. Rev. Lett. **86**, 5546 (2001)) that spin density functional potentials are not unique functionals of the spin densities, so that our values could miss the contribution due to the discontinuity of the exchange–correlation potential, and so show a problem similar to the underestimation of the band–gap in semiconductors occurring in the original (i.e. not spin–resolved) DFT.
29. P.J. Webster: J. Phys. Chem. Solids **32**, 1221 (1971)
30. A.G. Gavriliuk, G.N. Stepanov, and S.M. Irkaev: J. Appl. Phys. **77**, 2648 (1995)
31. S. Stadler, D.L. Harley, J.P. Craig, D.H. Minott, M.Khan, I. Dubenko, N. Ali, J. Dvorak, Y.U. Idzerda, D.A. Arena, and V.G. Harris: unpublished
32. P.J. Webster and K.R.A. Ziebeck. In: *Alloys and Compounds of d-Elements with Main Group Elements. Part 2.*, Landolt-Börnstein, New Series, Group III, vol 19c, ed by H.R.J. Wijn, (Springer, Berlin 1988) pp 75–184
33. Y.M. Yarmoshenko, M.I. Katsnelson, E.I. Shreder, E.Z. Kurmaev, A. Slebarski, S. Plogmann, T. Schlatholter, J. Braun, and M. Neumann: Eur. Phys. J. B **2**, 1 (1998); S. Plogmann, T. Schlatholter, J. Braun, M. Neumann, Y.M. Yarmoshenko, M.V. Yablonskikh, E.I. Shreder, E.Z. Kurmaev. A. Wrona, and A. Slebarski, Phys. Rev. B **60**, 6428 (1999)
34. J. Tersoff: J. Vac. Sci. Technol. B **4**, 1066 (1986) and references therein
35. J. Bardi, N. Binggeli, and A. Baldereschi: Phys. Rev. B **54**, 11102 (1996)
36. S. Picozzi, A. Continenza, S. Massidda, and A.J. Freeman: Phys. Rev. B **57**, 4849 (1998); *ibid.* **61**, 16736 (2000)
37. S.C. Erwin, S.-H. Lee, and M. Scheffler: Phys. Rev. B **65**, 205422 (2002)
38. S. Massidda, B.I. Min, and A.J. Freeman: Phys. Rev. B **59**, 144 (1987) and references therein

Magnetism and Structure of Magnetic Multilayers Based on the Fully Spin Polarized Heusler Alloys Co_2MnGe and Co_2MnSn

K. Westerholt, A. Bergmann, J. Grabis, A. Nefedov and H. Zabel

Institut für Experimentalphysik IV, Ruhr-Universität Bochum, D-44780 Bochum, Germany
kurt.westerholt@ruhr-uni-bochum.de

Abstract. Our Introduction starts with a short general review of the magnetic and structural properties of the Heusler compounds which are under discussion in this book. Then, more specifically, we come to the discussion of our experimental results on multilayers composed of the Heusler alloys Co_2MnGe and Co_2MnSn with V or Au as interlayers. The experimental methods we apply combine magnetization and magnetoresistivity measurements, x-ray diffraction and reflectivity, soft x-ray magnetic circular dichroism and spin polarized neutron reflectivity. We find that below a critical thickness of the Heusler layers at typically $d_{cr} = 1.5$ nm the ferromagnetic order is lost and spin glass order occurs instead. For very thin ferromagnetic Heusler layers there are pecularities in the magnetic order which are unusual when compared to conventional ferromagnetic transition metal multilayer systems. In [Co_2MnGe/Au] multilayers there is an exchange bias shift at the ferromagnetic hysteresis loops at low temperatures caused by spin glass ordering at the interface. In [Co_2MnGe/V] multilayers we observe an antiferromagnetic interlayer long range ordering below a well defined Néel temperature originating from the dipolar stray fields at the magnetically rough Heusler layer interfaces.

1 Introduction

In the main part of this article we will describe our experimental results on thin films and multilayers based on the Heusler phases Co_2MnGe and Co_2MnSn. However, before coming to this special topic, we will give a short review of the structural and magnetic properties of the Heusler compounds in general with special emphasis on the half metallic Heusler compounds. This is intended to give a general introduction to the experimental chapters of this book, where the focus of the investigations is on thin films of Heusler phases with theoretically predicted half metallic character. A knowledge of the metallurgical, structural and magnetic properties of bulk Heusler compounds will help the reader to estimate the problems one encounters when trying to grow thin films of the Heusler compounds optimized for spintronic applications.

The Heusler compounds have a long history in magnetism, starting more than 100 years ago with the detection of the ternary metallic compound Cu_2MnAl by F. Heusler [1]. That time the most interesting point with this phase was that it represented the first transition metal compound with ferromagnetically ordered Mn-spins. Later it turned out that a whole class of isostructural ternary metallic alloys with the general composition X_2YZ exists, where X denotes a transition metal element such as Ni, Co, Fe or Pt, Y is a second transition metal element, e.g. Mn, Cr or Ti and Z is an atom from 3rd, 4th or 5th group of the periodic system such as Al, Ge, Sn or Sb. More than 1000 different Heusler compounds have been synthesized until now, a comprehensive review of the experimental work until the year 1987 can be found in [2].

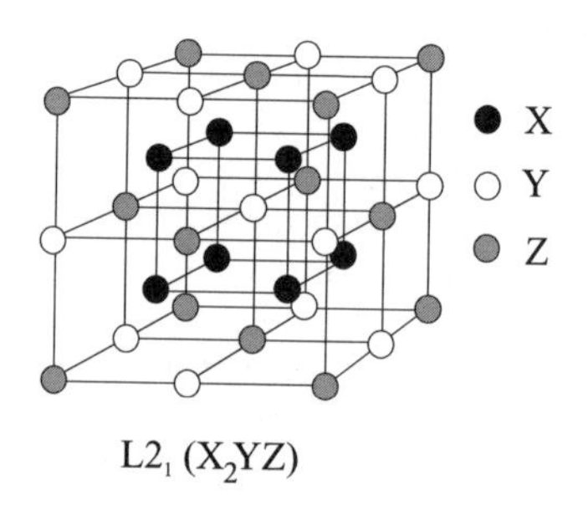

Fig. 1. Schematic representation of the $L2_1$-structure

The original Heusler phase Cu_2MnAl, which later gave its name to the whole class of ternary intermetallic compounds, can be considered as the prototype of all other Heusler compounds, including those with half metallic properties. It has been studied intensively over the past 100 years and we begin with describing its properties in some detail. The crystal structure of Cu_2MnAl is cubic with three different structural modifications: At high temperatures the crystal structure is bcc with random occupation of the Cu, Mn and Al atoms on the lattice sites of a simple bcc lattice with a lattice parameter $a = 0.297\,\mathrm{nm}$. At low temperatures the equilibrium crystal structure is fcc, space group $L2_1$, with the lattice parameter doubled to $a = 0.595\,\mathrm{nm}$ compared to the bcc high temperature phase. The $L2_1$-structure is most important for the predicted half metallic properties of the Heusler compounds discussed in this book, since the theoretical band structure calculations usually refer to this structure. The $L2_1$ unit cell is depicted in Fig. 1 and can be imagined to be combined of four interpenetrating fcc-sublattices occupied by Mn, Al and Cu atoms respectively and shifted along the space diagonal with the corner of the Al-sublattice at $(0, 0, 0)$, the first Cu-sublattice at $(1/4, 1/4, 1/4)$, Mn at $(1/2, 1/2, 1/2)$ and the second Cu-sublattice at $(3/4, 3/4, 3/4)$. There exists a third intermediate crystal structure for the Cu_2MnAl compound having B2-symmetry. It has the same lattice parameter as the $L2_1$ phase, but in the unit cell depicted in Fig. 1 the Mn- and Al-atoms are distributed at random on the Mn- and Al-sublattices whereas the two Cu-sublattices remain intact.

In Cu_2MnAl the B2 and the $L2_1$ phases develop from the high temperature bcc structure by chemical ordering of the atoms on the 4 fcc-sublattices in two steps via two second order structural phase transitions at 1040 K (bcc-B2) and 900 K (B2-$L2_1$) [3]. The structural order parameter of these transitions can be calculated by measuring the intensity of the superstructure Bragg reflection (200) for the bcc-B2 transition or the Bragg reflection (111) for the B2-$L2_1$-transition. Since in these order-disorder phase transitions diffusion of atoms is involved, long time annealing or slow cooling is necessary to achieve complete chemical ordering. In the majority of the Heusler compounds these or similar structural phase transitions exist, they can be hysteretic and of first order or reversible and of second order [2]. For several Heusler phases there in addition exist low temperature structural distortions of the cubic unit cell, the most prominent example being Ni_2MnGa with a martensitic phase transition at 175 K. Ni_2MnGa is the only ferromagnetic transition metal alloy which exhibits shape memory properties [4]. For Cu_2MnAl and other Heusler phases one often encounters problems with the metallurgical stability. Below $T = 900$ K the Cu_2MnAl phase tends to decompose into β-Mn, Cu_9Al_4 and Cu_3AlMn_2 [5]. Experimentally one can overcome this problem by suppressing the nucleation of the secondary phases by rapid quenching the Heusler alloys from high temperatures.

The magnetism of the Heusler alloys is very versatile and has been under continuous discussion all over the past hundred years. The majority of the Heusler alloys with a magnetic element on the Y-position order ferromagnetically, but several antiferromagnetic compounds also exists, Ni_2MnAl or Pd_2MnAl provide examples [6,7]. The main contribution to the magnetic moments in the Heusler phases usually stems from the atoms at the Y-position. If magnetic atoms also occupy the X-positions, their moment is usually quite small or even vanishing. In the above mentioned Ni_2MnAl compound e.g. the Ni-atoms are non-magnetic. Actually there exist a few Heusler compounds with rather large magnetic moments on both the X- and the Y-positions. In this case the ferromagnetic state is very stable and the ferromagnetic Curie temperatures T_c become exceptionally high. The best examples are provided by the Heusler phases Co_2MnSi and Co_2FeSi with a Co-moment of about 1 μB and Curie temperatures of 985 K [8] and 1100 K [9], respectively, the highest T_c-values known for the Heusler alloys. The reason for this very stable ferromagnetism is a strong next nearest neighbor ferromagnetic exchange interaction between the spins at the X- and the Y-position. If a non-magnetic element occupies the X-position, the dominant exchange interaction between the Y-spins is of weaker superexchange type, mediated by the electrons of the non-magnetic Z- and X-atoms [10].

Heusler compounds such as Cu_2MnAl with a magnetic moment only on the Y-position are considered as good examples of localized 3*d* metallic magnetism [10]. Since in the ideal $L2_1$ structure there are no Mn-Mn nearest neighbours, the Mn 3*d*-wave functions overlap only weakly and the magnetic moments remain essentially localized at the Mn-position. However, in the family of Heusler compounds there also exist beautiful examples of weak itinerant ferromagnetism with strongly delocalized magnetic moments. The compound Co_2TiSn with magnetic moments only on the X-positions belongs to this category [11]. As it is obvious from the crystallographic structure (Fig. 1), there are nearest neighbour X-atoms making the overlap of the 3*d*-wavefunctions and the delocalized character of the *d*-electrons much larger than in the case of only the atoms at the Y-position being magnetic. Replacing the Co-atoms in Co_2TiSn by Ni, this delocalization effect proceeds further, making the compound Ni_2TiSn a Pauli paramagnet [12]. Even more interesting, the Heusler compounds Fe_2TiSn and Fe_2VAl also do not order magnetically, but are marginally magnetic. They belong to the rare class of transition metal compounds exhibiting heavy Fermion like properties in the low temperature specific heat and the electrical resistivity, which attracted much interest in the literature in recent years [13–15].

Now, coming to the fully spin polarized Heusler compounds which are of main interest in the present book, the first Heusler compounds with a gap in the electron density of states for the spin down electron band at the Fermi level were detected in 1983 by electron energy band structure calculations [16]. The first two compounds with this spectacular property were NiMnSb and PtMnSb, so called semi Heusler alloys (space group $C1_b$) where in the Heusler unit cell of Fig. 1 one of the X-sublattices remains empty. The semi Heusler structure only exists if the Z-position is occupied by atoms with large atomic radii such as Sn or Sb. The NiMnSb and PtMnSb compounds have been dubbed half metals [16], since only for one spin direction there is metallic conductivity; for the other spin direction the conductivity is of semiconducting type. In a ferromagnetic transition metal alloy this half metallicity is a very rare property, since usually *s*- or *p*-bands with a small exchange splitting cross the Fermi energy and contribute states of both spin directions. For several years PtMnSb and NiMnSb remained the only ferromagnetic alloys with half metallic character, before starting from 1990 a second group of half metallic Heusler alloys, Co_2MnSi, Co_2MnGe and $Co_2Mn(Sb_xSn_{1-x})$ was detected [17–19].

The experimental proof of half metallicity in these Heusler alloys is a long and still ongoing controversial issue. The first attempts to prove the half metallicity used electron transport measurement to test the existence of a gap in the spin down electron band [20, 21]. Since in the half metal for temperatures small compared to the gap in the minority spin band there is only one spin direction at the Fermi level available, it is expected that electronic scattering processes involving spin flips and longitudinal spin wave excitations are inhibited. Thus one should expect an increasing electron mobility and a change of the power law describing the temperature dependence of the resistivity when the temperature becomes smaller than the gap for the minority spin band. Actually for NiMnSb this type of behavior for temperatures below 80 K has been detected, additionally in this temperature range the Hall coefficient exhibits an anomalous temperature dependence, strongly suggesting a thermal excitation of charge carriers across a gap coexisting with metallic conductivity [20,21]. More directly, positron annihilation experiments on bulk single crystals from the NiMnSb phase were found to be consistent with 100% spin polarization at the Fermi level [22]. Finally, analyzing the current-voltage characteristic below the superconducting gap of a point contact between a Nb-superconductor and a bulk PtMnSb sample, which is dominated by Andreev reflections at the ferromagnet/superconductor interface, the authors derived a spin polarization of 90% at the Fermi level [23].

For the Co_2Mn(Si,Ge,Sn) fully spin polarized group of Heusler compounds spin polarized neutron diffraction measurements on bulk samples have been employed to determine the degree of spin polarization at the Fermi level [24]. This methods probe the spatial distribution of the magnetization, details of which depend sensitively on the spin polarization. The main conclusion is that the spin polarization is large but not 100%. More recent superconducting/ferromagnetic point contact measurements on a Co_2MnSi single crystal gave a spin polarization of 55% [25]. Similarly, the degree of spin polarization determined from the analysis of spin resolved photoemission spectra was always found to be definitely below 100% [26, 27].

During the first years after the discovery of the half metallic character in the Heusler compounds they were considered as exotic ferromagnets of mainly academic interest. This attitude has changed completely with the development of new ideas of data storage and processing designed to use both, the charge and the spin degree of freedom of the conduction electrons, nowadays called spin electronics (spintronics) [28,29]. In the spintronic community there is a strong belief that in the future these new concepts have the perspective to complement or even substitute conventional Si technology. It was rapidly realized how valuable it would be for spintronic devices to have a ferromagnet available with only one conduction electron spin direction at the Fermi level. With an electrode possessing 100% spin polarization the generation of a fully spin polarized current for spin injection into semiconductors would be very easy [30], spin filtering and spin accumulation in metallic thin film systems would be most effective [31] and the tunneling magnetoresistance (TMR) [32] of a device prepared of two half metallic electrodes should have a huge magnetoresistance, since the tunneling probability is vanishing to first order if the two electrodes have an antiparallel magnetization direction.

Thus the novel concepts of spintronics started an upsurge of interest in ferromagnetic half metals in the literature. Principally there are three different classes of ferromagnetic half metals which are under intense discussion in the literature and regarded as possible candidates to be used in spintronic devices. These three groups combine semiconductors such as GaAs or ZnO doped by magnetic transition metal ions [33], magnetic oxides such as CrO_2, Fe_3O_4 or $(Sr_{1-x}La_x)MnO_3$ [34, 35] and, last but not least, the fully spin polarized Heusler compounds under discussion in this book.

Actually with each of these three essentially different groups of materials one encounters serious problems when thinking of possible applications. The question which group or which compound is the best is completely open. The magnetically doped semiconductors all share the problem of magnetic inhomogeneity and rather low ferromagnetic Curie temperatures, making them unsuitable for room temperature applications. For the ferromagnetic oxides, on the other hand, the main problem is that the preparation and handling of these compounds is completely incompatible with current semiconductor technology. In this respect the fully spin polarized Heusler compounds appear to be the better choice, since they can be prepared by conventional methods of metallic thin film preparation. A further advantage of the Heusler compounds, especially in comparison with the magnetically doped semiconductors, is their high ferromagnetic Curie temperature. However, as already stated above and frequently discussed in the following chapters of this book, the half metallicity of the Heusler compounds is a subtle property which is easily lost in a real sample.

Recent intense work by theoretical groups, several of them represented in this book, increased the number of Heusler compounds with predicted half metallic properties to more than 20, among them Rh_2MnGe, Fe_2MnSi and Co_2CrAl, to mention a few of the new compounds [36–40]. From the experimental side, most of these phases have not been studied thoroughly yet, thus leaving a vast field for future experimental investigations. The experimental work in the literature mainly concentrated on the classical Heusler half-metallic phases NiMnSb and PtMnSb, the alloys Co_2MnSi and Co_2MnGe and the newly discovered compound $Co_2(CrFe)Al$ [38].

When thinking of possible applications in spintronic devices, which at the moment is the main motivation behind the experimental research on the fully spin polarized Heusler compounds in the literature, it is obvious that one needs thin films of the Heusler compounds; bulk samples are not very useful. This is the reason why in the majority of the experimental chapters in this book thin film systems are studied. Additional detailed research on high quality bulk material, however, is highly welcome and urgently needed for purposes of comparison with the thin films and in order to elucidate the basic properties of the compounds. The recent XAFS and neutron scattering study of Co_2MnSi single crystals provides a beautiful example [41, 42]. It allowed a precise determination of the number of antisite defects in the crystal (Co sitting on the Mn position and vice versa), which would be very difficult in thin films, but is important for a realistic judgment of the perspective to reach full spin polarization in a Co_2MnSi thin films prepared under optimum conditions.

In the experimental chapters of this book it will become clear to the reader that using the predicted full spin polarization of the Heusler compounds in thin film structures is difficult and the technical development is still in its infancy. Thin film preparation in general and especially the preparation of thin film heterostructures, often imposes limits on the process parameters and this might severely interfere with the needs to have of a high degree of spin polarization. For obtaining a large spin polarization a perfect crystal structure with a small number of grain boundaries is important. This can best be achieved by keeping the substrate at high temperatures during the thin film deposition. However, most Heusler phases grow in the Vollmer-Weber mode (three-dimensional islands) at high temperatures, thus when using high preparation temperatures there might be a strong roughening of the surfaces which for spintronic devices is strictly prohibited. In addition, in thin film heterostructures combining different metallic, semiconducting or insulating layers with the Heusler compounds, high preparation temperatures are forbidden, since excessive interdiffusion at the interfaces must be avoided.

Interfaces of the Heusler compounds with other materials are a very delicate problem for spintronic devices. For spin injection into semiconductors or a tunneling magnetoresistance the spin polarization of the first few monolayers at the interfaces is of utmost importance. A large spin polarization in the bulk of a Heusler compound does not guarantee that it is a good spintronic material, unless it keeps its spin polarization down to the first few monolayers at the interfaces. Thin films and TMR-devices have first been investigated systematically using the semi Heusler compounds PtMnSb and NiMnSb [42]. The performance was rather disappointing e.g. the maximum value for the TMR achieved was of the order of 12% only. Later, TMR devices based on Co_2MnSi was shown to work much better [44] (see article by J. Schmalhorst in this book), but still the calculated spin polarization is definitely smaller than 100%.

This experience leads to the suspicion that at least for a few monolayers at the interfaces the full spin polarization is somehow lost. One possible explanation for this could be modifications in the electronic energy band structure at surfaces and interfaces, since the theoretical predictions of a full spin polarization in a strict sense only hold for bulk crystals. Recent band structure calculations taking the surfaces explicitly into account revealed that for certain crystallographic directions the full spin polarization at the surfaces is actually lost [45, 46]. Even more seriously, it seems that atomic disorder of the Heusler compound at surfaces and interfaces has the tendency to severely affect the full spin polarization. Theoretical band structure calculations predicting the half metallicity assume $L2_1$-symmetry with perfect site symmetry of all four X, Y and Z sublattices. As mentioned above, the existence of some antisite disorder, however, cannot be completely avoided even in well annealed single crystals [41, 42] .

An essential question is, which type of disorder is most detrimental for the spin polarization of the Heusler compounds. Model calculations taking different types of point defects into account are very illustrative in this respect [39,47] (see also the article by S. Picocci in this book). These calculations show that for the Co_2MnGe and the Co_2MnSi Heusler phase the electronic states of Co-antisite atoms sitting on regular Mn-sites fill the gap in the minority spin band. The magnetic moment of these Co antisite defect remains virtually unchanged and couples ferromagnetically to the surrounding spins. A Mn-antisite atom sitting on a regular Co-position, on the other hand, does not introduce electronic states in the minority spin gap, however has its magnetic moment coupled antiferromagnetically to the surrounding spins. This leads to a drastic reduction of the saturation magnetization, which frequently has been observed for ferromagnetic Heusler phases not prepared under optimum conditions.

Introducing now the present article which reviews our work during the last four years on thin films and multilayers based on the Heusler alloys Co_2MnGe and Co_2MnSn, our main interest was not on spintronic devices for the moment, although we will present examples of GMR devices in Sect. IV. We, instead, go one step backwards and study first the basic structural and magnetic properties of very thin films of these Heusler compounds. The magnetism of a ternary metallic compound is complex and little is known of what happens with the magnetic order in the limit of very thin films or at interfaces. Then we present our results on the magnetic order in multilayers of the Heusler alloys using V and Au as interlayers, which can be grown with high quality. Neutron and x-ray scattering provide well established methods to get insight into the structural and magnetic properties of the interfaces. And, by the way, we follow the interesting and until now unsolved question, whether an oscillating interlayer exchange coupling (IEC) between the Heusler layers can be observed when varying the thickness of the V- and Au-layers systematically.

In detail, our article below is organized as follows: In Sect. 2 we discuss the magnetic and structural properties of single films of the Co_2MnGe and Co_2MnSn phase for a thickness range down to about 3 nm. In Sect. 3 we discuss the properties of multilayers prepared by combining these Heusler layers with Au and V interlayers. In this section we also review in some detail a novel type of antiferromagnetic interlayer long range order which we have observed in $[Co_2MnGe/V]_n$ multilayers and an unidirectioal exchange anisotropy originating from the interface magnetic order in the $[Co_2MnGe/Au]_n$ multilayers. In Sect. 4 we present results of our resistivity and magnetoresistivity measurements on the single films, multilayers and GMR devices, before we summarize and draw conclusions in Sect. 5.

2 Single Co_2MnGe and Co_2MnSn Thin Films

2.1 Structural Properties

All films of the present study were grown by rf-sputtering with a dual source HV sputtering equipment on single crystalline Al_2O_3 (11–20) surfaces (sapphire a-plane). The base pressure of the system was 5×10^{-8} mbar after cooling of the liquid nitrogen cold trap. We used pure Ar at a pressure of 5×10^{-3} mbar as sputter gas, the targets were prepared from stoichiometric alloys of the Heusler phases. The sputtering rates during the thin film preparation were 0.06 nm/s for the Co_2MnSn phase and 0.04 nm/s for the Co_2MnGe phase, the Au or V seed layers were deposited at a sputtering rate of 0.05 and 0.03 nm/s respectively. The sapphire substrates were cleaned chemically and ultrasonically after cutting and immediately before the deposition they were additionally etched by an ion beam for 300 s in order to remove any residual surface contaminations. There are two conditions to achieve good structural and magnetic quality of the Heusler layers. First, the substrate temperature must be high, in our optimized procedure the substrate temperature was 500°C. Second, seed layers of a simple metal with a good lattice matching to the Heusler compounds are important to induce epitaxial or textured growth with a flat surface. Growing the films directly on sapphire results in polycrystalline films and surface roughening. In our previous investigations we tested different possible seed layers such as Cr, Nb and Ag, later we concentrated on Au and V which gave the best structural results [66, 67]. The surfaces of the Heusler compounds oxidize rapidly and must be covered by a protection layer before handling in air. If nothing else is stated we use an amorphous Al_2O_3 film of 2-nm thickness as a cap layer.

The structural quality of the samples was studied by small angle x-ray reflectivity and high angle out-of-plane Bragg scattering using Cu K_α radiation. Figure 2 show an out of plane x-ray Bragg scan of a Co_2MnSn film with a 2-nm thick Au seed layer and Co_2MnGe film with a 5-nm V seed layer. Only the (220) and the (440) Bragg peaks of the Heusler phases are observed, proving perfect (110) texture out-of-plane. In plane the films are polycrystalline. Table 1 summarizes the relevant structural data derived from the x-ray scans. The lattice parameters virtually coincide with those of the bulk phases. The half width of the rocking curve (width in 2Θ at half maximum) characterizing the mosaicity of the crystallite is 1.5° for the Co_2MnSn phase and 0.8° for the Co_2MnGe phase. The half width of the radial Bragg scans corresponds to the experimental resolution of the spectrometer implying that the total thickness of the Heusler layer is structurally coherent in the growth direction. Examples of a small angle x-ray reflectivity scan for films of the same phases are shown in Fig. 3. Total thickness oscillations are observed up to $2\theta = 10°$, indicatiove of a very flat surface morphology. additional oscillations from the buffer layer are also observed. A simulation of the curves with the Parratt formalism gives a typical rms roughness for the film surface of

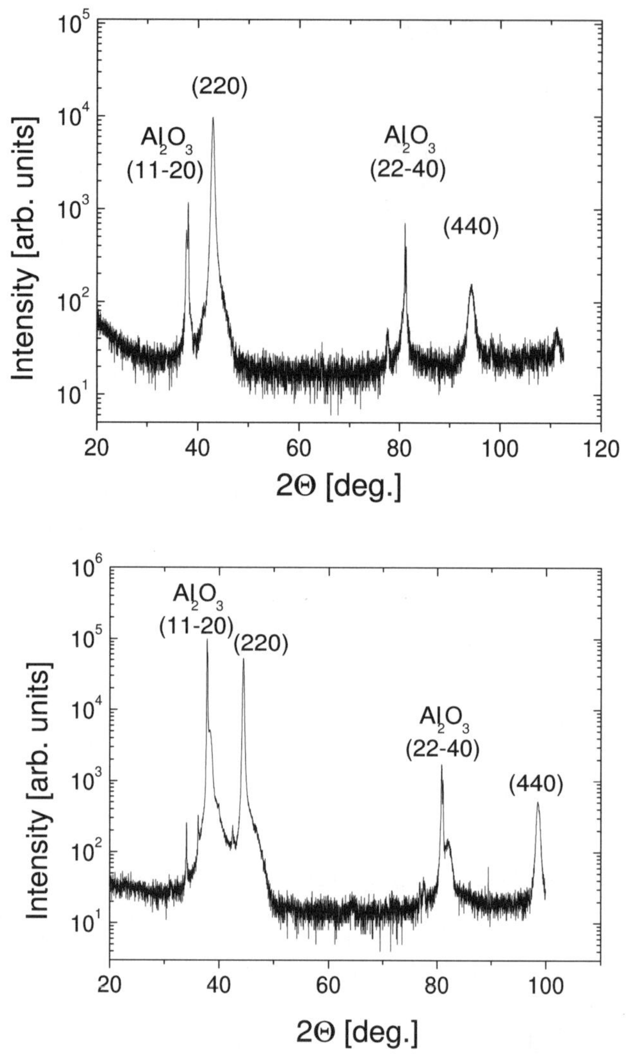

Fig. 2. Upper panel (**a**): Out of plane Bragg scans of a Au(3 nm)/ Co_2MnSn(100 nm) film measured with Cu K_α radiation. Bottom panel (**b**): same for a V(3 nm)/Co_2MnGe(100 nm) film

about 0.3 nm. This is corroborated by atomic force microscopic (AFM) pictures of the surface which reveal a very smooth surface morphology. We also have grown films at lower preparation temperatures down to a substrate temperature $T_{sub} = 100°$C. The structure of films prepared below $T_s = 500°$C is still perfect (110) texture, however with a rocking width slightly increasing with decreasing preparation temperature.

Table 1. Lattice parameters, saturation magnetization and saturation magnetic moments (per formula unit) of the pure, thick Heusler films in comparison to the bulk values. The bulk values have been taken from [2]

Phase	Lattice Parameter [nm] Bulk	Lattice Parameter [nm] Film	Sat. Magnetization $[\frac{emu}{g}]$ Bulk	Sat. Magnetization $[\frac{emu}{g}]$ Film	Sat. Moment $[\mu_B]$ Film
Co_2MnGe	0.5743	0.575	111	114	5.02
Co_2MnSn	0.600	0.6001	91	92	4.95

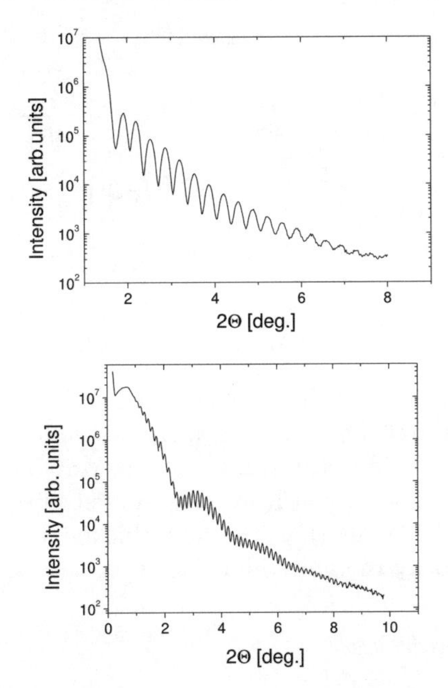

Fig. 3. X-ray reflectivity scans of Au(3 nm)/Co_2MnSn(30 nm) in the upper panel (**a**) and V(4 nm)/Co_2MnGe(60 nm) in the bottom panel (**b**)

Since for the multilayers discussed in the next section the typical layer thickness of the Heusler layers is of the order of 3 nm only and T_s is limited to 300°C in order to avoid strong interdiffusion at the interfaces, it seems worthwhile to study the structural and magnetic properties of single very thin Heusler layers prepared under the same conditions. Figure 4 shows an out-of- plane Bragg scan of a trilayer V(5 nm)/Co_2MnGe(4 nm)/V(5 nm). One observes a very broad (220) Bragg peak with a half width of $\Delta 2\Theta = 2°$ at $2\Theta = 43°$ from the Co_2MnGe phase. Using the Scherrer formula for the Bragg reflection of small particles, which correlates the coherence length L_c of the particles with the half width of the Bragg peaks

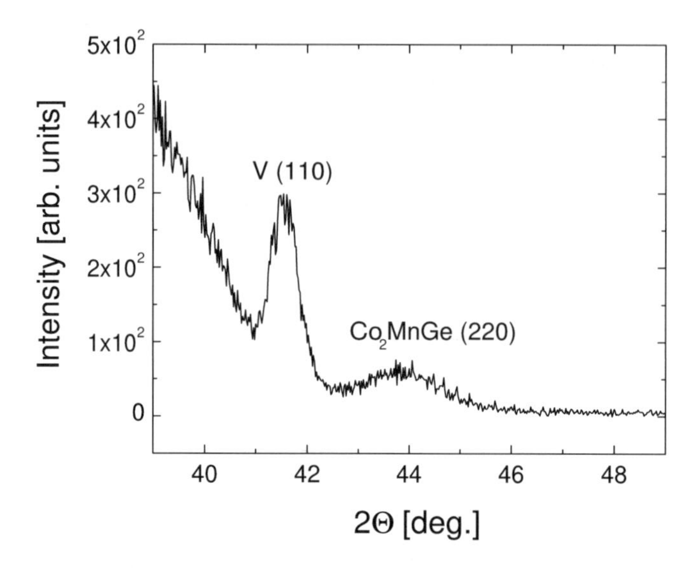

Fig. 4. Out of plane Bragg scan of a V(5 nm)/Co_2MnGe(4 nm)/V(5 nm) trilayer

$$L_c = 0.89\lambda/[\Delta(2\Theta) \cdot \cos(\Theta)] \quad (1)$$

we derive $L_c = 3$ nm i.e. a value slightly smaller than the layer thickness. This is an indication that the mean size of the crystalline grains in the film is slightly smaller than the film thickness. For Co_2MnGe layers of 4-nm thickness between two Au layers, which we also have studied, the grain size was found to be definitely larger than the film thickness, since the correlation length is only limited by the finite film thickness.

2.2 Magnetic Properties

The dc magnetization of our films was studied by a commercial superconducting quantum interference device (SQUID) based magnetometer (Quantum Design MPMS system). Examples of magnetic hysteresis loops of the Co_2MnGe and the Co_2MnSn films are presented in Fig. 5. The films possess a growth induced weak, uniaxial anisotropy with an anisotropy field H_K of about 50 Oe. For the measurements in Figs. 5 and 6 the external field axis was directed parallel to the magnetic easy axis, thus the hysteresis loops are rectangular. The coercive field for the Co_2MnGe phase is $H_c = 20$ Oe, for the Co_2MnSn film we get $H_c = 50$ Oe at room temperature. The saturation magnetization at 5 K is in good agreement with the values measured in bulk samples within the experimental error bars (see Table 1). The saturation magnetic moment per formula unit calculated from the saturation magnetization is 4.95 μ_B for Co_2MnSn and 5.02 μ_B for Co_2MnGe. These values agree with those derived from the theoretical band structure calculations for

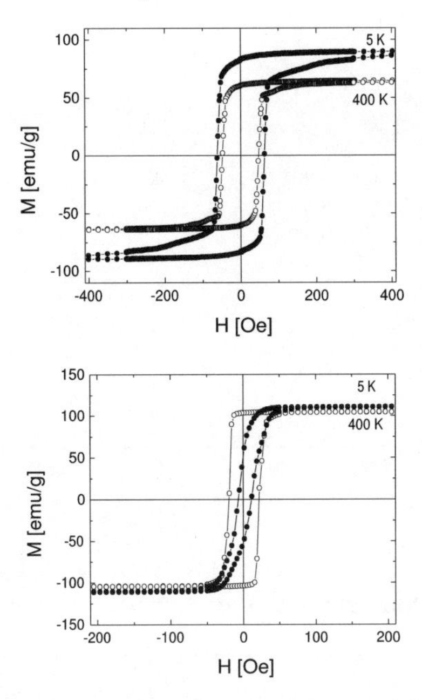

Fig. 5. Hysteresis loops of a Co_2MnSn in the upper panel (**a**) and a Co_2MnGe film in the bottom panel (**b**) from Fig. 2 measured at 5 K and 400 K

perfect $L2_1$ type of order [36], indicating that the films have a high degree of metallurgical order. Thus we can conclude that thick films of the Co based Heusler phases can be grown at high temperatures with a quality comparable to bulk samples.

However, when decreasing the substrate temperature while keeping the thickness of the Co_2MnGe layer and all other parameters constant, we observe a continuous decrease of the saturation magnetization down to about 70% at $T_s = 100°C$ (Fig. 6). This decrease of the magnetization is accompanied by an increase of the lattice parameter of about 1%. For the Co_2MnSn phase and Cu_2MnAl the degradation of the ferromagnetic saturation magnetization with decreasing T_s is even faster (Fig. 6). It seems plausible to attribute the decrease of the saturation magnetization to an increasing number of antisite defects in the $L2_1$ structure. This effect is well known in Cu_2MnAl, where the disordered B2 phase which can be prepared by quenching from high temperatures has a very low saturation magnetization [73]. For films with a small thickness of the Heusler phase of the order of a few nm the situation becomes even worse. In this case a preparation at 500°C is prohibited, since then interdiffusion at the seed layer/Heusler interface is too strong. A practical limit for the substrate temperature for avoiding excessive interdiffusion is 300°C. As a first example of a very thin film, Fig. 7 depicts the hysteresis loop measured at 5 K of a 4-nm thick Co_2MnGe film grown directly on sapphire

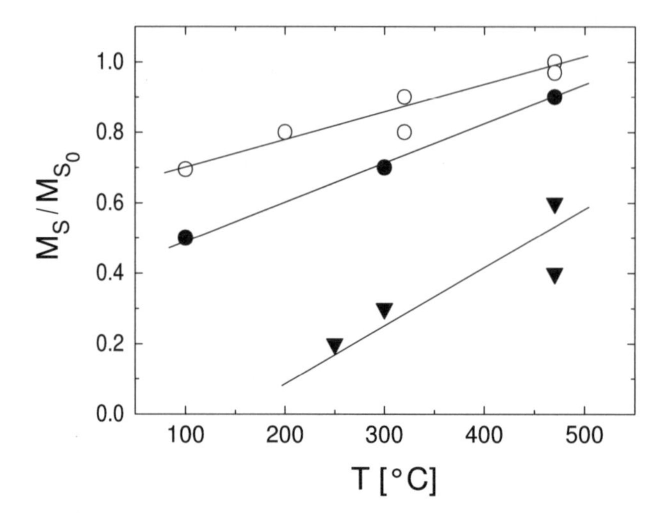

Fig. 6. Saturation magnetization of Co_2MnGe (*open circles*), Co_2MnSn (*solid circles*) and Cu_2MnAl (triangles) versus the substrate temperature during preparation

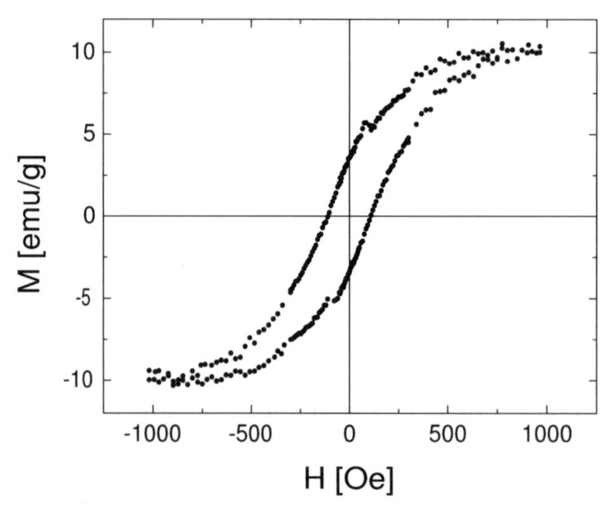

Fig. 7. Hysteresis loop of a 4-nm thick Co_2MnGe film grown directly on sapphire a-plane at a substrate temperature of 300°C measured at 5 K

a-plane at $T_s = 300$°C. The x-ray structural analysis showed no resolvable Bragg peak, thus the crystalline structure seems to be polycrystalline with very small grains. The film has a very low saturation magnetization of only about 10% of the bulk value, showing that the ferromagnetic properties are completely different from those of the Co_2MnGe phase in the $L2_1$ structure. Growing the same film thickness for Co_2MnGe on a V seedlayer, a (220) Bragg peak can be observed (see Fig. 4) and about 50% of the ferromagnetic

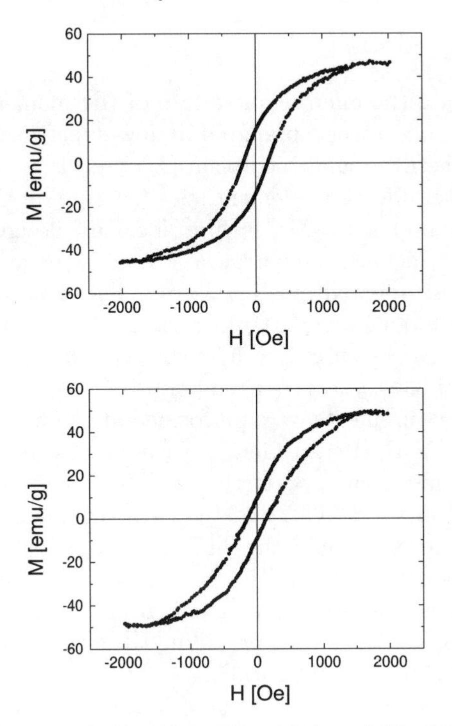

Fig. 8. Hysteresis loops measured at 5 K of V(3 nm)/Co_2MnGe(4 nm)/V(3 nm) with the field applied parallel to the sample plane in the upper panel (**a**) and perpendicular to the sample plane in the bottom panel (**b**)

saturation magnetization of the Co_2MnGe phase is recovered (Fig. 8). The magnetization is isotropic in the film plane with a magnetic remanence of only about 30% of the saturation magnetization.

It is interesting to note that for a field direction perpendicular to the film plane (Fig. 8b) the hysteresis curves are very similar to those observed in a parallel field, thus the strong anisotropy of the demagnetising field characterizing a homogeneous magnetic thin film is absent. This clearly shows that the magnetization of the film in Fig. 8 is not homogeneous, but breaks up into weakly coupled small magnetic particles pointing in their own magnetically easy direction given by the geometric shape and the crystal magnetic anisotropy. For a 4-nm thick Co_2MnGe film grown on Au the small particle character of the hysteresis loop also exists, but the magnetic anisotropy of a homogenous thin film is partly recovered. For this film the ferromagnetic saturation magnetization is also strongly reduced compared to the bulk value of Co_2MnGe. We would attribute the different magnetic behaviour of Co_2MnGe grown on V and Au to much smaller crystalline grains in the case of a V seed layer, where the single grains seem to be nearly decoupled magnetically.

2.3 XMCD on Co_2MnGe

In order to elucidate the microscopic origin of the moment reduction in the Co_2MnGe Heusler alloy when prepared at low substrate temperatures (see Fig. 6) x-ray magnetic circular dichroism (XMCD) is a very suitable experimental method as it allows an element specific study of the magnetism [74]. We therefore prepared a Co_2MnGe film specially designed for an XMCD study with 16-nm thickness grown on a V seed layer and with a 2-nm Au cap layer, all layers prepared at $T_s = 300°C$. The saturation magnetization of the film as measured by SQUID magnetometry was found to be slightly smaller than the value given in Fig. 5, it corresponds to a magnetic moment per formula unit of 2.3 μ_B at room temperature and 2.98 μ_B at 4 K.

The XMCD measurements were performed at the bending magnet beamline PM3 at BESSY II (Berlin Germany) using the new ALICE chamber for spectroscopy and resonant scattering [75]. The measurements were taken by the total electron yield (TEY) method. At an angle of incidence of 40° saturation effects are small and the TEY is proportional to the absorption coefficient to a good approximation. During the experiment the helicity of the photons was fixed whereas the magnetization of the sample was switched by a magnetic field of ± 0.1 T thus providing the electron yield with the magnetization parallel (Y_+) and antiparallel (Y_-) to the photon helicity. The Y_+ and Y_- scans measured at the L_3 edge of Mn and Co are normalized to the flux of the incoming photon beam. The XMCD spectrum (Y_+–Y_-) at the $L_{2,3}$ edges of Mn and Co measured at room temperature is plotted in Fig. 9. The XMCD spectra contain quantitative information on the spin and orbital magnetic moments which can be extracted via the sum rule analysis [71].

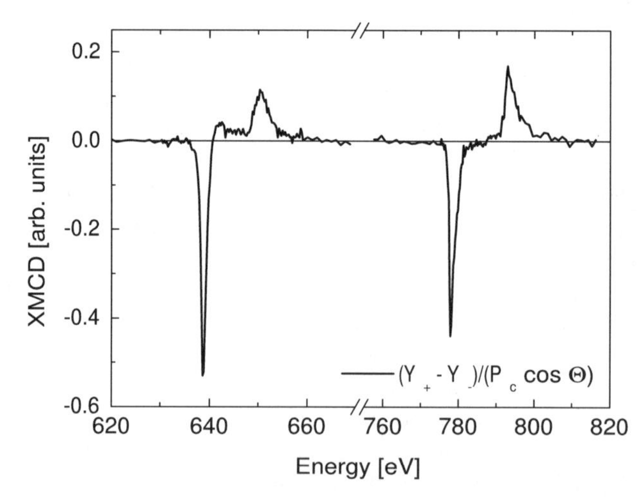

Fig. 9. XMCD spectra of a 11-nm thick Co_2MnGe film at the Mn $L_{3,2}$ and at the Co $L_{3,2}$ edge

There are several sources of systematic errors in this analysis which might affect the absolute values for the magnetic moment questionable. This is first the number of holes in the d−band, which we can precisely take from electronic band structure calculations. Second, it is the neglecting of magnetic dipolar interactions in the model, which in our case seems justified because of the cubic symmetry of the Co_2MnGe phase. Third, for the Mn atom there might be a mixing of the L_3 and the L_2 levels by the relatively strong $2p-3d$ Coulomb interactions. The correction factor x taking this effect into consideration has been calculated ranging from $x = 1$ for negligible jj-mixing to $x = 1.5$ for strong jj-mixing [76]. Keeping these reservation in mind, the sum rule analysis yields for the case of the Co atom $m_{spin} = 0.55\ \mu_B$ for the spin magnetic moment and $m_{orb} = 0.028\ \mu_B$ for the orbital magnetic moment. For the case of the Mn atom the analysis yields $m_{spin} = 0.98\ \mu_B$ $(1.47\ \mu_B)$ and $m_{orb} = 0.056\ \mu_B$, where for the first value it is assumed that $x = 1.0$ holds, for the value given in brackets $x = 1.5$ is assumed. Summing up all values for the atomic magnetic moments and extrapolating to 4 K we determine a saturation magnetic moment of $m = 0.75\ \mu_B$ for Co and $m = 1.36\ \mu_B$ $(1.97\ \mu_B)$ for Mn [71]. The moment for Co agrees reasonable with the theoretical value from band structure calculations, for the Mn atom the theoretical calculations give $m = 3.6\ \mu_B$ i.e. a much higher value than the experiment, irrespective of the exact value of the correction factor x. The magnetization data yielded a saturation magnetic moment of 2.98 μ_B per Co_2MnGe formula unit, the XMCD results yield 2.83 μ_B $(3.46\ \mu_B)$, i.e. within the uncertainty range of the XMCD result the agreement is satisfactory.

As mentioned in the introduction, the theoretical model calculations [39, 47] show that antisite disorder in the $L2_1$ structure severely affects the Mn magnetic moments, since a Mn spin sitting on a regular Co position has a spin direction antiparallel to the other Mn and Co spins. This will strongly reduce the experimentally determined mean Mn moment. The value of the Co moment when sitting on a Mn position remains essentially unaffected. Thus the plausible hypothesis formulated above that a lower preparation temperatures causes site disorder in the $L2_1$ structure and concomitantly a lowering of the saturation magnetization finds strong support from the XMCD results and the theoretical model calculations.

Summarizing this chapter, we have shown that with optimized preparation conditions high quality thick films of the Co_2MnGe and the Co_2MnSn phase can be grown. But if experimental constraints are imposed when preparing devices such as limits for the substrate temperature, non applicability of seed layers or if in devices very thin Heusler layers are needed, one faces problems. Site disorder in the interior of the films and mixing and disorder at interfaces have the tendency to lower the ferromagnetic magnetization. The magnetic behaviour of very small Co_2MnGe grains as e.g. the nearly complete loss of ferromagnetism for Co_2MnGe deposited on bare sapphire or the typical small particle magnetic behaviour in very thin Co_2MnGe grown on V suggest that the grain boundaries are only weakly ferromagnetic or even non ferromagnetic.

3 Multilayers $[Co_2MnSn(Ge)/Au(V)]_n$

3.1 Structural and Magnetic Properties

Multilayers of the two Heusler phases Co_2MnGe and Co_2MnSn with V and Au as interlayers have been prepared by the same dual source rf-sputtering equipment described in Sect. 2. For the multilayer preparation the substrate holder is swept between the two targets, if nothing else is stated, the number of bilayers prepared is $n = 30$ starting with either V or Au, and the films are protected by a Al_2O_3 cap layer. The substrate temperature was held fixed at $T_S = 300°C$ for all multilayers, the deposition rates of the materials were the same as given in Sect. 2 for the single layers. Although the structural quality of the Heusler layers improve at higher substrate temperatures, $T_S = 300°C$ turned out to be the upper limit when strong interdiffusion at the interfaces must be avoided.

Using the natural gradient of the sputtering rate, the simultaneous preparation of up to 10 samples with the thickness of either the magnetic layer or the non magnetic layer altered is possible. The thickness can be varied by a factor of three and we exploit this feature e.g. for the preparation of several series of multilayers for the investigation of the thickness dependence of the magnetic interlayer coupling. In our previous investigations we have also tested the growth of Heusler multilayers with several other combinations of materials, including the Heusler compounds Co_2MnSi and Cu_2MnAl and Cr as interlayers [65, 67]. Here we concentrate on Co_2MnGe and Co_2MnSn with Au and V as interlayers which can be grown with best structural quality. The x-ray characterization of the multilayers was done by a standard thin film x-ray spectrometer or using synchrotron radiation at the Hasylab in Hamburg, Germany.

Figure 10a shows a low angle x-ray reflectivity scan measured on a $[Co_2MnGe(3\,nm)/V(2\,nm)]_{50}$ multilayer (the number in round brackets denotes the nominal thickness of the single layers, $n = 50$ is the number of bilayers) by synchrotron radiation at a wavelength of 0.177 nm. One observes superlattice reflections up to the 4th order revealing a smooth layered structure. From a fit with the Parratt formalism shown by the displaces thin line one derives a thickness of $d_V = 2.2\,nm$ and $d_{Heusler} = 2.8\,nm$ and a roughness parameter of $\sigma = 0.5\,nm$ for V and $\sigma = 0.7\,nm$ for Co_2MnGe. Thus at the interfaces there is interdiffusion and/or topological roughness on a scale of about 0.6 nm.

Figure 10b depicts a scan across the (220)/(110) Bragg reflection of the same multilayer. There is a rich structure with satellites up to 3rd order proving the growth of coherent V and Co_2MnGe layers. The out of plane

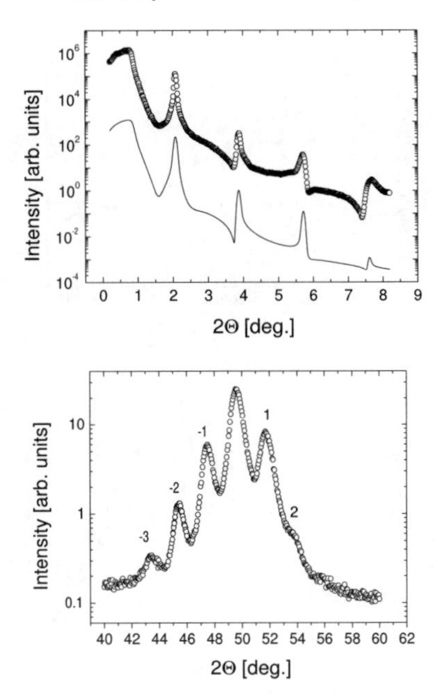

Fig. 10. Upper panel (**a**): X-ray reflectivity scan of a $[Co_2MnGe(3\,nm)/V(2\,nm)]_{50}$ multilayer measured at a wavelength of the synchroton radiation $\lambda = 1,77$ nm. The open symbols are the measured intensity, the straight line shows a simulation. Bottom panel: (**b**) Out of plane Bragg scan of the same multilayer at $\lambda = 1,77$ nm. The numbers in the figure denote the order of the superlattice reflections and the order of the satellites

coherence length derived from the width of the Bragg peak is about 13 nm i.e. comprises about 3 bilayers. Figure 11a shows the small angle reflectivity scan for a nominal $[Co_2MnGe(3\,nm)/Au(2.5\,nm)]_{50}$ multilayer. The fit with the Parratt formalism gives a thickness of 3.0 nm and 2.3 nm for the Heusler and Au layer respectively. As already evident from the narrower superlattice reflection peaks, the quality of the layered structure is even better than for the $[Co_2MnGe/V]$ multilayer of Fig. 10a. This is corroborated by the simulation which gives a roughness parameter $\sigma_{Au} = 0.5$ nm and $\sigma_{Heusler} = 0.3$ nm. The Bragg scan (Fig. 11b) at the (111)/(220) Bragg position also reveals an out of plane coherent crystalline lattice with a coherence length of 14 nm.

We have also grown successfully $[Co_2MnSn/V]$ and $[Co_2MnSn/Au]$ multilayers with comparable quality [66] i.e. with an out of plane coherence length and a roughness parameter similar to the $[Co_2MnGe/V]$ and $[Co_2MnGe/Au]$ multilayers. Multilayers of similar quality can be grown in the thickness range of down to 1 nm for the Heusler compounds and the interlayers, for a smaller thickness the layered structure and the crystalline coherence is gradually lost.

The magnetic properties of the multilayers were studied by measurements of the magnetic hysteresis loops. Figures 12 and 13 show examples of hys-

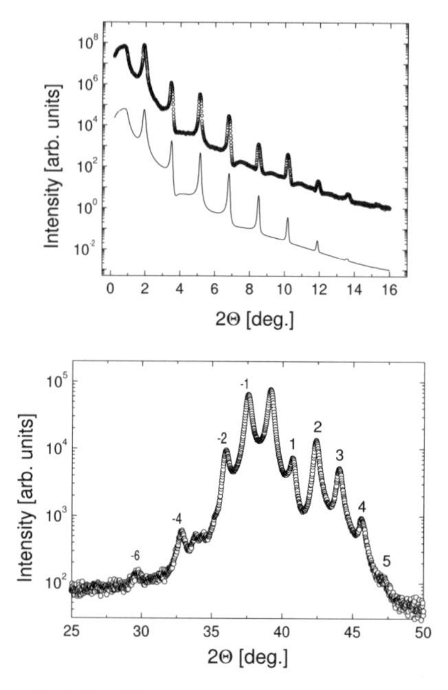

Fig. 11. Upper panel (**a**): X-ray reflectivity scan of a $[Co_2MnGe(3\,nm)/Au(2.5\,nm)]_{50}$ multilayer measured at $\lambda = 1,54\,nm$. The open symbols are the measured intensity, the straight line shows a simulation. Bottom panel (**b**): Out of plane Bragg scan of the same multilayer at $\lambda = 1,54\,nm$

teresis loops from the $[Co_2MnGe/Au]$, the $[Co_2MnSn/Au]$, the$[Co_2MnGe/V]$ and the $[Co_2MnSn/V]$ multilayer systems at different temperatures. Several important features should be noted.

The saturation magnetization measured at 5 K is smaller than the value for the bulk compound, the reduction is comparable to the one which we observed in single films of the same systems in Sect. 2. The coercive field is strongly increasing at low temperatures from values of typically $H_c = 50$ Oe at 60 K to several hundred Oe at 4 K. This effect is shown in more detail in Fig. 14 where the coercive force drastically increases below 50 K. This feature is often observed in magnetically inhomogeneous films and indicates that thermal activation plays an important role in the remagnetization processes at higher temperatures. Interestingly for the multilayers with V in Fig. 13 there is no observable magnetic remanence at higher temperature, the magnetization curve is completely reversible for temperatures above about 150 K. On the other hand for the multilayers with Au (Fig. 12) there is a hysteresis with a finite remanent magnetization up to the ferromagnetic Curie temperature, as it should be for a normal ferromagnetic compound. Most of our multilayers possess a growth induced uniaxial magnetic anisotropy similar to the thick films discussed in Sect. 2, but with a definitely smaller amplitude of the order of 20 Oe for the anisotropy fields. For the hysteresis curve

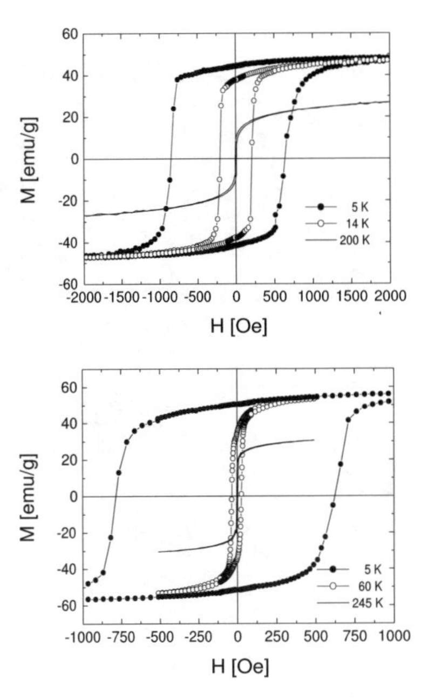

Fig. 12. Hysteresis loops of the multilayers [Co_2MnGe(2.2 nm)/Au(3 nm)]$_{30}$ in the upper panel (**a**) and [Co_2MnSn(3 nm)/Au(1.5 nm)]$_{30}$ in the bottom panel (**b**) measured at different temperatures given in the figure

measurements in Fig. 12 the field is applied along the magnetic easy axis, thus the vanishing remanent magnetization cannot simply be explained by a magnetic anisotropy field perpendicular to the field direction.

We prepared series of 10 multilayer samples with either the thickness of the Heusler compound or the thickness of the V or Au layer varied between 1 nm and 3 nm and systematically studied the hysteresis curves. These results are summarized in the next three figures. In Fig. 15 we have plotted the relative remanent magnetization for [Co_2MnGe/V] and [Co_2MnSn/Au] multilayers determined at a temperature of 200 K as a function of the thickness of the non magnetic interlayers. We find no systematic variation with the thickness, for the multilayers with Au the remanent magnetization is about 60% of the saturation magnetization independent of the thickness of the Au layer, for [Co_2MnGe/V]$_n$ the remanent magnetization vanishes for all thicknesses above $d = 1.5$ nm. Thus there is no indication of an oscillation of the interlayer exchange interaction in [Co_2MnSn/Au]$_n$ which in the case of an antiferomagnetic coupling should give a lowering of the remanent magnetization. Similarly in [Co_2MnGe/V]$_n$ the vanishing remanent magnetization could indicate an af coupling, but then the af coupling would be independent of the interlayer thickness, which does not at all fit into the scheme of an IEC mechanism [72]. We will come back to this question in Sect. 3.3 below. In Fig. 16 we have plotted the ferromagnetic saturation magnetization measured

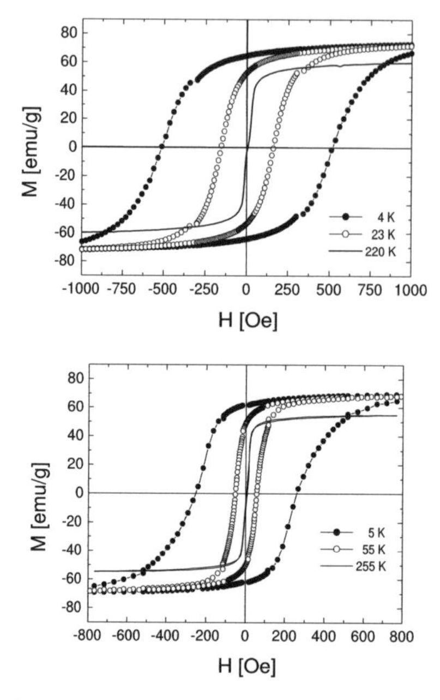

Fig. 13. Hysteresis loops measured at different temperatures for the samples [Co_2MnGe(3 nm)/V(3 nm)]$_{30}$ in the upper panel (**a**) and [Co_2MnSn(3 nm)/V(3 nm)]$_{30}$ in the bottom panel (**b**)

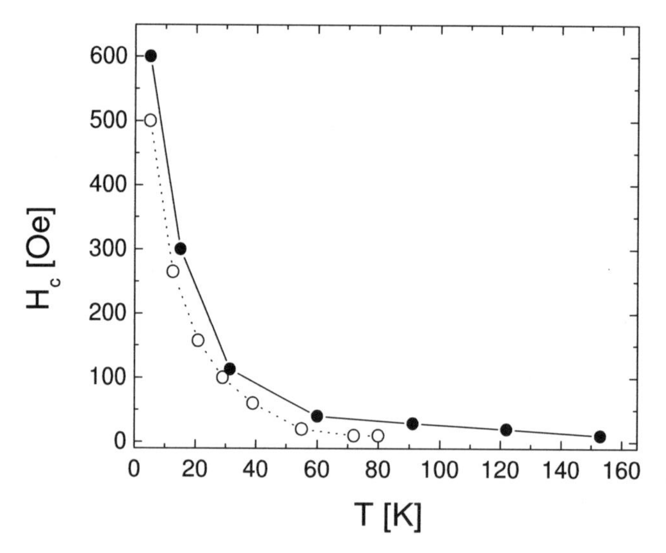

Fig. 14. Coercive field versus temperature for the multilayers [Co_2MnSn(3 nm)/Au(1.5 nm)]$_{30}$ (*solid circles*) and [Co_2MnGe(3 nm)/V(4 nm)]$_{50}$ (*open circles*)

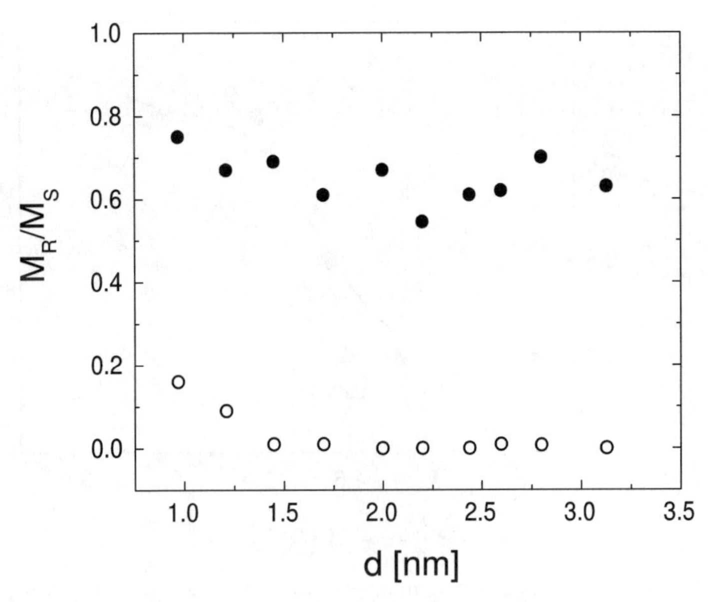

Fig. 15. Relative remanent magnetization for $[Co_2MnSn(3\,nm)/Au(d)]_{30}$ (*solid circles*) and $[Co_2MnGe(3\,nm)/V(d)]_{30}$ (*open circles*) multilayers determined at $T =$ 200 K versus thickness of the non magnetic interlayer d

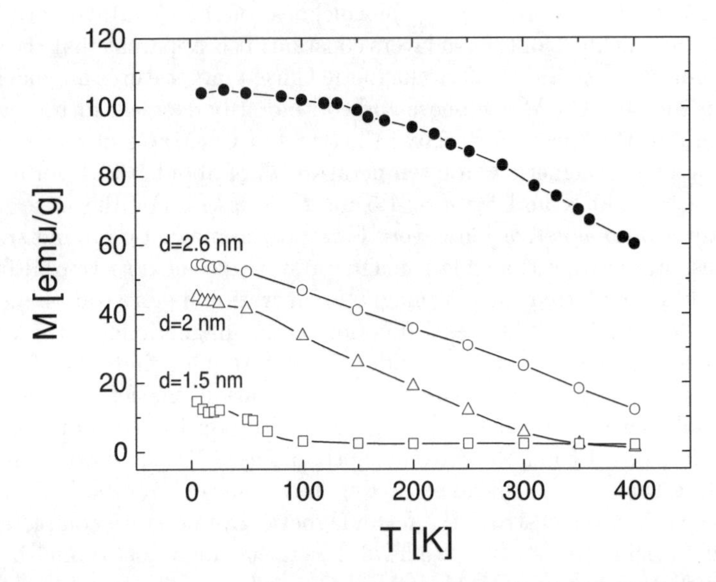

Fig. 16. Ferromagnetic saturation magnetization of bulk Co_2MnGe (solid circles) and $[Co_2MnGe(d)/Au(3\,nm)]_{30}$ multilayers (*open symbols*) with varied Co_2MnGe layer thickness d (given in the figure) versus temperature

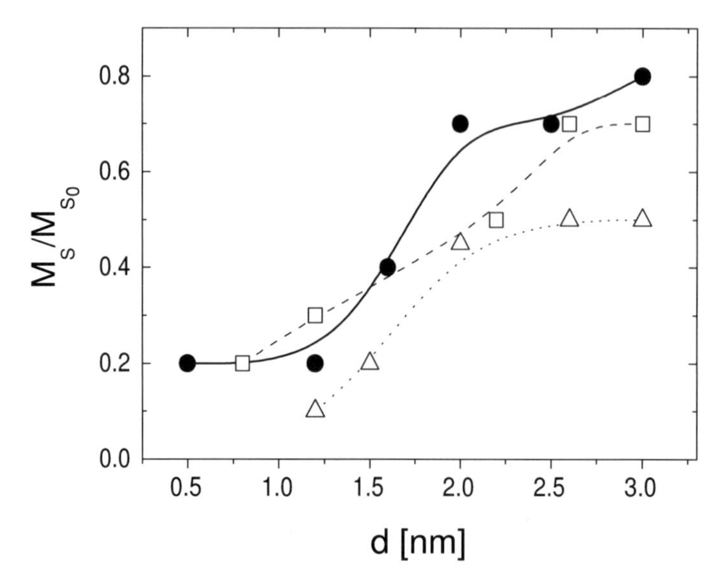

Fig. 17. Relative saturation magnetization versus the thickness of the Co_2MnGe and Co_2MnSn layer for multilayers $[Co_2MnSn(d)/V(3\,nm)]_{30}$ (circles), $[Co_2MnGe(d)/V(3\,nm)]_{30}$ (squares) and $[Co_2MnSn(d)/Au(3\,nm)]_{30}$ (*triangles*)

in a field of 2 kOe when varying the thickness of the Co_2MnGe layers while keeping the thickness of the Au layers constant. It is apparent that the saturation magnetization and the ferromagnetic Curie temperatures are much lower than for the bulk Co_2MnGe phase and continuously decrease further with decreasing film thickness. For a layer thickness of Co_2MnGe of $d = 2.6\,nm$ we estimate a ferromagnetic Curie temperature T_c of about 500 K, for $d = 2\,nm$ we get $T_c \sim 340$ K and for $d = 1.5\,nm$ $T_c \sim 90$ K. For the latter sample the saturation magnetization is very low and the magnetic ground state is a spin glass state rather than a ferromagnetic state (see next section). Finally in Fig. 17 we have plotted the ferromagnetic saturation magnetization measured at $T = 5$ K and $H = 2$ kOe as a function of the magnetic layer thickness for the different multilayer systems under study here. One finds that for a layer thickness of $d = 3\,nm$ the saturation magnetization reaches only 50 to 80% of the bulk saturation magnetization, depending on the combination. Typically below $d = 1.5\,nm$ the ferromagnetic moment breaks down completely, suggesting that at this thickness mixing and disorder from both sides of the ferromagnetic layer destroys the ferromagnetic ground state completely.

Summarizing this section, we have shown that high quality multilayers of the Heusler compounds Co_2MnGe and Co_2MnSn can be grown on sapphire a-plane. The ferromagnetic saturation magnetization is definitely lower than the bulk value but similar to what we have observed for the very thin single films. The reduction of the magnetization has two different sources. First there are many antisite defects since the multilayers are prepared at 300°C and not at the much higher optimum temperature. Second, there is an additional drastic breakdown of the saturation magnetization at the interfaces due to strong disorder and intermixing here. From the investigations of the multilayers with very thin magnetic films we may conclude that typically 0.7 nm of the Heusler layers at the interfaces contribute only a small magnetization and are not ferromagnetic. This is supported by the results of the small angle x-ray reflectivity, from which we derived an interface roughness of the same order of magnitude. When varying the thickness of the non magnetic interlayers Au and V we find no indication of an oscillating interlayer exchange coupling, but for the case of the V interlayers there is an interesting anomalous behaviour, namely a vanishing remanent magnetization over a broad thickness range which will be the subject of the Sect. 3.3.

3.2 Exchange Bias in [Co_2MnGe/Au] Multilayers

Apart from the more general magnetic properties of the Co-based Heusler multilayers presented in the previous section, there are intriguing peculiarities in the magnetic order of these multilayers which we want to discuss in this and the next section.

After cooling in a magnetic field most of the Co-based Heusler multilayers which we have studied exhibit asymmetric magnetic hysteresis loops shifted along the magnetic field axis, evidencing the existence of an unidirectional exchange anisotropy, nowadays dubbed "exchange bias" [77]. This is an effect known to occur at interfaces between an antiferromagnet (af) and a ferromagnet (f), the classical example being the Co/CoO interface [77]. The exchange bias phenomenon attracted considerable new interest in recent years and there is a much deeper understanding of it now than it was ten years before, although a quantitative theoretical model is still lacking. The exchange bias sets in at a blocking temperature T_B slightly below the Néel temperature T_N of the antiferromagnet and there is general agreement in the literature that the spins at the f/af interface which are coupled to both, the ferromagnet and the antiferromagnet, cause the exchange bias effect. The coupling to the ferromagnet in saturation defines the direction of the exchange bias field H_{EB}. Upon cooling through T_B the interface spins are blocked and keep the direction of H_{EB} fixed. Theoretical models developed for the exchange bias effect in recent years focus on different aspects of the complex problem. The domain state model [78] assumes that the domain formation in the antiferromagnet is the most essential point, other models start from the frustrated, weakly coupled spins at the interface or at domain walls of the antiferromagnet [79,80].

We have observed an exchange bias shift of the hysteresis loop in multilayers of Co_2MnSn and Co_2MnGe when combined with Au, Cr or Cu_2MnAl [65]. Only with V interlayers it does not occur. Since certainly there is no antiferromagnetism in these multilayers, the existence of H_{EB} is puzzling at the first glance.

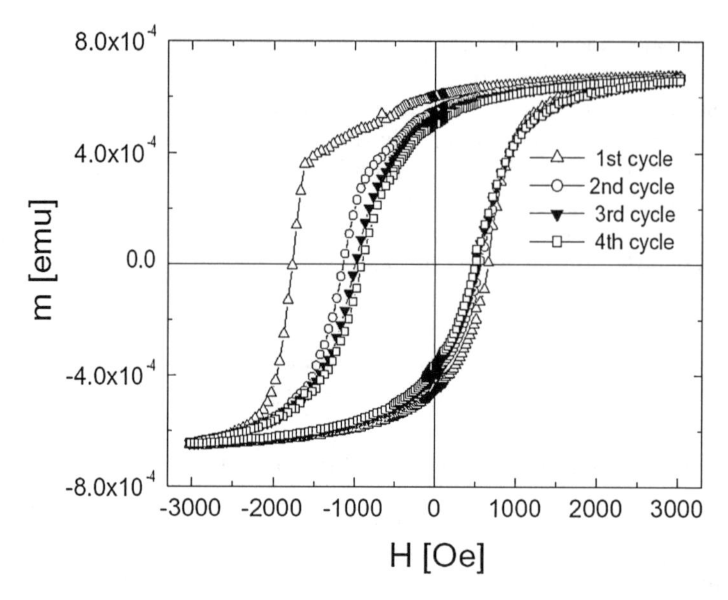

Fig. 18. Sequence of magnetic hysteresis loops of the multilayer [Co_2MnGe (2.6 nm)/Au(3 nm)]$_{30}$ measured at 2 K after field cooling in $H = 2$ kOe

Before coming back to this question we shortly introduce the main experimental results for the example of the [Co_2MnGe/Au] multilayer system. In Fig. 18 we show the hysteresis loop measured at 2 K after field cooling in an applied field of 2 kOe. The loop is found to be shifted by a rather large exchange bias field $H_{EB} = 500$ Oe. Similar to many other exchange bias systems [77] there is a strong relaxation of H_{EB} when sweeping the magnetic field several times (see sweep nb. 1 to 4 in Fig. 18). The exchange bias effect sets in rather sharply at a blocking temperature of 15 K (Fig. 19), showing that the magnetic exchange interactions defining T_B are rather weak. We have also tested the dependence of H_{EB} on the thickness of the magnetic layer d_m and found $H_{EB} \sim 1/d_m$ as usual for exchange bias systems and proving that the exchange anisotropy originates from interface spins.

There is one enlightening experiment which is difficult to realize in conventional f/af exchange bias systems and which we show in Fig. 20. When sweeping the external field to -4 T at 5 K, one can nearly eliminate H_{EB}. Simultaneously the excess magnetization (asymmetry of the loop along the magnetization axis) vanishes, directly proving the intimate correlation of the excess positive magnetization and H_{EB}. It seems quite natural to conclude that interface spins determining H_{EB} are rotated out of their metastable orientation when sweeping the field up to 4 T. Now coming back to the question concerning the origin of the exchange bias shift in our multilayers, the mere existence of H_{EB} shows that at the Co_2MnGe/Au interface antiferromagnetic exchange interactions must be present. The excess positive

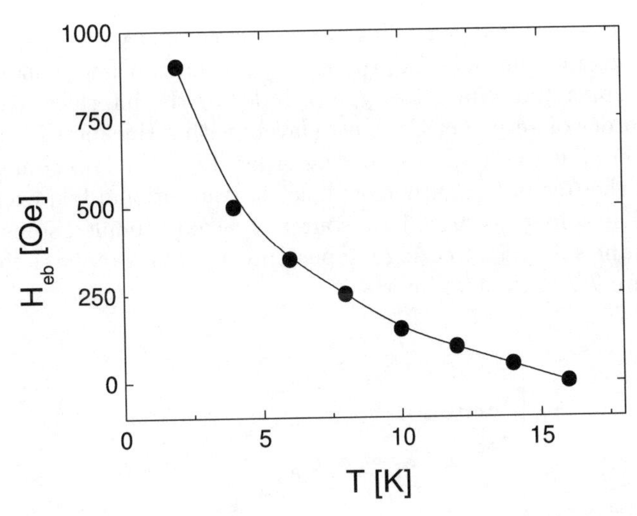

Fig. 19. Exchange bias field versus temperature for the multilayer [Co_2MnGe (2.6 nm)/Au(3 nm)]$_{30}$ for a cooling field $H = 200$ Oe

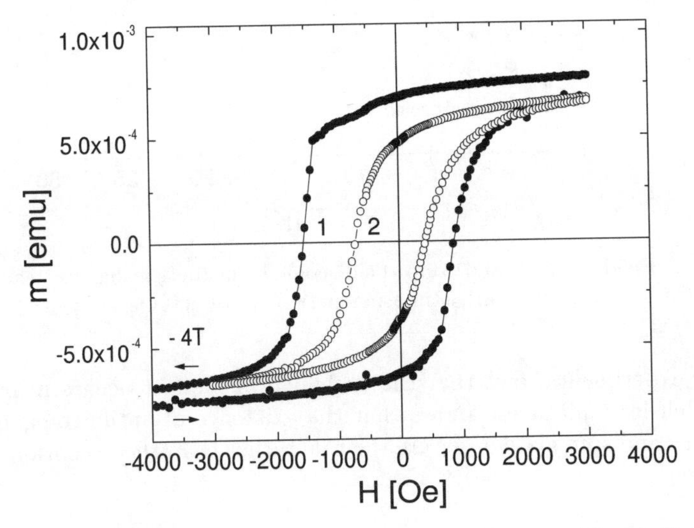

Fig. 20. Magnetic hysteresis loops of the multiayer [Co_2MnGe(2.6 nm)/ Au(3 nm)]$_{30}$ measured at 2 K after field cooling in $H = 40$ kOe. The labels in the figure denote the number of the field cycles. The field has been driven to -40 kOe during the first field cycle

magnetization which vanishes after magnetizing to high negative fields shows that there are spins at the interface with a component antiparallel to the ferromagnet. Adopting the point of view of localized models for magnetic exchange interactions, there seems to be an antiferromagnetic superexchange type of interaction at the interface e.g. an af Co-Au-Co superexchange or an af Co-Au-Mn superexchange. This interaction competing with the ferromagnetic exchange of the pure Co_2MnGe phase creates a few monolayers with frustrated spins and spin glass type of order at the interface. Actually the magnetic order of very thin Co_2MnGe layers with a thickness of the order of 1 nm exhibit all ingredients of spin glass order (Fig. 21). This spin glass phase takes over the role of the antiferromagnet in conventional f/af exchange bias systems. The spin glass provides a source of weakly coupled spins which are blocked at the spin glass freezing temperature T_f, which replaces the blocking temperature T_B of the f/af interface.

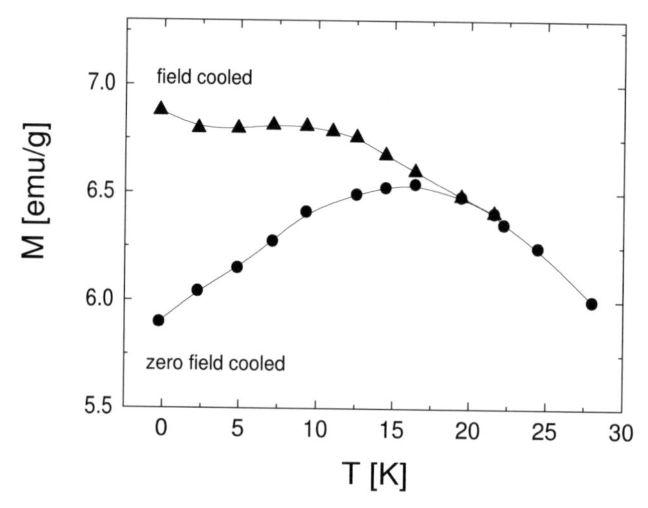

Fig. 21. Field cooled and zero field cooled magnetization for the sample $[Co_2MnGe(1.2\,nm)/Au(3\,nm)]_{30}$ in a magnetic field of 500 Oe

The experimental fact that the exchange bias effect occurs in many Co based Heusler multilayers shows that the existence of antiferromagnetic interactions and spin glass order are the rule rather than the exception in these multilayers.

3.3 Antiferromagnetic Interlayer Coupling in $[Co_2MnGe/V]$ Multilayers

It is an interesting principal question whether in the Co Heusler based multilayers an oscillatory interlayer exchange coupling (IEC) of the type existing in multilayers combined of the 3*d* transition element ferromagnets and normal metals [72] can be observed, too. Thus we carefully inspected the magnetic hysteresis loops of the Co-based Heusler multilayers from Sect. 3.1 in order to find indications for an IEC mechanism. We were successful in the

[Co_2MnGe/V] multilayer system, where a vanishing remanence at room temperature (see Fig. 13a) is seen, giving a first indication of an antiferromagnetic IEC.

We then started a systematic investigation of the magnetic order in this system by magnetic neutron reflectometry. The experiments were carried out at the ADAM reflectometer at the ILL in Grenoble. This instrument is equipped with neutron spin analysers so that spin polarized neutron reflectivity (PNR) measurements can be performed. PNR gives additional and important information about the magnetization direction in the film [81–83], since a magnetization direction parallel (or antiparallel) to the neutron spin polarization causes scattering with the polarization direction unchanged (up-up (+,+) and down-down (−,−) channels) only, whereas a magnetization component perpendicular to the spin quantization axis causes spin flip scattering appearing in the down-up (−,+) and up-down (+,−) channels.

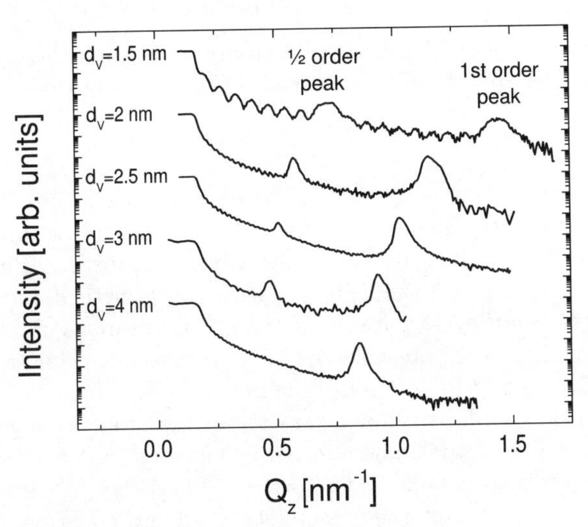

Fig. 22. Specular unpolarized neutron reflectivity scans of [Co_2MnGe(3 nm)/ $V(d)]_n$ multilayers. The V layer thickness d_V is given in the figure. For $d_V = 1.5$ nm the number of bilayers is $n = 20$; in all other cases $n = 50$

Figure 22 displays (unpolarized) neutron reflectivity scans for a series of [Co_2MnGe/V] multilayers with different thickness of the V layer d_V ranging from 1.5 nm to 4 nm. The observation of the half order peak clearly proves the existence of an antiferromagnetic (af) interlayer coupling. The af coupling is observed for all multilayers with an interlayer thickness d_V smaller than 3 nm, we find no indication of an oscillatory IEC in this thickness range. We also studied other, intermediate V thicknesses by magnetization measurements and found that the coupling is always antiferromagnetic. At this point serious doubts arise whether actually the IEC mechanism is responsible for

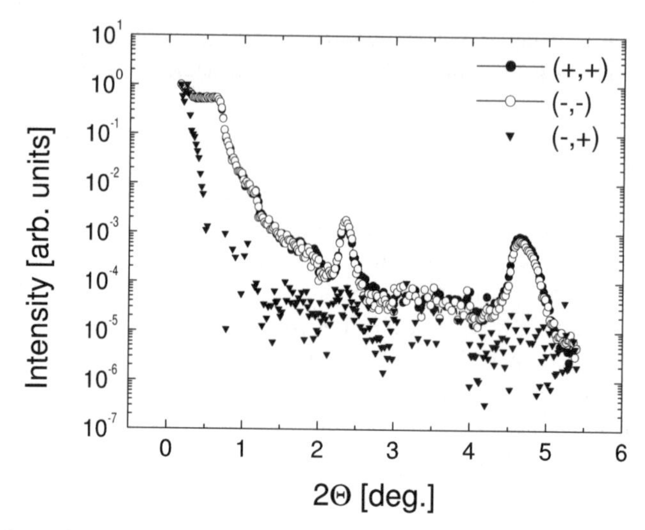

Fig. 23. Specular polarized neutron reflectivity scans for non spin flip ((+,+) and (−,−)) and for spin flip (−,+) channels of the multilayer [Co_2MnGe(3 nm)/ V(2 nm)]$_{50}$

the af interlayer coupling, since the oscillatory character of this coupling is an intrinsic property of the geometry of the Fermi surface of the interlayers [72].

In order to characterize the af interlayer ordering further, PNR measurements of the [Co_2MnGe/V] multilayers with the growth induced magnetic easy axis of the multilayers parallel to the small neutron guiding field were taken (Fig. 23). Figure 23 presents the intensity measured for the spin channels ((+,+), (−,−) and (+,−)); only the non spin flip channels show a finite intensity. Thus the sublattice magnetization of the af lattice is pointing parallel or antiparallel to the magnetic field. The first order structural peak in Fig. 23 has the same intensity for the (+,+) and the (−,−) channel. This proves that the magnetic interlayer superstructure has no ferromagnetic component which could occur e.g. by some canting of the af coupled layers. From this we can conclude that the af interlayer structure of the multilayer is well defined.

The measurement of PNR scans as a function of an applied field is shown in Fig. 24. The af peak intensity is rapidly suppressed at higher magnetic fields, simultaneously the splitting of the (−,−) and the (+,+) scattering intensities increases until at the ferromagnetic saturation field there is only a finite scattering intensity in the (+,+) channel. The ferromagnetic saturation field derived from these measurements is 50 Oe at 200 K and 300 Oe at 4 K. Thus the interlayer coupling interaction in this system is very weak. This holds for all af ordered multilayers of the [Co_2MnGe/V] system.

There are three additional features in Fig. 24 worth mentioning: First, at intermediate fields there is no observable spin flip scattering i.e. the

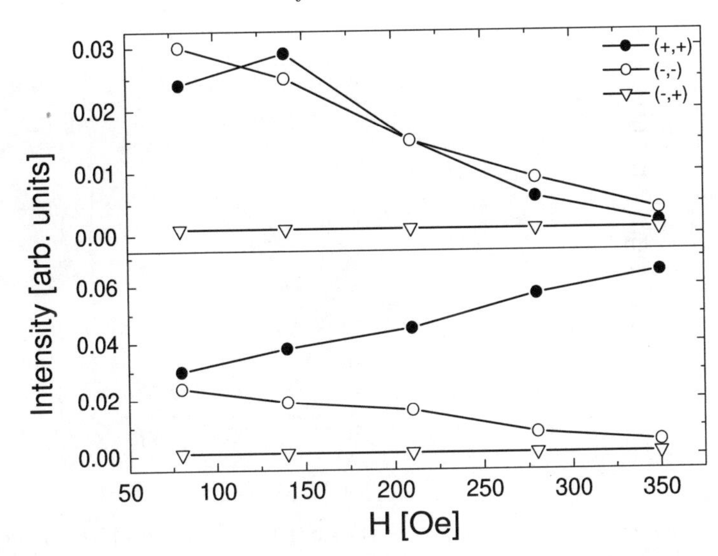

Fig. 24. Field dependence of the af peak intensity (*upper panel*) and 1st order structural peak (*lower panel*) for the spin flip (−,+) and non spin flip ((+,+) and (−,−)) channels for the multilayer $[Co_2MnGe(3\,nm)/V(2\,nm)]_{50}$

magnetization reversal of the af state takes place by domain wall movements within the single ferromagnetic layers and not by coherent rotations [84]. Second, the af structure is irreversibly lost when magnetizing the sample to the ferromagnetic saturation field at low temperatures. Only at high temperatures above about 100 K the af order is partly restored. This is also clearly observed in the series of magnetic hysteresis curves measured for $T < T_N$ shown in Fig. 13a. At low temperatures one observes typical ferromagnetic hysteresis loops with a remanent magnetization of up to 90% of the saturation magnetization. Third, we note that in ferromagnetic saturation in Fig. 24 the intensity in the (+,+) channel vanishes completely i.e the neutron reflected from the multilayer are completely spin polarized in the up direction. This originates from the fact that in the Co_2MnGe layer the magnetic and structural neutron scattering cross section have the same magnitude so that for one spin direction in ferromagnetic saturation the magnetic and the structural scattering length just cancel.

The temperature dependence of the af peak intensity as measured after cooling in zero-field and after cooling in 1000 Oe and then switching off the field at the measuring temperature is displayed in Fig. 25. After zero-field cooling the af peak intensity develops below 270 K in a phase transition like fashion, reaches a maximum at about 200 K and decreases slightly towards lower temperatures. Cooling in a high field there is no detectable intensity below 100 K, but approaching the phase transition at 270 K the af order recovers after switching off the field and close to the transition temperature

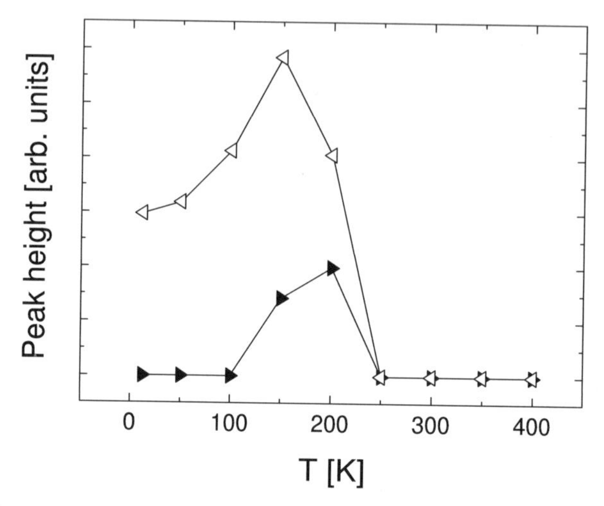

Fig. 25. Temperature dependence of the af peak intensity of the multilayer $[Co_2MnGe(3\,nm)/V(3\,nm)]_{30}$ measured after field cooling (*solid* triangles) and zero field cooling (open triangles)

the peak intensity coincides with that measured after zero-field cooling. Since the half order peak intensity is proportional to the squared sublattice magnetization in an antiferromagnet, this behaviour clearly reveals that there is a reversible antiferromagnetic phase transition at 270 K.

The transition can also easily be detected in the dc magnetic susceptibility (Fig. 26a) which exhibits a peak at 270 K, similar to a conventional bulk antiferromagnet. An antiferromagnetic phase transition in the $[Co_2MnGe/V]$ system at 270 K is rather surprising, since from the study of the single $V/Co_2MnGe/V$ trilayers with the same thickness of the Co_2MnGe layer we estimated a ferromagnetic ordering temperature T_c of 600 K (see Sect. 2). In multilayers coupled by antiferromagnetic IEC the Néel temperature of the af ordering between the layers and the ferromagnetic Curie temperature of the single layers should coincide. Thus the low value for T_N provides further evidence that the af interlayer ordering in $[Co_2MnGe/V]$ is not explicable by the conventional IEC mechanism.

The key for the understanding of the af interlayer magnetic ordering in $[Co_2MnGe/V]$ lies in the magnetic granular behaviour of very thin Co_2MnGe layers as discussed in Sect. I. By mixing and disorder at interfaces and grain boundaries the films break up into small weakly interacting ferromagnetic clusters and exhibit the typical behaviour of a small particle magnet. If the clusters are small and the cluster interactions weak, it is expected that the clusters are superparamagnetic at high temperatures [85]. In this case there is no ferromagnetic phase transition at a ferromagnetic Curie temperature T_c but a freezing of the clusters at a superparamagnetic blocking temperature $T_B < T_c$. Actually this has not been observed in our single Co based

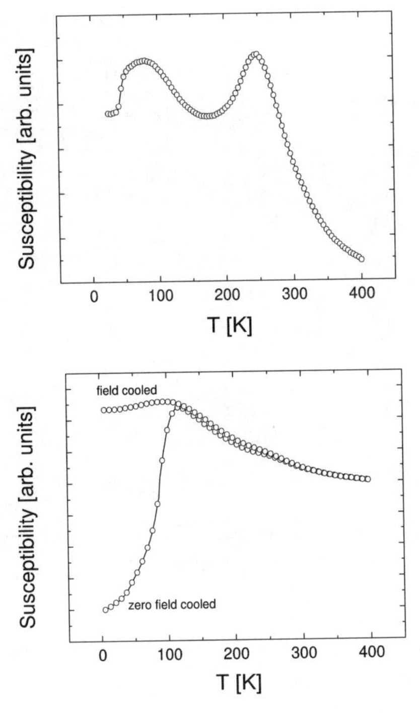

Fig. 26. The dc magnetic susceptibility of the $[Co_2MnGe(3\,nm)/V(3\,nm)]_{30}$ multilayer (*upper panel*) and the $[Co_2MnGe(3\,nm)/V(4\,nm)]_{30}$ multilayer (*lower panel*)

Heusler layers from Sect. 2 at least up to the maximum experimental temperature of 400 K. But it may well be that at even higher temperatures a corresponding phenomenon might exist. Now, in multilayers composed of the Co_2MnGe layers with the same thickness superparamagnetic behaviour and cluster blocking clearly appear below 400 K, probably due to a change of the microstructure and the additional influence of the dipolar fields from the neighbouring layers.

The superparamagnetic behaviour is shown in Fig. 27, where magnetization curves for the af ordered sample from Fig. 23 are plotted for temperatures above T_N. The $M(H)$ curves are completely reversible and saturate at fields of the order of 1 kOe which is characteristic for a superparamagnet with a huge magnetic moment. We applied the classical formula for the superparamagnetic magnetization as a first approximation

$$M(H,T) = N_c \mu_c \mathcal{L}\left(\frac{\mu_c H}{k_b T}\right) \tag{2}$$

with the number of clusters N_c and the cluster moment $\mu_c = N\mu_B$. $\mathcal{L}$ is the Langevin function. Fitting this formula to our results in Fig. 27 at 400 K

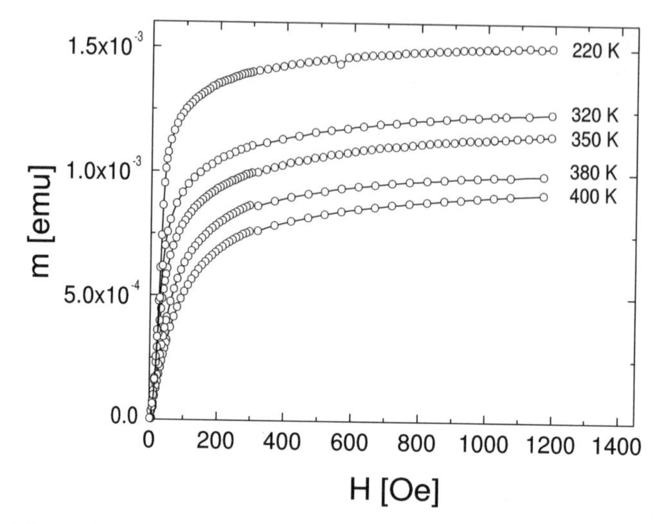

Fig. 27. Magnetization curves of the af coupled multilayer [Co_2MnGe(3 nm)/V(3 nm)]$_{50}$ measured at different temperatures $T > T_N$

we find a cluster moment of 1.6×10^5 μ_B corresponding to a cluster with a lateral diameter of about 70 nm.

These experimental findings suggest that at T_N of the [Co_2MnGe/V] multilayers thermally rotating ferromagnetic clusters within each Co_2MnGe layer undergo a second order phase transition with simultaneous intrta-plane ferromagnetic order and inter-plane antiferromagnetic order. This is a rather spectacular phenomenon which to the best of our knowledge has never been observed in multilayer systems before. An interesting question now arises concerning the type of magnetic order for the multilayer in Fig. 22 with a thickness of the V layer of 4 nm, which obviously does not exhibit af long range order. The answer is given in Fig. 26b, where the zero field cooled and the field cooled magnetization of this sample is compared. There is an onset of a strong magnetic irreversibility at about 150 K, reminiscent of a cluster blocking or a cluster glass magnetic ordering [85]. Actually in the hysteresis loops measured at low temperatures for this sample one finds features consistent with this type of transition, namely a temperature dependence of the coercive force and a magnetic remanence following a power law behaviour $H_c \sim T^{-a}$.

The magnetic order in the [Co_2MnGe/V] multilayer system for different thickness of the V layer is summarized in a magnetic phase diagram in Fig. 28. There is an antiferromagnetic range for $d_V \leq 3$ nm with the Néel temperature increasing with decreasing d_V up to $T_N = 380$ K for $d_V = 1.5$ nm. Then a range with cluster blocking or cluster glass freezing follows with the magnetic transition temperature nearly independent of d_V and extending up to the maximum thickness we have studied $d_V = 10$ nm.

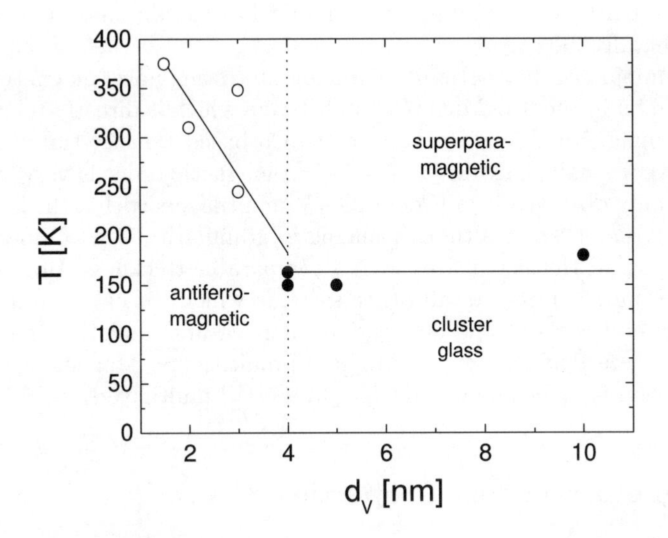

Fig. 28. Magnetic phase diagram for multilayers $[Co_2MnGe(3\,nm)/V(d)]_{50}$ with different vanadium thickness d_V. There is an antiferromagnetic range for $d_V \leq 3\,nm$ and a range with cluster blocking for $d_V > 4\,nm$

Concerning the microscopic origin of the ordering phenomenon observed in the $[Co_2MnGe/V]$ multilayers here, we would like to emphasize that the IEC coupling plays no essential role. It is either too weak or cancels out because of thickness fluctuations. The remaining other candidate which can produce an interlayer coupling and af long range order are interlayer magnetic dipolar fields. Finite dipolar stray fields at the surface of each single Co_2MnGe layer are inevitably present, since the layers possess an internal granular magnetic structure. Topologically the magnetic surface is rough on the length scale of the dimension of the clusters i.e. on a scale of about several 10 nm.

Theoretical model calculations show that the dipolar coupling field seen by the neighbouring layers depends on the topological structure of the surface and can be ferromagnetic or antiferromagnetic, depending on details of the roughness [86]. Quantitative calculations with a surface structure mimicking typical thin film systems lead to an estimation of a coupling field strength of the order of 100 Oe i.e. the order of magnitude which we observed here. Experimentally there is the well known Néel type af coupling [87] between thin films separated by a very thin nonmagnetic interlayer, which is a disturbing factor in some spin valve devices [88]. Originally this coupling, which is of dipolar origin, has been proposed by Néel [87]. Dipolar fields can also produce antiferromagnetic long range order in multilayers, as has been shown for [Co/Cu] multilayers with a thickness of the Cu layer of 6 nm [89]. But in this system the af long range order is only observed in the virgin state and is lost irreversibly after once magnetizing the sample. In [Nb/Fe] multilayers the existence of af interlayer dipolar interaction has also been shown experimentally [90].

In summary of this section, we find an interesting novel magnetic ordering phenomenon in the [Co_2MnGe/V] multilayers which is directly related to the granular magnetic microstructure of the Co based Heusler thin films in the limit of very small thickness. The antiferromagnetic order is very subtle and we have only observed it in [Co_2MnGe/V] multilayers with a thickness of the Heusler layer of 3 nm, although magnetic granularity is a common feature of all Co based Heusler phases with a comparable thickness. However, when combined into multilayers all other systems which we have studied remain ferromagnetic without the cluster blocking features and the af long range order characteristic for the [Co_2MnGe/V] multilayers. Thus in this sense the phenomenon is rather unique for [Co_2MnGe/V] multilayers.

4 Magnetotransport Properties

For possible technical applications such as GMR devices based on the fully spin polarized Heusler compounds, the electrical transport properties are of utmost importance. Thus we summarize here our results on the electrical conductivity, the Hall conductivity and the magnetoresistance of the pure Heusler films and the Heusler multilayers. The electrical conductivity was measured by a 4 point technique with silver painted electrical contacts, for the measurement of the Hall conductivity the films were patterned into the conventional Hall bar geometry with five contacts produced by ion beam etching. Figure 29 displays the electrical resistivity versus temperature for the optimized 100-nm thick pure films of the Co based Heusler alloys. The

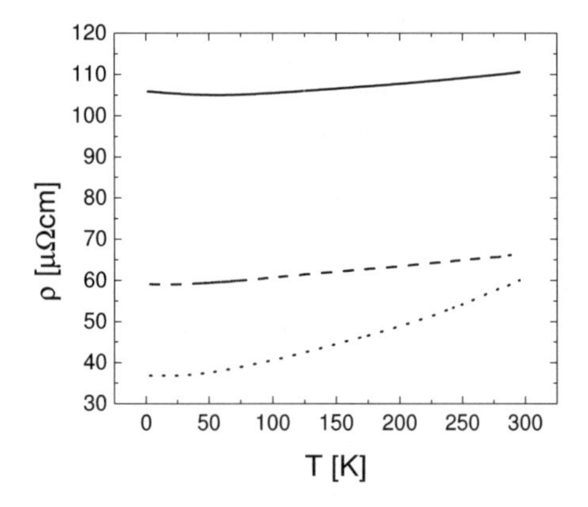

Fig. 29. Electrical resistivity versus temperature for V(3 nm)/Co_2MnSi(100 nm) (*solid line*), Au(3 nm)/Co_2MnSn(100 nm) (*dashed line*) and V(3 nm)/Co_2MnGe (100 nm) (*dotted line*)

residual resistivity is rather high, especially for the Co_2MnSi film (taken from reference [67]) with a residual resistivity ratio (RRR) (defined as the ratio of the room temperature resistivity and the resistivity at 4 K) of $RRR = 1.07$. The Co_2MnSn and the Co_2MnGe films have a higher electrical conductivity and a higher RRR ($RRR = 1.15$ for the Co_2MnSn phase and $RRR = 1.55$ for the Co_2MnGe phase). Actually the RRR of 1.55 for Co_2MnGe is the highest value reported for thin films of the Co based fully spin polarized Heusler compounds in the literature until now.

It holds generally true for thin films of the Co based Heusler compounds that there is strong electron scattering at defects such as grain boundaries, antisite defects, interstitials and voids. We determined the Hall coefficient R_0 by fitting the Hall voltage measured up to a magnetic field $\mu_0 H = 3$ T by the formula $U_H = R_0 H + R_H 4\pi M$ with the magnetization M and the anomalous Hall coefficient R_0 and derived $R_0 = 2 \times 10^{-10}$ m^3/As for the Co_2MnSi phase and $R_0 = 2.2 \times 10^{-10}$ m^3/As for the Co_2MnSn phase. Combining with the electrical conductivity and assuming parabolic conduction bands we estimate an electron mean free path l_e at 4 K of $l_e = 1.3$ nm for the Co_2MnSi phase and 2.2 nm for the Co_2MnSn phase. Assuming that for the Co_2MnGe layer we have a similar value for R_0 we estimate $l_e = 5.5$ nm for this thin film in Fig. 29. These remarkably low values for l_e stresses the strong defect scattering in these compounds.

In Fig. 30 we show resistivity measurements of several multilayers from Sect. 3. We find high absolute values for the resistivity, the resistivity ratio is below $RRR = 1.1$ for Co_2MnGe and Co_2MnSn with V and Au interlayers. In these multilayers l_e is very short namely of the order of 1 nm. If one inspects

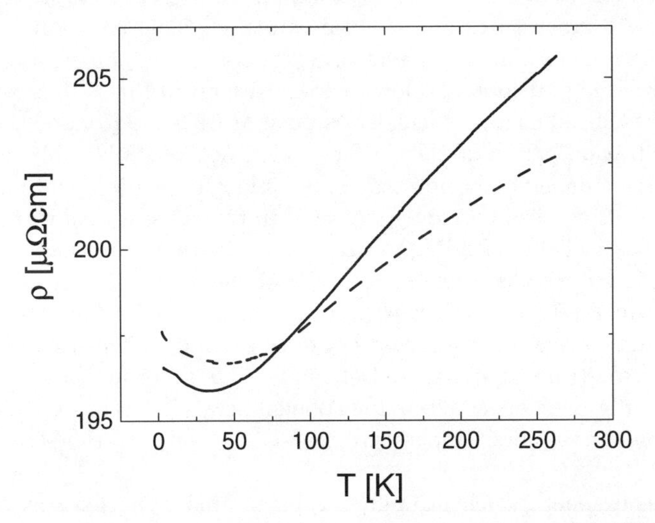

Fig. 30. Resistivity versus temperature for the multilayers [Co_2MnSn(3 nm)/V(3 nm)]$_{30}$ (*dashed line*) and [Co_2MnGe(3 nm)/V(3 nm)]$_{50}$ (*straight line*)

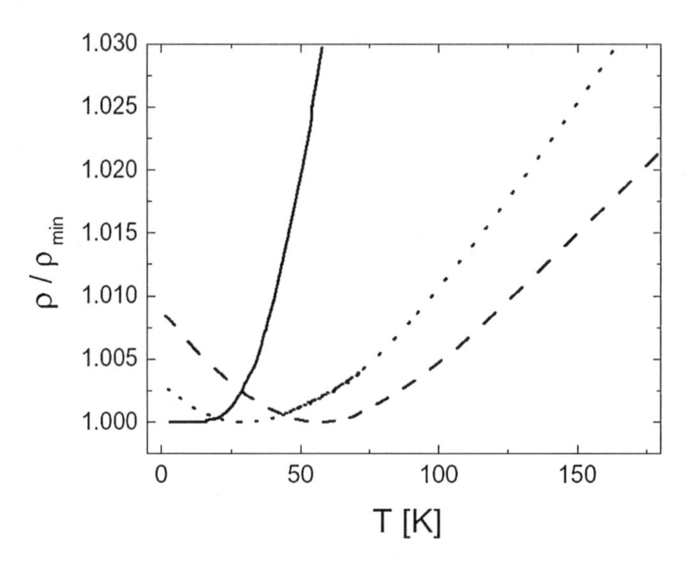

Fig. 31. Low temperature resistivity of Co_2MnSi (*dashed line*), Co_2MnSn (*dotted line*) and Co_2MnGe (*straight line*) films. The Co_2MnSi and the Co_2MnSn phase show an upturn of the resistivity towards lower temperatures

the resistivity curves in Figs. 29 and 30 in more detail, one observes a shallow minimum in $\rho(\mathrm{T})$ at temperatures between 20 K and 50 K for all samples with an RRR below about $RRR = 1.2$. This is a common feature for the Co based Heusler thin films. The low temperature resistivity $\rho(\mathrm{T})$ is plotted on an enlarged scale in Fig. 31. The amplitudes of the resistivity upturn towards low temperatures and the temperature of the resistivity minimum increases with increasing defect scattering, thus the low temperature anomaly is clearly related to disorder. A low temperature upturn in $\rho(\mathrm{T})$ is well known for metallic thin films and is usually associated with the Kondo effect [91] or weak localization [92]. But these effects seem not very plausible here, since in a strong ferromagnet Kondo scattering is hardly possible and measurable weak localization effects are only expected in two dimensional metals. There is a third classical effect leading to a low temperature upturn in $\rho(\mathrm{T})$ which is less well known. This is a renormalization of the electronic density of states at the Fermi level in metals with strong disorder [93]. This essentially is an electronic correlation effect and becomes relevant if the scattering of the conduction electrons is strong and l_e small. This renormalization effect has experimentally been observed in amorphous metals [94]and gives a plausible explanation for the low temperature anomaly in $\rho(\mathrm{T})$ of the Heusler films, too.

A measurement which strongly supports this hypothesis is shown in Fig. 32, where for the example of a [Co_2MnSn/Au] multilayer it is shown that a strong magnetic field of 4 T applied in the direction parallel and

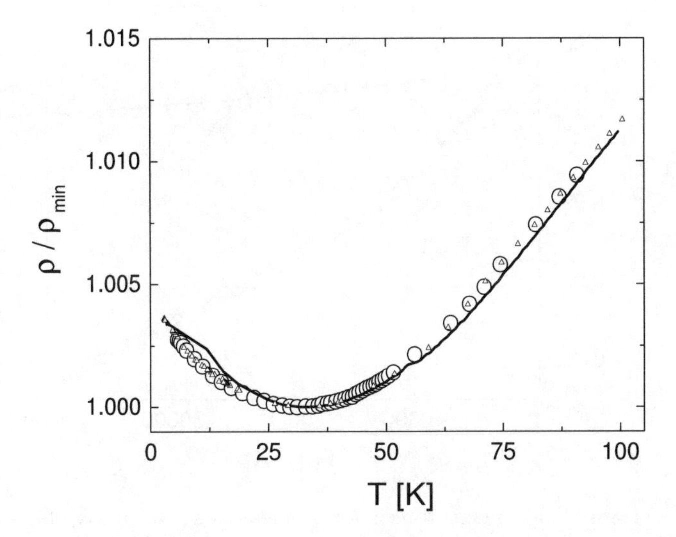

Fig. 32. Electrical resistivity versus temperature for the multilayer [Co_2MnGe (3 nm)/Au(3 nm)]$_{50}$ measured in zero field (*straight line*) and with an applied field of 4 T parallel to the film (*open triangles*) and perpendicular to it (*open dots*)

perpendicular to the film does not change the low temperature upturn in ρ(T). This is expected for the renormalization effect, whereas for Kondo scattering and also in the case of weak localization the upturn should be suppressed.

Coming to the magnetoresistance (MR) of the pure thin films and the multilayers we first discuss the high field MR i.e. the resistance for fields above the ferromagnetic saturation field where the anisotropic magnetoresistance (AMR) and the Giant magnetoresistance (GMR) play no role. This high field magnetoresistance measured at 4 K is plotted for the pure Heusler films in Fig. 33 and for several multilayers in Fig. 34. The high field magnetoresistance for the pure Heusler films is negative and isotropic. The total decrease of the resistance for fields up to 4 T is about 0.16% for the Co_2MnSi and the Co_2MnSn phase and definitely higher, namely 0.8% for the Co_2MnGe single crystalline film. We interpret the high field MR as a reduction of spin disorder scattering in a ferromagnet with non perfect ferromagnetic alignment of all spins. This interpretation is supported by the results of the high field MR on the multilayers in Fig. 34. For the [Co_2MnSn/Au] multilayer where the exchange bias effect discussed in Sect. 3.2 gives clear evidence for the existence of magnetic frustration and canted spins at the interfaces, we find an unusually large MR value with an amplitude of 3% up to a field of 3 T, whereas for the [Co_2MnGe/V] multilayer showing no indication of magnetic frustration the MR is only 0.3%.

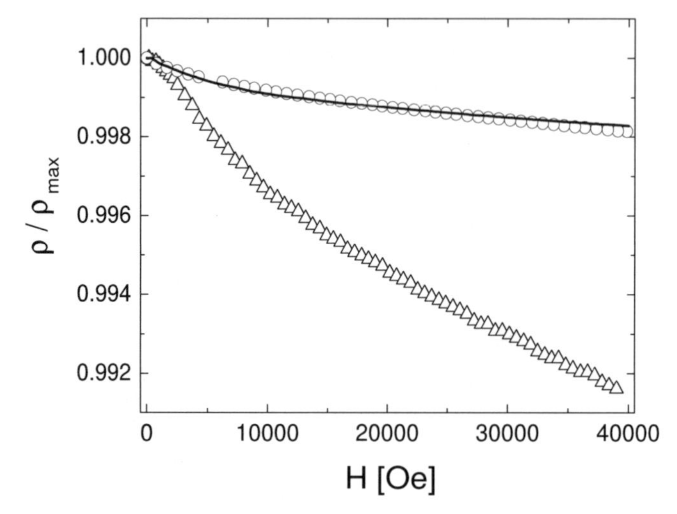

Fig. 33. High field magnetoresistance measured at 4 K for V(3 nm)/ Co_2MnSi(100 nm) (*straight line*), Au(3 nm)/Co_2MnSn(100 nm) (*open dots*) and V(3 nm)/Co_2MnGe(100 nm) (*open triangles*)

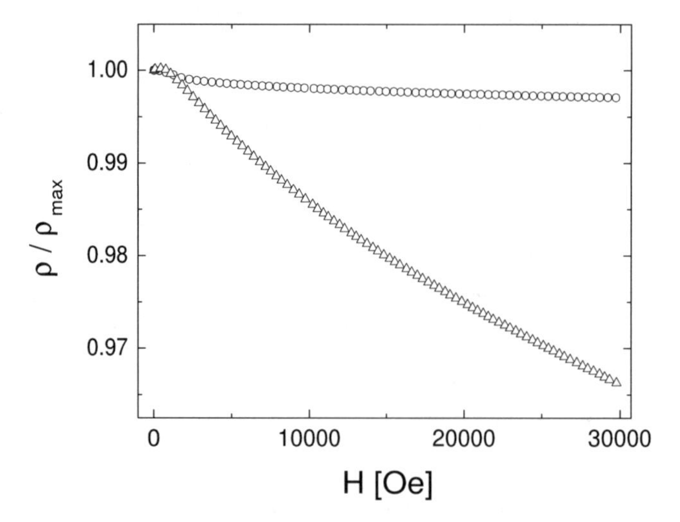

Fig. 34. High field magnetoresistance measured at 4 K for the multilayers $[Co_2MnSn(3\,nm)/V(3\,nm)]_{30}$ (*open dots*) and $[Co_2MnSn(3\,nm)/Au(3\,nm)]_{30}$ (*straight line*)

In the low field MR measurements i.e. the MR for fields up to the coercive force H_c we observed the standard AMR of a ferromagnetic film, however with a very small relative difference of the resistance for the field applied parallel and perpendicular to the current of the order $\Delta\rho/\rho = 4 \times 10^{-4}$. For the

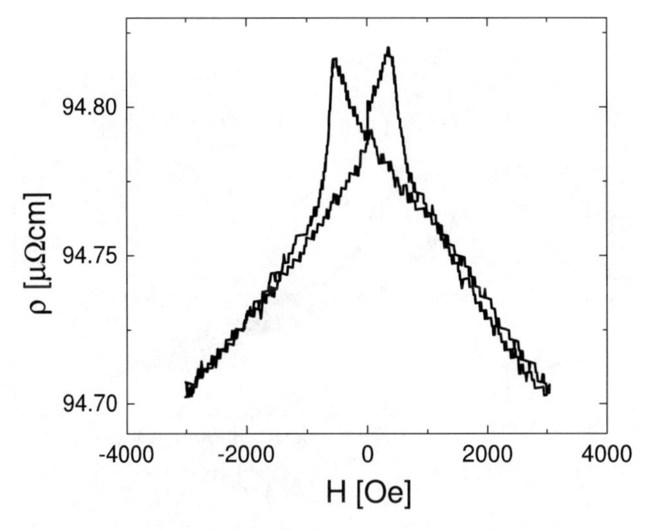

Fig. 35. Low field magnetoresistance measured at 4 K of a Co_2MnGe(3 nm)/ V(3 nm)/Co_2MnGe(3 nm)/CoO(2 nm) trilayer

multilayers the AMR is even smaller i.e. below the experimental resolution limit for $\Delta\rho/\rho$ of about 10^{-5}.

In Figs. 35 and 36 we present low field MR measurements related to the GMR effect. Figure 35 displays the MR at 4 K of a Co_2MnGe/V/Co_2MnGe/ CoO trilayer, where an exchange bias with CoO is used to pin the magnetization direction of the upper Co_2MnGe layer [77]. One finds a small jump in the MR at H_c with an amplitude $\Delta\rho/\rho = 5 \times 10^{-4}$ which could be associated with a GMR effect. In Fig. 36 we show the low field MR of a [Co_2MnGe/V] multilayer measured at 4 K and starting from the antiferromagnetically ordered state of the multilayer obtained after zero field cooling (see Sect. 3.3). Driving the antiferromagnetically coupled Co_2MnGe layers to ferromagnetic saturation and returning to zero field, gives an irreversible change of the MR $\Delta\rho/\rho = 5 \times 10^{-4}$, which is the upper limit for the GMR effect.

This section should demonstrate the difficulties one encounters when trying to use the fully spin polarized Co based Heusler alloys in magnetotransport applications such as GMR or AMR. The electrical conductivity and the electron mean free path are small, pointing towards the existence of lattice defects with a high conduction electron scattering efficiency. Our results suggest that these are the same antisite defects in the $L2_1$ structure which have the tendency to destroy the half metalicity. The rather low values for the GMR in the spin valve device and the antiferromagnetically coupled multilayers are not really surprising since there are three detrimental factors which acting together are able to nearly eliminate the GMR effect: First,non ferromagnetic interlayers are present introducing strong spin independent scattering thereby reducing the GMR. Second, for the observation of a sizable GMR ef-

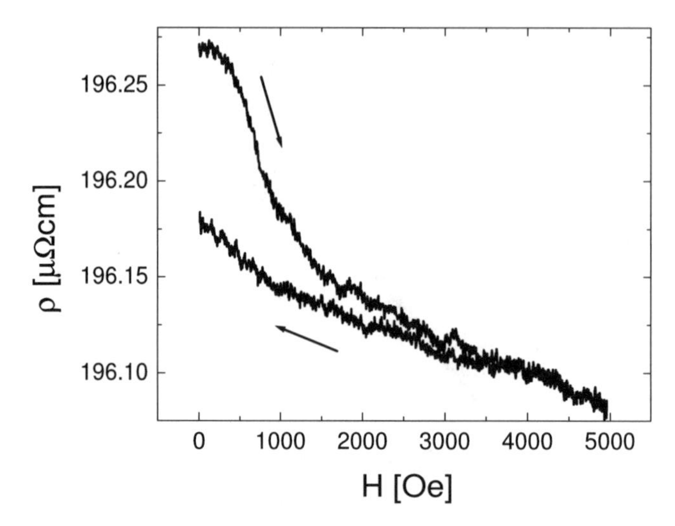

Fig. 36. Low field magnetoresistance measured at 4 K of a [Co_2MnGe(3 nm)/ V(3 nm)]$_{50}$ multilayer, starting at the antiferromagnetic state after zero field cooling (arrows indicate the field cycle)

fect, the conduction electron mean free path should be larger than the double layer thickness [55], which is certainly not fulfilled in our samples. Last but not least, the high spin polarization expected for the ideal Co based Heusler compounds is most certainly lost in very thin Co based Heusler layers with a saturation magnetization of typically only 50% of bulk samples.

5 Summary and Conclusions

We have shown that thick single films of the Heusler compounds Co_2MnGe and Co_2MnSn can be grown with a saturation magnetization and structural quality comparable to bulk samples. However, when decreasing the thickness to the order of 3 nm the films show a definitely lower value of the saturation magnetization and exhibit properties reminiscent of small particle ferromagnets. We interpret this behaviour by a magnetic decoupling of the Heusler films at the grain boundaries of the very small grains forming the film. Structural disorder and a high concentration of antisite defects at the grain boundaries weaken the ferromagnetism drastically.

As we have shown by the XMCD measurements on the Mn and the Co $L_{2,3}$ absorption edges, it is only the Mn atom in the Heusler compounds which reduces the saturation magnetization. Theoretical model calculations suggest that this originates from an antiparallel spin orientation of antisite Mn spin rather than from a loss of the atomic magnetic moment of Mn. In the multilayers a similar phenomenon occurs at the interfaces of the Heusler alloys with other metals. For a thickness range of about 0.5 to 0.7 nm at the interfaces the Heusler alloys are only weakly ferromagnetic or exhibit a spin glass type of order. The exchange bias shift of the hysteresis loops observed at low temperatures gives clear evidence for this.

In [Co_2MnGe/V] multilayers we have detected a peculiar antiferromagnetic interlayer magnetic ordering with a well defined Néel temperature far below the ferromagnetic Curie temperature of a single Heusler layer of the same thickness. The antiferromagnetic ordering exists in the thickness range of the V interlayer between 1 nm and 3 nm, very unlike the antiferromagnetic interlayer order in multilayer systems coupled by interlayer exchange mechanism. The antiferromagnetic order is directly related to the granular ferromagnetic structure of very thin Heusler layers. The small ferromagnetic particles defined by the weak magnetic coupling at the grain boundaries exhibit superparamagnetic behaviour above the Néel temperature. The interlayer dipolar interactions between the superparamagnetic particles cause a reversible magnetic phase transition with antiferromagnetic order between the layers and ferromagnetic order within the layers at a well defined Néel temperature.

Finally, the magnetotransport properties revealed that the electron mean free path in the single layers and in the multilayers is very small, indicative of the presence of large amounts of very effective electron scattering centers. Simultaneously there exists an isotropic negative spin disorder magnetoresistance up to very high fields, which shows that the magnetic ground state is not a perfect ferromagnet but possesses abundant magnetic defects. We think that the main source of defects responsible for the small mean free path and the magnetic disorder scattering are the antisite defects in the $L2_1$ structure.

Coming back to the issue discussed in the introduction, namely the perspective of the Heusler compounds as full spin polarized ferromagnetic layers in magnetoelectronic devices, our results indicate the severe difficulties which one inevitably encounters. The constraints in the preparation of thin film heterostuctures in many cases prohibit the optimum preparation conditions for the Heusler phase thus rendering it difficult to get the $L2_1$ structure with a high degree of atomic order. At the interfaces additional problems arise since they tend to be strongly disordered and weakly ferromagnetic. What one ideally would need to overcome these problems is a Heusler phase with a more robust spin polarization at the Fermi level which principally might already exist among the many new Heusler half metals which have been predicted theoretically. In any case, the evolving field of magnetoelectronics motivated the magnetism community to extend the experimental activities in thin film preparation towards complex intermetallic ternary alloys, which is certainly much more difficult than the preparation of simple films but offers new and interesting perspectives for basic research and applications.

References

1. F. Heusler: Verh. Dtsch. Phys. Ges. **5**, 219 (1903)
2. P.J. Webster and K.R.A. Ziebeck. In: *Alloys and Compounds of d-Elements with Main Group Elements. Part 2.*, Landolt-Börnstein, New Series, Group III, vol 19c, ed by H.R.J. Wijn, (Springer, Berlin 1988) pp 75–184
3. J. Soltys: Phys. Stat, Sol. (a) **66**, 485 (1981)
4. A. Zheludev, S.M. Shapiro, P. Wochner, A. Schwartz, M. Wall, and L.E. Tanner: Phys. Rev. B **51**, 113109 (1995)
5. R. Kosubski, and J. Soltys: J Mat. Science **17**, 1441 (1982)
6. K.R.A. Ziebeck, and P.J. Webster: J. Phys. F **5**, 1756 (1975)
7. P.J. Webster, and R.S. Tebbel: J. Appl. Phys **39**, 471 (1968)
8. P.J. Webster: J. Phys. Chem. Sol. **32**, 1221 (1971)
9. K.H.J. Buschow, and P.G. van Engen: J. Magn. Mag. Mat. **25**, 90 (1983)
10. J. Kübler, A.R. Williams, and C.B. Sommers: Phys. Rev. B **28**, 1745 (1983)
11. K.H.J. Buschow, and M. Erman: J. Magn. Magn. Mat. **30**, 374 (1983)
12. J. Pierre, R.V. Skolozdra, and Yu. V. Stadnyk: J. Magn. Magn. Mat. **128**, 93 (1993)
13. Y. Nishino, M. Kato, S. Asano, K. Soda, M. Hayasaki, and U. Mizutani: Phys. Rev. Lett. **79**, 1909 (1997)
14. A. Siebarski, M. B. Maple, E.J. Freeman et al.: Phys. Rev. B **62**, 3296 (2000)
15. H. Okamura, J. Kawahara, T. Nanba et al.: Phys. Rev. Lett. **84**, 3674 (2000)
16. R. de Groot and P. van Engen: Phys. Rev. Lett. **50**, 2024 (1983)
17. S. Fujii, S. Sugimura, S. Ispida, and S. Asano: J. Phys.: Cond. Matter **2**, 8583 (1990)
18. S. Ishida, S. Kashiwagi, S. Fujii, and S. Asano: Physica B **210**, 140 (1995)
19. S. Ishida, S. Fujii, S. Kashiwagi, and S. Asano: J. Phys. Soc. Jpn. **64**, 2154 (1995)
20. K.E.H.M Hansen, P.R. Mijnarends, L.P.L.M. Rabou, and K.H.J. Buschow: Phys. Rev. B **42**, 1533 (1990)
21. M.J. Otto, R.A.M. van Woerden, P.J. van der Valk, J. Wijngaard, C.F. van Bruggen, and C. Haas: J. Phys.: Cond. Matter **1**, 2351 (1989)
22. C. Hordequin, J. Pierre, and R. Currat: J. Magn. Magn. Mat. **162**, 75 (1996)
23. R.J. Soulen, J.M. Byers, M.S. Osofsky, B. Nadgony et al.: Science **282**, 85 (1998)
24. P.J. Brown, K.U. Neumann, P.J. Webster, and K.R.A. Ziebeck: J. Phys.: Cond. Matter **12**, 1827 (2000)
25. L. Ritchie, G. Xiao, Y. Ji, T.Y. Chen, M. Zhang et al.: Phys. Rev. B **68**, 104430 (2003)
26. D. Ristoiu, J.P. Nozieres, C.N. Borca, T. Komesu, H.-K. Jeong, and P.A. Dowben: Europhys. Lett. **49**, 624 (2000)
27. W. Zhu, B. Sinkovic, E. Vescovo, C. Tanaka, and J. S. Moodera: Phys. Rev. B**64**, 060403 (2001)
28. G.A. Prinz: Science **282**, 1660 (1998)
29. D. Awschalom and J. Kikkawa: Physics Today **52**, 33 (June 1999)
30. G. Schmidt, D. Ferrand, L.W. Molenkamp, A.T. Filip, and B.J. van Wees: Phys. Rev. B **62**, R4790 (2000)
31. C. Cacho, Y. Lassailly, H.-J. Drouhin, G. Lampel, and J. Peretti: Phys. Rev. Lett. **88**, 066601 (2002)

32. J. Moodera, R. Kinder, L.T. Wong, and R. Meservey: Phys. Rev. Lett. **74**, 3273 (1998)
33. Eun-Cheol Lee, and K.J. Chang, Phys. Rev. B **69**, 85205 (2004)
34. K. Schwarz: J. Phys. F **16**, L211 (1986)
35. J.M.D. Coey, M. Viret, and S. von Molnar: Adv. Phys. **48**, 167 (1999)
36. I. Galanakis, P.H. Dederichs, and N. Papanikolaou: Phys. Rev. B **66**, 174429 (2002)
37. I. Galanakis, P.H. Dederichs, and N. Papanikolaou: Phys. Rev. B **66**, 134428 (2002)
38. C. Felser, B. Heitkamp, F. Kronast, D. Schmitz, S. Cramm, H.A. Dürr, H-J. Elmers G.H. Fecher, S. Wurmehl, T. Block, D. Valdaitsev, S.A. Nepijko, A. Gloskovskii, G. Jakob, G. Schonhense, and W. Eberhardt: J. Phys.: Condens. Matter **15**, 7019 (2003)
39. S. Picozzi, A. Continenza, and A.J. Freeman: Phys. Rev. B **69**, 094423 (2004)
40. M.P. Raphael, B. Ravel, Q. Huang, M.A. Willard, S.F. Cheng, B.N. Das, R.M. Stroud, K.M. Bussmann, J.H. Claassen, and V.G. Harris: Phys. Rev. B **66**, 104429 (2002)
41. B. Ravel, M.P. Raphael, V.G. Harris, and Q. Huang: Phys. Rev. B. **65**, 164431 (2002)
42. C. Hordequin, J.P. Nozieres, and J. Pierre: J. Magn. Magn. Mat. **183**, 225 (1998)
43. C. Tanaka, J. Novak, and J.S. Moodera: J. Appl. Phys **81**, 5515 (1997)
44. J. Schmalhorst, S. Kämmerer, M. Sacher, G. Reiss, A. Hütten, and A. Scholl: Phys. Rev. Lett. **70**, 024426 (2004)
45. I. Galanakis: J. Phys.: Condens. Matter **14**, 6329 (2002)
46. S. Ishida, T. Masaki, S. Fujii, and S. Asano: Physica B **245**, 1 (1998)
47. D. Orgassa, H. Fujiwwara, T.C. Schulthess, and W.H. Butler: Phys. Rev. B **60**, 13237 (1999)
48. R.W. Overholser, M. Wuttig, and D.A. Neumann: Scripta Marterialia **40**, 1095 (1999)
49. G. Fritsch, V.V. Kokorin, and A. Kempf: J. Phys.: Condens. Matter **6**, L107 (1994)
50. J. Tobola and J. Pierre: J. Alloy Comp. **196**, 234 (2000)
51. E. DiMasi, M.C. Aronso, and B.R. Cole: Phys. Rev. B **47**, 1430 (1993)
52. J. Pierre, R.V. Skolozda, and M.A. Kouacou: Physica B **206**, 844 (1995)
53. J. Tobola, J. Pierre, S. Kapryzyk, R.V. Skolozda, and M.A. Kouacou: J. Magn. Magn. Mater. **159**, 192 (1996)
54. J.F. Gregg, I. Petey, E. Jougelet, and C. Dennis: J. Phys. D: Appl. Phys. **35**, R121 (2002)
55. B. Dieny: J. Magn. Magn. Mater. **136**, 335 (1994)
56. K.A. Kilian, and R.H. Victora: J. Appl. Phys. **87**, 7064 (2000)
57. V. Yu Irkhin, and M.I. Katsnelśon: Physics-Uspekhi **37**, 659 (1994)
58. W.E. Picket: Phys. Rev. Lett. **77**, 3185 (1996)
59. S. Picozzi, A. Continenza, and A.J. Freeman: Phys. Rev. B **66**, 094421 (2002)
60. S. Picozzi, A. Continenza, and A.J. Freeman, IEEE Trans. Magn. **38**, 2895 (2002)
61. I. Galanakis, S. Ostanin, M. Alouani, H. Dreyssé, and J.M. Wills,: Phys. Rev. B **61**, 4093 (2000)
62. M.C. Kautzky, F.B. Mancoff, J.-F. Bobo, B.M. Clemens, et al. : J. Appl. Phys. **81**, 4026 (1997)

63. K. Inomata, S. Okamura, R. Goto, and N. Tezuka: Jap. J. Appl. Phys. Part 2-Lett. **42**, L419 (2003)
64. G.A. de Wijs and R.A. deGroot: Phys. Rev. B **64**, 020402 (2001)
65. K. Westerholt, U. Geiersbach, and A. Bergmann: J. Magn. Magn. Mater. **257**, 239 (2003)
66. U. Geiersbach, A. Bergmann, and K. Westerholt: Thin Sol. Films **425**, 226 (2003)
67. U. Geiersbach, A. Bergmann, and K. Westerholt,: J. Magn. Magn. Mater. **240**, 546 (2002)
68. H. Bach, K. Westerholt, and U. Geiersbach: J. Cryst. Growth **237**, 2046 (2002)
69. A. Bergmann, J. Grabis, V. Leiner, M. Wolf, H. Zabel, and K. Westerholt: Superlattices and Microstructures **34**, 137 (2003)
70. A. Nefedov, J. Grabis, A. Bergmann, K. Westerholt, and H. Zabel: Physica B **345**, 250 (2004)
71. J. Grabis, A. Bergmann, A. Nefedov, K. Westerholt, and H. Zabel: Phys. Rev. B *submitted*
72. P. Bruno: Phys. Rev. B **52**, 411 (1995)
73. M. Goto, T. Kamimori, H. Tange, K. Kitao, S. Tomioshi, K. Ooyama, and Y. Yamaguchi: J. Magn. Magn. Mater. **140–145**, 277 (1995)
74. J. Stöhr: J. Magn. Magn. Mater. **200**, 470 (1999)
75. J. Grabis and A. Nefedov, and H. Zabel: Rev. Sci. Instr. **74**, 4048 (2003)
76. H.A. Dürr, G. van der Laan, D. Spanke, F.U. Hillebrecht, and N.B. Brookes: Phys. Rev. B **56**, 8156 (1997)
77. J. Nogues, and I.K. Schuller: J. Magn. Magn. Mater. **192**, 203 (1999)
78. P. Miltenyi, M. Gierlings, J. Keller, B. Beschoten, G. Güntherodt, U. Novak, and K.D. Usadel: Phys. Rev. Lett. **84**, 4224 (2000)
79. A.P. Malozemoff: Phys. Rev. B **35**, 3679 (1987)
80. D. Mauri, H.C. Siegmann, P.S. Bagus, and E. Kay: J. Appl. Phys. **62**, 3047 (1987)
81. S.J. Blundell and J.A.C. Bland: Phys. Rev. B **46**, 3391 (1992)
82. H. Zabel: Physica B **198**, 156 (1994)
83. H. Zabel, and K. Theis-Bröhl: J. Phys.: Condens. Matter **15**, S505 (2003)
84. K. Theis-Bröhl, V. Leiner, A. Westphalen, and H. Zabel: Physica B **345**, 161 (2004)
85. B.D. Culity: *Introduction to Magnetic Materials*, (Addison-Wesley, 1972)
86. D. Altbir, M. Kiwi, Ramírez, and I.K. Schuller: J. Magn. Magn. Mater. **149**, L246 (1995)
87. L. Neél: C.R. Acad. Sci. **255**, 1545 (1962)
88. D. Wang, J.M. Daughton, Z. Qian, C. Nordman, M. Tondra, and A. Pohm: J. Appl. Phys. **93**, 8558 (2003)
89. J.A. Borchers, J.A. Dura, J. Unguris, D. Tulchinsky, M.H. Kelley, C.F. Maijkrzak et al.: Phys. Rev. Lett. **82**, 2796 (1999)
90. Ch. Rehm, D. Nagengast, F. Klose, H. Maletta, and A. Weidinger: Europhys. Lett. **38**, 61 (1997)
91. Melvin D.Daybell: The Kondo Problem, in: *Magnetism V*, ed. by G.T. Rado and H. Suhl (Academic Press, New York and London, 1973) p. 121
92. P.A. Lee and T.V. Ramakrishnan: Rev. Mod. Phys. **57**, 287 (1985)
93. B.L. Altshuler and A.G. Aronov: Sov. Phys. JETP **50**, 968 (1979)
94. W.L. McMillan and Jack Mochel: Phys. Rev. B **46**, 556 (1981)

The Properties of $Co_2Cr_{1-x}Fe_xAl$ Heusler Compounds

Claudia Felser[1], Hans-Joachim Elmers[2], and Gerhard H. Fecher[3]

[1] Institut für Anorganische Chemie und Physikalische Chemie, Johannes Gutenberg-Universität Mainz, 55099 Mainz, Germany
felser@uni-mainz.de
[2] Institut für Physik, Johannes Gutenberg-Universität Mainz, 55099 Mainz, Germany
elmers@uni-mainz.de
[3] Institut für Anorganische Chemie und Physikalische Chemie, Johannes Gutenberg-Universität Mainz, 55099 Mainz, Germany
fecher@uni-mainz.de

Abstract. The classical concept of band structure tuning as used for semiconductors by partly replacing one atom by a chemical neighbor without altering the structure is applied examplarily to the half-metallic ferromagnetic Heusler compound $Co_2Cr_{1-x}Fe_xAl$. Band structure calculations are presented for ordered and disordered compounds. We present experimental and theoretical results. The connection between specific site disorder and the band structure is shown explicitly with particular emphasis on the half-metallic properties. Experimentally observed deviations from the ideal Heusler structure and from the simple Slater-Pauling rule for the magnetization are discussed in close relation to theoretical models. It has been found that the orbital magnetic moment and hence the spin-orbit coupling is important for the understanding of the half-metallicity. Experimental techniques which can be used to determine the electronic properties are described.

1 Introduction

Magnetoelectronics or spintronics, an electronics which uses the spin degrees of freedom, is currently attracting great interest due to its potential for applications in magnetic sensors, magnetic random access memories, and magnetologic [1]. A precondition for successful development of the new devices seems to be a high degree of spin polarization at the Fermi energy. A group of materials, the Heusler compounds, shows particular features in their electronic structure that may cause a high spin polarization. Guided by striking features in the electronic structure of several magnetic compounds we have identified $Co_2Cr_{1-x}Fe_xAl$ as a possible candidate. Our approach for identifying appropriate materials is described in the following.

Usually, new materials are found first experimentally followed by a theoretical explanation. Our starting point is different. We study the electronic structure of known relevant compounds systematically, try to understand their electronic structure, and extract similarities and differences. This idea is not new; the names which are related with this viewpoint are R. Hoffmann, J. Burdett and others [2–4]. The approach used here for finding new compounds goes further by trying to formulate a recipe for specific electronic properties like magnetoresistance. This approach is not arbitrary. We recognized that in a few structure types interesting properties are found more frequently compared to others. As an example, we recently suggested that, comparing the electronic structures of layered manganites and cuprates, the nesting features in the Fermi surface (a van Hove singularity) might be suggestive for magnetism as well as for superconductivity [5].

In a systematic search for magnetic analogues (from the electronic structure viewpoint) of superconductors, our studies have led us to examine GdI_2 [6], a layered d^1 compound that is isostructural with and nominally isoelectronic to the superconductors $2H$-TaS_2 and $2H$-$NbSe_2$. GdI_2 is known to undergo a ferromagnetic transition around 300 K [7]. However, the presence of the half-filled 4f band results in the compound being ferromagnetic. The systematic seen in the layered manganites and cuprates, led us to search for colossal magnetoresistance effects in GdI_2. Indeed, we found a colossal magnetoresistance effect in GdI_2 with a significant magnitude, approximately 70% at 7 T close to the room temperature [8]. However, GdI_2 is very sensitive to air and a compound with a Curie temperature at room temperature is expected to have a high spin polarization only at low temperatures. Therefore, we have focused our search on magnetic Heusler and half-Heusler compounds with a high Curie temperature. Large magnetoresistance effects have already been found in doped Fe_2VAl [9], Fe_2VGa [10], $Fe_{3-x}V_xSi$ [11] and in $CoV_{0.6}Mn_{0.4}Sb$ [12]. A high spin polarization is clearly favorable for both colossal magnetoresistance effect and spin polarized tunneling.

Co-based Heusler compounds are of particular interest because they show comparatively high Curie temperatures. Half-metallic Heusler and half-Heusler compounds show a Slater-Pauling behavior, and the saturation magnetization scales with the number of valence electrons following a simple electron counting scheme [13–17]. Paramagnetic semiconducting Heusler compounds are formed for electron counts of 24 valence electrons per formula unit.

Our band structure calculations predict that the ordered Heusler compounds Co_2CrAl and $Co_2Cr_{0.6}Fe_{0.4}Al$ should exhibit a complete spin polarization at the Fermi level [18]. In $Co_2Cr_{0.6}Fe_{0.4}Al$ the density of states reveals a half-metallic ferromagnet (HMF) with a van Hove singularity in the vicinity of the Fermi energy in the majority spin channel and a gap in the minority spin channel. This feature turns $Co_2Cr_{0.6}Fe_{0.4}Al$ into a potential candidate to show high magnetoresistance effects. In pressed powder pellets, we have indeed found a magnetoresistive effect of 30% in a small magnetic field of 0.1 T at room temperature [18]. This observation has motivated us to look closer into this interesting system.

Our initial findings have started a considerable effort in order to exploit the electronic properties of $Co_2Cr_{1-x}Fe_xAl$ compounds. Recent band structure calculations [14, 17, 19–22] confirmed the half-metallic character for Cr-rich compounds. Zhang et al. [20] found that Co_2CrAl is a true half-metallic ferromagnet with a magnetic moment of $3\mu_B$/f.u.. Reproducible magnetic moments were obtained in agreement with the thumb rule if Cr is partly replaced by Fe. Co_2FeAl is no half-metall due to a small overlap of the conduction band with the valence band at the X point. Ishida et al. [22] have investigated $Co_2Cr_{1-x}Fe_xAl$ for varying values of x by superstructure calculation. For $x < 0.625$ the half-metallicity is destroyed. The replacement of Cr by Fe may be seen as a doping with electrons. Band structure calculations for the disordered variant also confirm a strongly reduced magnetic moment (about $1\mu_B$/f.u.) as well as a loss of the half-metallic character. Miura et al. [21] investigated the effects of disorder on the electronic and magnetic properties of $Co_2Cr_{1-x}Fe_xAl$ using a coherent potential approximation. It was found that the compound preserves the high spin polarization also in the disordered B2-structure. Only disorder between Co and Cr or Fe reduces the spin polarization. In addition to bulk properties, Galanakis [19] investigated the influence of surface states on the half-metallic properties in Heusler compounds and reported that Co_2CrAl with a CrAl-terminated surface behaves different from all other Heusler compounds, the pseudogap is preserved in the minority bands.

Inomata et al. [23] prepared a spin valve type tunneling junction with a $Co_2Cr_{0.6}Fe_{0.4}Al$ Heusler compound film which shows a tunneling magnetoresistance of 16% at room temperature. Clifford et al. [24] found a spin polarization of up to 80% using point contact spectroscopy.

The review is organized as follows: In Sect. 2 we present band structure calculations carried out at different levels. The obtained band structures are compared with existing literature data and used as a base for detailed comparison to experimental results. The experimental findings for $Co_2Cr_{1-x}Fe_xAl$ alloys are summarized in Sect. 3 for bulk materials and in Sect. 4 for thin films. We put a focus on recent results obtained with soft X-ray photoemission and absorption spectroscopy.

2 Theory

Electronic structure calculations were carried out at different levels of sophistication. Ordered compounds were calculated by means of full potential methods. Using supercells, such methods may also be used to calculate compounds with particular fractional mixtures between Cr and Fe in the doped case or to estimate the influence of disorder. This results, however, in an artificial periodicity leading to distinct deviations in comparison to experiments. To overcome this problem, the coherent potential approximation was used to determine the properties of doped compounds and randomly disordered alloys. Finally, in order to determine spectroscopic properties correctly, fully relativistic calculations were performed for ordered and disordered compounds. In the following, the calculations and their results are presented and compared with the work of other groups.

2.1 Calculations for Ordered Compounds

First, self – consistent band structure calculations were carried out for well ordered structures using the full potential linearized augmented plane wave method (FLAPW) provided by Blaha et al. [25, 26] (Wien2k 04). In this method only core-states are treated relativistically. The exchange - correlation functional was taken within the generalized gradient approximation (GGA) in the parameterization of Perdew et al [27]. For comparison, calculations were also performed using the linear muffin-tin orbital (LMTO) method [28] provided by Savrasov [29] (LMTART 6.5) on different levels of sophistication from simple atomic sphere approximation (ASA) to full potential plane wave representation (FP-LMTO-PLW).

The properties of the pure Cr or Fe containing compounds where calculated in $Fm\bar{3}m$ symmetry using the experimental lattice parameter ($a = 10.82a_{0B}$, $a_{0B} = 0.529$Å) as determined by X-ray powder diffraction. All muffin tin radii where set to nearly touching spheres with $r_{MT} = 2.343a_{0B}$ in both full potential methods. The overlapping spheres where set to $r_{MT} = 2.664a_{0B}$ for the ASA calculations. Ordered compounds are derived from the supercell method by replacing certain Y = Cr atoms of the full $X_8Y_4Z_4$ cubic cell by Y' = Fe atoms. Table 1 gives the symmetry and positions of the atoms in the different compositions used.

Table 1. Space group of ordered $Co_2Cr_{1-x}Fe_xAl$ and Wyckoff positions of the constituents for $x = 0, 1/4, 1/2, 3/4, 1$

Compound		Co	Cr	Fe	Al
Co_2CrAl	$Fm\bar{3}m$	8c	4a	−+	4b
$Co_8Cr_3FeAl_4$	$Pm\bar{3}m$	8g	1a	3c	1b, 3d
$Co_4CrFeAl_2$	$P4/mmm$	4i	1a	1d	1b, 1c
$Co_8CrFe_3Al_4$	$Pm\bar{3}m$	8g	3c	1a	1b, 3d
Co_2FeAl	$Fm\bar{3}m$	8c	–	4a	4b

Results for Ordered Compounds

The electronic structure of the pure and doped compounds will be discussed in the following. First, the band structure and the density of states of the ordered

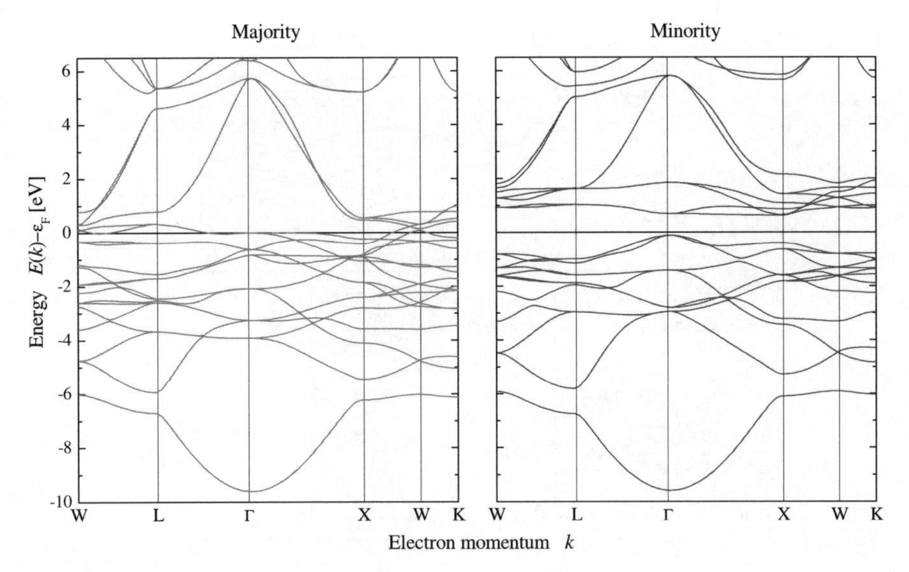

Fig. 1. Band structure of Co_2CrAl.
The energy scale is referenced to the Fermi energy (ϵ_F)

compounds are presented. This is followed by a more specific discussion of the magnetic properties on hand of measured and calculated magnetic moments.

The self-consistent FLAPW band structure of Co_2CrAl is shown in Fig. 1.

From the spin-resolved bands, it is seen that the majority bands cross or touch the Fermi energy (ϵ_F) in almost all directions of high symmetry. On the other hand, the minority bands exhibit a gap around ϵ_F, thus confirming a HMF character. For Co_2CrAl, the width of the gap is given by the energies of the highest occupied band at the Γ-point and the lowest unoccupied band at the Γ or X-point. The smaller value is found between Γ and X, thus it is an indirect gap. However, there is only a small difference in the maxima at the X and Γ points and details of the calculation methods may change this observation [22].

We will restrict the following comparison of the doped compounds to the Δ-direction being parallel to [001]. The Δ-direction possesses in all cases the C_{4v} symmetry. It has the advantage that the compound with $x = 1/2$ can be compared directly to the others even if it is calculated for tetragonal symmetry where the corresponding Λ-direction is between Γ and Z.

The Δ-direction is perpendicular to the Co_2 (100)-planes, playing an important role for the understanding of the HMF character of Heusler compounds. This was first pointed out by Öĝüt and Rabe [30].

The band structures in Δ-direction of the pure ($x = 0, 1$) and the doped ($x = 1/2$) compounds are compared in Fig. 2 for energies above the Heusler gap. In general, the doped compounds exhibit much more bands compared

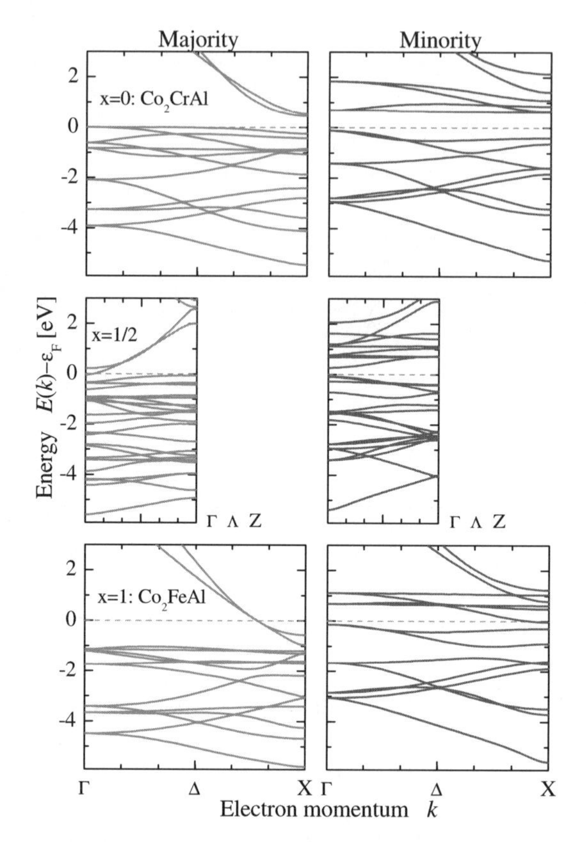

Fig. 2. Δ,Λ-bands of $Co_2Cr_{1-x}Fe_xAl$.
(The high symmetry points are X for $x = 0, 1$ and Z for $x = 1/2$. Note that the absolute values of k at X are: 1.1Å^{-1} for $x = 0, 1$ and only 0.55Å^{-1} at Z for $x = 1/2$)

to the pure ones as a result of the lowered symmetry. Therefore, results are shown only for the mixed compound with equal Fe and Cr concentration.

First, we compare the majority bands of the two pure compounds. The Fermi-energy is higher in the Fe case compared to Cr, as expected from the larger number of d-electrons. The indirect gap of the Δ-direction (clearly seen for the Cr based compound) is not only shifted below ϵ_F but nearly closes for the Fe based compound. This gap is also nearly closed in the majority bands of the mixed compound with $x = 1/2$ as well as those with $x = 1/4, 3/4$ (not shown here). This observation questions any rigid band model. In that case the bands would be simply filled with increasing number of d-electrons, leaving the shape of the bands unchanged.

More interesting is the behavior of the minority bands as those determine the HMF character of the compounds. Comparing Cr and Fe based compounds, one finds that the energies of the states at Γ are nearly the same below ϵ_F. The shapes of the bands close to Γ are similar, too. The situation is different at X where the unoccupied states are shifted towards ϵ_F in Co_2FeAl compared to Co_2CrAl.

The first unoccupied minority band of Co_2FeAl just touches ϵ_F at X. Therefore, any temperature above 0 K will immediately destroy the HMF gap. Even if accounting for small numerical deviations, this compound may not be a good candidate for spin-injection devices.

The mixed compounds are described by P lattices. Therefore, the Brillouin zone of these compounds is generally smaller compared to the F lattice. This results in a seemingly back-folding of bands from the larger F-Brillouin zone into a smaller one. This effect is accompanied with some additional splitting (removed degeneracies) at points of high symmetry. Due to the manifold of bands in the mixed compounds, it is not easy to compare the results directly, therefore we concentrate on the width of the gap in the minority bands.

The minority gap is mainly characterized by the bands in the Δ-direction as found from the band structure for all directions of high symmetry (not shown here). The width of the gap in the minority bands is shown in Fig. 3. The direct band gap at the Γ-point becomes successively smaller with increasing Fe content x and ranges from 750 meV at $x = 0$ to 200 meV at $x = 0.75$. The direct gap at Γ is much wider in Co_2FeAl, therefore this compound is characterized by the indirect gap only. In contrast to our calculation Ishida et al. [22] found that the gap is closed for compositions $Co_2Cr_{1-x}Fe_xAl$ with $x < 5/8$.

The character of the gap changes from indirect to direct if comparing pure and mixed compounds, respectively. This change of the character of the gap in the Δ-direction is a consequence of the smaller Brillouin zone in the mixed compounds that leads to a so called back-folding of bands.

The total density of states (DOS) is shown in Fig. 4 for varying Fe content x. The gap at the Fermi energy is clearly seen in the DOS of our calculation of the minority states for all compounds.

The majority DOS at the Fermi energy decreases with increasing Fe concentration x. The density of majority electrons at ϵ_F is a crucial point for spectroscopic methods investigating the spin polarization, like spin-resolved photoemission. A complete spin polarization may be only detectable if there is a high majority density. The same may be true for spin injection systems where one is interested in a high efficiency.

It is also seen that the minority DOS seems to be much less effected by the Fe doping compared to the majority DOS. Mainly the unoccupied part of the DOS above ϵ_F changes its shape but not the occupied part.

In summary, it is seen that doping of the compound with Fe changes mainly the occupied majority and the unoccupied minority DOS. Doping by

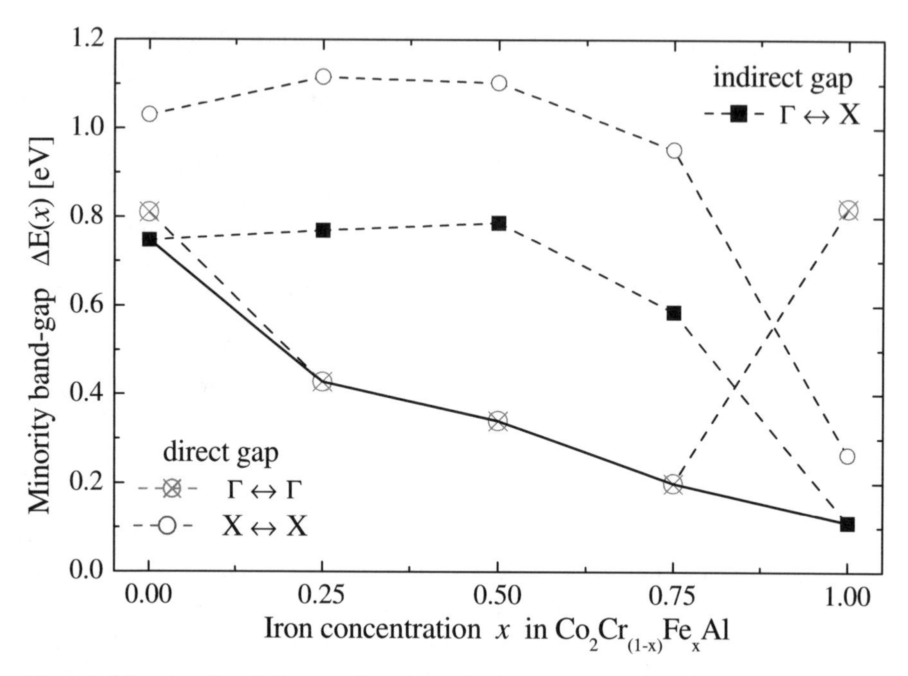

Fig. 3. Minority Band Gap in $Co_2Cr_{1-x}Fe_xAl$ for $x = 0, 1/4, 1/2, 3/4, 1$. (Lines are drawn to guide the eye. The full drawn line follows the limit for the HMF-gap)

Fe results thus not just simply in a shift of the DOS as would be expected from a rigid band model.

More details of the change in the DOS and electronic structure can be extracted if analyzing the partial DOS (PDOS), that is the atom type resolved density of states. The PDOS of the pure compounds is compared in Fig. 5 to the PDOS of the mixed compound with equal Cr and Fe content ($x = 1/2$).

From Fig. 5, it is seen that the high majority DOS at the Fermi-energy emerges from Cr. Both, Co and Fe exhibit only a small majority PDOS at ϵ_F. Overall, the decrease of the majority DOS of $Co_2Cr_{1-x}Fe_xAl$ around ϵ_F can be clearly attributed to the increasing amount of Fe with respect to chromium. The minimum in the minority DOS around ϵ_F is mainly restricted by the shape of the Co PDOS. This indicates that the HMF like behavior is mainly characterized by Co. This statement is in agreement with the results found by Galanakis et al. [14, 15] also for other Co_2YZ Heusler compounds.

Doping with Fe does not only change the total DOS but also the PDOS of Co and Cr. In particular, the slight shift of the Cr PDOS to lower energies causes an additional decrease of majority states at ϵ_F. This shift increases with increasing Fe concentration as was found from the PDOS for $x = 1/4$ and 3/4 (not shown here). The aluminium PDOS stays rather unaffected from the Cr or Fe concentration. It exhibits only small energy shifts.

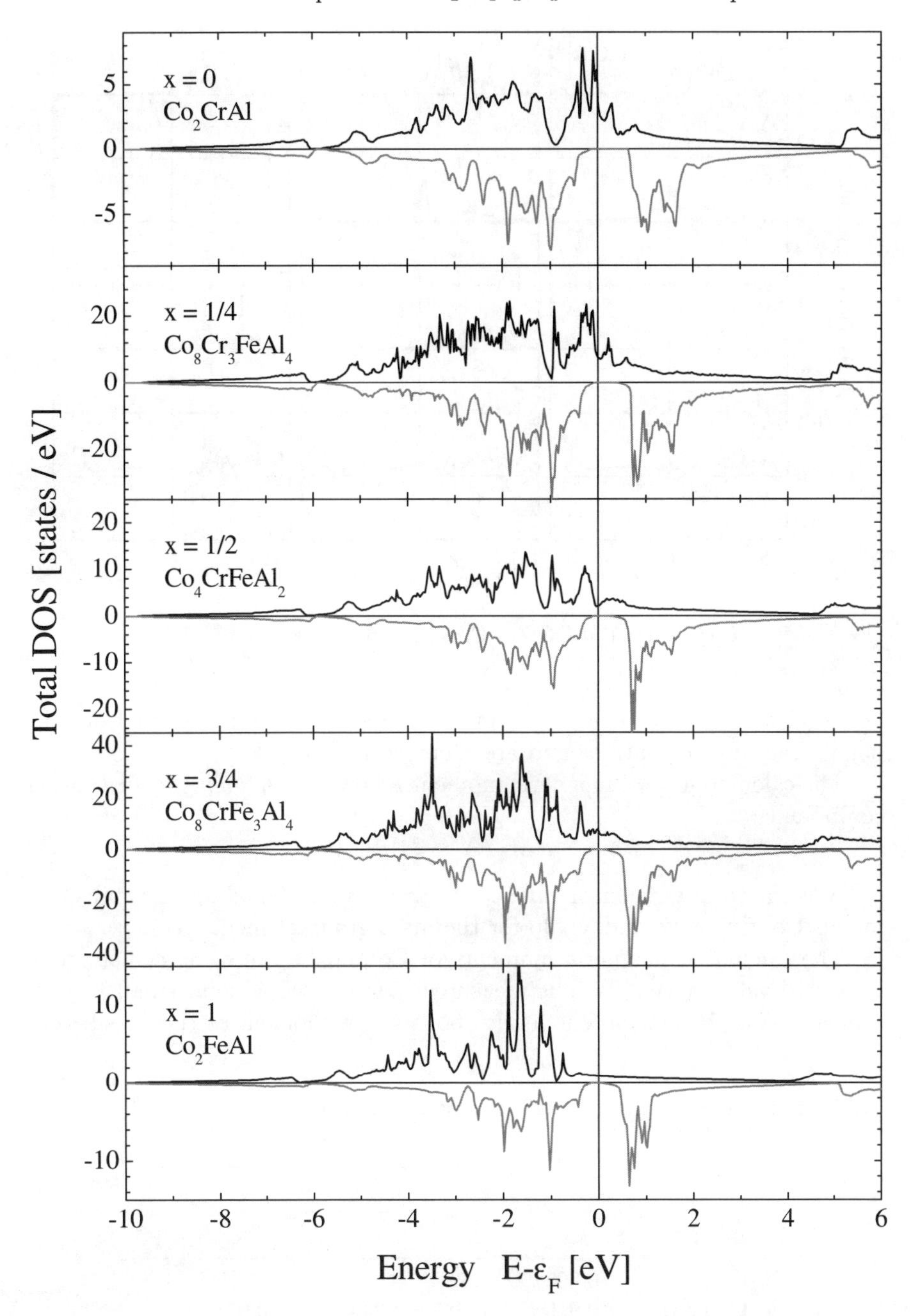

Fig. 4. Total DOS of ordered $Co_2Cr_{1-x}Fe_xAl$ for $x = 0, 1/4, 1/2, 3/4, 1$

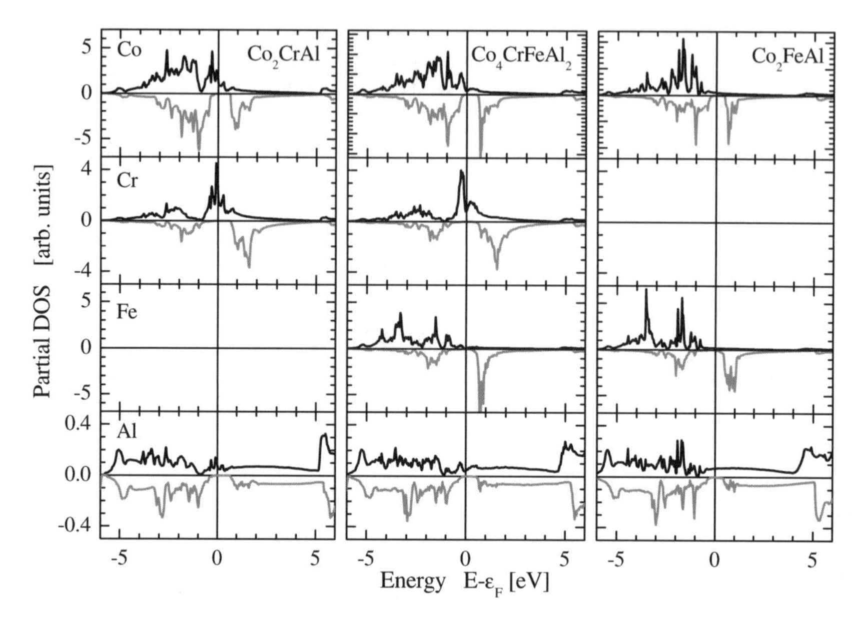

Fig. 5. Partial DOS of $Co_2Cr_{1-x}Fe_xAl$ ($x = 0, 1/2, 1$)

Total and element specific magnetic moments were extracted from the band structure calculations and are reported in Table 2.

The calculated total magnetic moment follows the *thumb-rule* for Heusler compounds:

$$\mu_s = (N - 24)\mu_B \tag{1}$$

N is the total number of valence electrons (here: 4s, 3d for the transition metals Co, Cr, or Fe and 3s, 3p for the main group element Al).

The calculated magnetic moments of Co and Fe are in agreement with measured values (see 3.2). The measured, lower total value at small Fe concentration can be attributed clearly to a too low moment at the Cr sites.

Table 2. Total and element specific magnetic moments of ordered $Co_2Cr_{1-x}Fe_xAl$ for $x = 0, 1/4, 1/2, 3/4, 1$

	$m_s[\mu_B]$				
Compound	Co	Cr	Fe	Al	tot
Co_2CrAl	0.805	1.524	–	–0.07	3
$Co_8Cr_3FeAl_4$	0.916	1.509	2.802	–0.071	3.5
$Co_4CrFeAl_2$	1.014	1.552	2.777	–0.081	4
$Co_8CrFe_3Al_4$	1.1	1.69	2.783	–0.075	4.5
Co_2FeAl	1.22	–	2.796	–0.081	5

Most evidently, the calculated magnetic moment of the Co_2CrAl compound does not agree with the measured one. The experimental value is only about $1.3\mu_B$ whereas theory predicts a value of about $3\mu_B$ per formula unit. The value found here is in rough agreement to the ground state magnetic moment of $1.55\mu_B$ per formula unit reported earlier by Buschow [31]. It was previously considered that mainly the Co atoms carry the magnetic moment, whereas the contribution of Cr and Al atoms remains negligible [32]. However, this does not explain the physics behind the lowering.

One may assume that the high magnetic moment of Cr is an artifact of a particular calculation scheme. Therefore, different calculation schemes were used to check a probable occurrence of such effects. The results are summarized in Table 3. The results derived from KKR-calculations (see Sect. 2.2) are shown for comparison. All values found here for the pure compounds are in the same order of those calculated by Galanakis et al. [14] using also the KKR-method.

Table 3. Calculated magnetic moments in Co_2CrAl.
Element specific (per atom) and total magnetic moments (per formula unit) calculated by different methods (for details of KKR and SPRKKR see 2.2 and 2.3, respectively)

	$m_s[\mu_B]$		
Method	Co	Cr	tot
FLAPW (GGA)	0.81	1.52	3.00
FP-LMTO-PLW	0.73	1.61	3.01
KKR (GGA)	0.6	1.66	3.01
KKR (VWN-ASA)	0.79	1.51	3.00
KKR (VWN-MT)	0.71	1.66	3.00
LMTO (ASA)	0.67	1.72	2.98
LMTO (GGA-ASA)	0.70	1.63	2.98
SPRKKR	0.77	1.53	3.01

It is seen from Table 3 that all values stay comparable within a few percent. In particular, the total moment is $3\mu_B$, and integer within the accuracy of the numerical integration. Therefore, the deviation from the experiment cannot be attributed to the peculiarities of one or another theoretical method. The Cr moment is about $(1.6 \pm 0.1)\mu_B$ rather independent of the method of calculation.

A reduction of the observed Cr moment may be caused by site-disorder. Part of the Cr atoms may become anti-ferromagnetically ordered either among each other or with respect to the Co atoms. A mixture of ordered and disordered crystallites will then result in a measured value being too small compared to the calculated moment.

To roughly estimate the influence of disorder between Co and Cr atoms, the band structure was also calculated for the X-structure ($F\bar{4}3m$) [33]. The exchange of one of the X atoms with a Y type atom leads to the X structure with symmetry $F\bar{4}3m$. The X atoms (X = Co) occupy 4a (0,0,0) and 4c (1/4,1/4,1/4) sites and the Y (Y = Cr, or Fe) and Z (Z = Al) atoms occupy 4b (1/2,1/2,1/2) and 4d (3/4,3/4,3/4) sites, respectively. This structure was also considered among several phases with random anti-site disorder by Feng et al. [34] investigating Fe_2VAl. It is similar to the $C1_b$ Half-Heusler phase but with the vacancies filled up in a different way compared to the $L2_1$ structure. The X-structure consists of successive CoCr and CoAl (001)-planes.

The calculated, element specific and total magnetic moments are compared in Table 4. For Co the sum of both atoms is given. The calculation reveals for the X-structure an anti-ferromagnetic ordering of the Cr atoms with respect to Co and thus a reduced overall magnetic moment.

Table 4. Structural dependence of the magnetic moments in Co_2CrAl calculated by FLAPW

Structure		m [μ_B]			
		Co_2	Cr	Al	tot
$L2_1$	$Fm\bar{3}m$	1.42	1.63	−0.05	2.98
X	$F\bar{4}3m$	1.84	−0.83	−0.02	0.91

A more detailed analysis of the X-structure shows not only an anti-ferromagnetic order of Cr but also an enhancement of the Co magnetic moments. Their values are $0.94\mu_B$ in CoAl and $0.76\mu_B$ in CoCr (001)-planes. Evidently the shortest distance between two Co atoms is smaller in the X-structure (2.48Å) compared to the $L2_1$-structure (2.86Å). The shortest Co-Cr and Cr-Cr distances stay the same in both structures.

The gap in the minority band structure is closed in the X-structure, that means that the alloy under study is no-longer a half-metallic ferromagnet. The vanishing of the gap was also found for Co_2FeAl in the X-structure, but with the Fe atoms aligned ferromagnetically with respect to Co. It is interesting to note that the $cP16$ structure (mixing of Cr and Al in the (100)-planes) still exhibits the gap for both pure compounds. From this observation it can be concluded that the existence of the gap is directly related to the occurrence of Co_2– and mixed CrAl (001)-planes and that the $L2_1$-structure is not the only possibility for the presence of a HMF gap in Heusler-like compounds X_2YZ.

This finding can be understood easily considering symmetry. The magnetization will reduce the symmetry. Applying any special orientation along one of the principal axes (e.g.: [001]) will reduce the symmetry from $Fm\bar{3}m$ to $I4/mmm$, at least. The properties of the electron spin will cause a further reduction to $I4/m$, as vertical mirror operations would change the sign of the spin. In particular, the Γ-point of the ferromagnetic Heusler compounds will belong to the $D_{4h}(C_{4h})$ color group and not longer to O_h like in the paramagnetic state. The point group in brackets assigns the *magnetic* symmetry with removed vertical mirror planes. The X-point becomes Z $(D_{4h}(C_{4h}))$ and the point group symmetry of the Λ-direction (formerly Δ) is reduced to $C_{4v}(C_4)$ in the ferromagnetic case.

The Γ and Z points of the $cP16$ structures, as representatives for Cr-Al disorder, have in the ferromagnetic state *also* $D_{4h}(C_{4h})$ symmetry. Therefore, this structures is expected to behave *similar* like the $L2_1$-structure. The Γ and Z points of the X-structure, as representative for Co-Cr anti-site disorder, have in the ferromagnetic state the *lower* $D_{2d}(S_4)$ point group symmetry. Therefore, that structure is expected to behave *different*. Indeed, the most pronounced difference is the local environment of the atoms in the (001)-planes.

2.2 Calculations for Random Alloys

Self-consistent band structure calculations were carried out for random alloys using the Korringa-Kohn-Rostocker method (KKR) with a code provided by Akai [35,36]. The random alloys were calculated within the coherent potential approximation (CPA). The program allows to use various types of the exchange correlation functional, that are for example: v. Barth and Hedin [37], Moruzzi, Janak, and Williams [38], Vosko, Wilk, and Nussair (VWN) [39,40], or the generalized gradient approximation (GGA) in the parameterization of Perdew et al. [27, 41], or Engel and Vosko [42]. The program allows also to work in both approximations, muffin tin (MT) and atomic spheres (ASA).

The properties of the ordered $L2_1$ and X compounds where calculated for comparison (see Sect. 2.1). To account for a random distribution of Cr and Fe in $Co_2Cr_{1-x}Fe_xAl$, the CPA method was used. Co and Al where placed on 8c (1/4,1/4,1/4) and 4a (0,0,0) sites, respectively. Disorder between Cr and Fe was accounted by setting the 4b (1/2,1/2,1/2) site occupations to 0.6 for Cr and 0.4 for Fe.

Several types of disorder have been considered. The exchange of one of the X atoms with a Y type atom leads to the X structure as already described in Sect. 2.1. This is indeed still a well ordered structure, whereas the following structures arise from some random distribution of one or another of the atoms. Such types of disorder are described in detail by Bacon and Plant [33] for rationally fractional occupancies with 1/2 and 1/4.

Disorder between X (X = Co) and Y (Y = Cr, Fe) or X and Z (Z = Al) atoms leads to the DO_3 structure, still with symmetry $Fm\bar{3}m$, for disorder between the transition metal atoms. This type of disorder is found by setting the site occupations of the 8c and 4a sites to 2/3 for X and 1/3 for Y keeping Z on 4b sites.

Disorder in the Y and Z (Y=Cr, Fe, Z=Al) (111)-planes but keeping the Z (Z=Co) (111)-planes leads to the $B32$ structure. The symmetry is $F\,d\bar{3}m$ with X on 8a sites and Y and Z equally distributed on 8b sites.

Disorder in the YZ (Y=Cr, Fe, Z=Al) (100)-planes leads to the $B2$ structure with symmetry $P\,m\bar{3}m$. This type of disorder is found by setting the $F\,m\bar{3}m$ site occupations of the 4a and 4b sites equally to 1/2 for Y and Z. The resulting structure with reduced symmetry was used for the calculations. The lattice parameter is $a_2 = a/2$. X was placed at the 1b site (center of the cube) and the site occupations for Y and Z at 1a (origin of the cube) were set equally to 1/2.

Complete disorder between all atoms results in the $A2$ structure with symmetry $Im\bar{3}m$ ($a_2 = a/2$). In that case the site occupations of 2a (1/2,1/2,1/2) were set to 1/2 for X and 1/4 for Y and Z.

For the mixed Fe-Cr systems the site occupation factors for Y and Y' have to be weighted by x and $(1 - x)$, respectively. Also, Miura et al. [21] investigated this system by KKR-CPA methods. They compared the ordered $L2_1$ structure with $B2$ type disorder and Co-Cr type disorder.

Results for the magnetic moment of the $L2_1$ Co_2CrAl are reported above in Sect. 2.1. In the following we concentrate on the disordered alloys.

Results for Random Alloys

A band structure in the usual definition has no meaning for alloys with random disorder, due to the lack of periodicity. Therefore, only the integrated quantities will be discussed here. Indeed, the magnetic moments depend more directly on the DOS rather than the particular form of the dispersion of the electronic bands.

Table 5 shows the results for the magnetic moment in the disordered $Co_2Cr_{1-x}Fe_xAl$ alloy in comparison to the ordered compounds.

Obviously, Co_2FeAl shows already in the perfect $L2_1$ structure a non-integer value of the total moment, at least for the lattice parameter used in the calculation. That means it is not in a HMF ground state. It should be

Table 5. Calculated (KKR – VWN) magnetic moments in disordered $Co_2Cr_{1-x}Fe_xAl$, $x = 0, 0.4, 1$

Structure	Co	Cr	Co_2CrAl	Co	Cr	Fe	$Co_2Cr_{0.6}Fe_{0.4}Al$	Co	Fe	Co_2FeAl
$L2_1$	0.71	1.66	3.00	0.88	1.62	2.78	3.77	1.13	2.7	4.8
X	1.12	−0.9	0.88	1.14	−1.03	2.17	2.27	1.63	2.06	4.59
DO_3	1.92	0.21	2.69	1.92	0.19	2.74	3.56	1.86	2.69	4.81
$B32$	0.15	−0.03	0.26	0.8	−0.18	2.26	2.5	1.34	2.42	4.94
$B2$	0.71	1.8	3.1	0.88	1.64	2.87	3.68	1.07	2.82	4.76
$A2$	1.2	0.27	2.6	1.33	0.31	2.15	3.6	1.47	2.29	5.04

noted that the integer value of $5\mu_B$ was only found by FLAPW and SPRKKR (see below in Sect. 2.3), if using the experimental lattice parameter. The magnetic moment of correctly ordered $Co_2Cr_{0.6}Fe_{0.4}Al$ is close to the value of $3.8\mu_B$ expected for a HMF ground state. The total magnetic moment of the disordered compounds is in nearly all cases lower compared to the $L2_1$ structure. The only exceptions are the $B32$ and $A2$ alloys of Co_2FeAl.

The interesting point is that $Co_2Cr_{0.6}Fe_{0.4}Al$ shows in the X-structure also an anti-ferromagnetic alignment of the Cr atoms with respect to Co (and Fe) as was before observed for pure Co_2CrAl. Inspecting the DOS shows that this behavior destroys the minority gap. The Cr moments are nearly vanishing for DO_3 or $B32$ type disorder. All DOS exhibited the destruction of the minority gap in the cases of random disorder used for the calculations in Table 5. Our results are in agreement with the results found by Miura et al. [21].

2.3 Relativistic Calculations for Ordered Compounds and Random Alloys

Self-consistent band structure calculations were carried out using the spin polarized, full relativistic Korringa-Kohn-Rostocker method (SPRKKR) provided by Ebert [43]. The exchange correlation functional was taken within the VWN parameterization [39, 40]. Disorder and random alloys were calculated within CPA.

The properties of the $L2_1$ compounds and random alloys where calculated using the experimental lattice parameter. Co and Al where placed on 8c (1/4,1/4,1/4) and 4a (0,0,0) sites, respectively. Disorder between Cr and Fe was accounted by setting the 4b (1/2,1/2,1/2) site occupations to $1-x$ for Cr and x for Fe. Several types of disorder have been considered (for details see Sects. 2.1 and 2.2).

Results for Ordered Compounds and Random Alloys from Relativistic Calculations

Figure 6 shows the magnetic moments calculated in the full relativistic approach. The Fe concentration was varied in steps of 0.1.

The calculated total spin magnetic moment follows the *thumb-rule* Equation 1 for Heusler compounds (full drawn line). The calculated total magnetic moments of Co and Fe are in rough agreement with the measured values. The calculated moments of Co increase with increasing Fe concentration, whereas the Fe moment decrease and the Cr moments are nearly constant. This behavior can be explained by an energy shift of the partial densities of Co, Cr and Fe with change of the Fe content. The calculated Al spin moment is negative (see also results from FLAPW in Table 2), independent of the Fe concentration. This points on an *anti-ferromagnetic* order of the Al moments with respect to the transition metal moments. The measured, lower

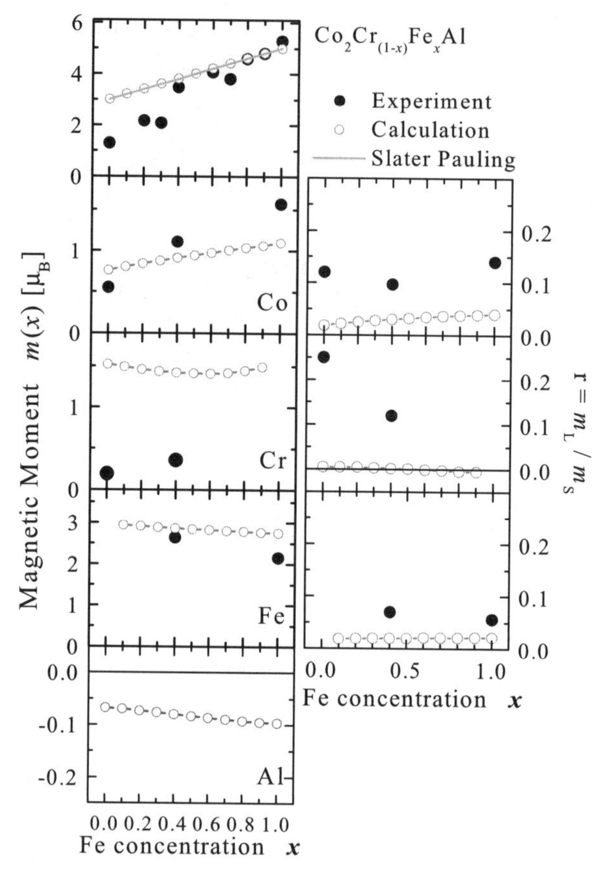

Fig. 6. Total spin magnetic moment (*top left panel*) and atom-resolved ones (*left panel*) for $Co_2Cr_{1-x}Fe_xAl$ as a function of the Fe concentration. In the right panel the ratio of the orbital magnetic moment over the spin magnetic moment for the transition metal atoms.
Measured values are from SQUID magnetometry and XMCD (see Sect. 3.2), calculations were performed with SPRKKR-CPA

total value at small Fe concentration can be attributed clearly to a too low moment at the Cr sites.

Most evidently the calculated ratios between orbital m_l and spin m_s magnetic moments are much smaller than the experimental values. For Cr, a vanishing of the orbital momentum is expected, as observed in the calculation. On the other hand, the determination of the Cr moments at the $L_{2,3}$ edges is complicated due to the partial overlap of the lines, what may cause the observation.

The too low values of m_l at Co and Fe sites cannot be explained that way. However, it is clear that the usual Hamiltonian used in LSDA contains directly a spin dependent coupling only. The orbital part of the moment thus arises only via spin-orbit interaction, or in words of the full relativistic case from the coupling of the large and the small components of the wave functions. This leads already in the case of pure Co and Fe metal to an underestimation of the orbital moments. One way to overcome that problem is inclusion of the Brooks orbital polarization [44] term (OP) in the Hamiltonian [45]. The results are compared in Table 6.

Table 6. Orbital magnetic moments in μ_B per atom

Method	VWN			VNW+OP			Exp.		
System	Co	Cr	Fe	Co	Cr	Fe	Co	Cr	Fe
Co_2CrAl	0.01	0	–	0.02	0	–	0.06	0.04	–
$Co_2Cr_{0.6}Fe_{0.4}Al$	0.03	0	0.06	0.04	0	0.08	0.11	0.05	0.18
Co_2FeAl	0.04	–	0.06	0.06	–	0.08	0.22	–	0.13

The inclusion of the OP term increases the magnitude of the orbital moments slightly, but evidently, the calculated values of the orbital magnetic moments are still smaller by a factor of 3 to 4 compared to the experimental values.

Both, experiment and calculation, exhibit the same trend for m_l as was already seen from the ratio $r = m_l/m_s$ in Fig. 6. The orbital magnetic moment at the Co sites increases with increasing Fe concentration, whereas the orbital magnetic moments at the Cr and Fe sites stay constant.

Calculations for the disordered system like in Sect. 2.2 did not show larger values for m_l. Thus an enhancement of m_l, as observed in the experiment by disorder can be excluded.

The full relativistic calculation (SPRKKR) of the density of states of $Co_2Cr_{0.6}Fe_{0.4}Al$ in $L2_1$ structure with random substitution of Cr by Fe is shown in Fig. 7.

It is seen that both, DOS and PDOS, of $Co_2Cr_{0.6}Fe_{0.4}Al$ are very similar to the ones of the supercell compound $Co_2Cr_{1/2}Fe_{1/2}Al$. Indeed, as mentioned in Sect. 2.2, the density of states and magnetic moments do not depend too much on peculiarities of the band structure. Thus it is not surprising that the DOS of the two systems with nearly equal stoichiometry are similar even the $Cr_{1/2}Fe_{1/2}$ compound has an artificial periodicity.

The minority gap is, however, less pronounced. The minority DOS does not become zero within the gap. This is mainly caused by the integration algorithm used in KKR methods [46]. It should be noted, however, that a distinction in purely spin up and spin down bands is not longer possible in the full relativistic approach. The spin-orbit interaction will mix states of opposite pure spin character. This causes a non-vanishing DOS within the minority-spin band gap formerly seen in the non-relativistic approach. Mavropoulos et al. [47] report calculations that result in a reduction of the

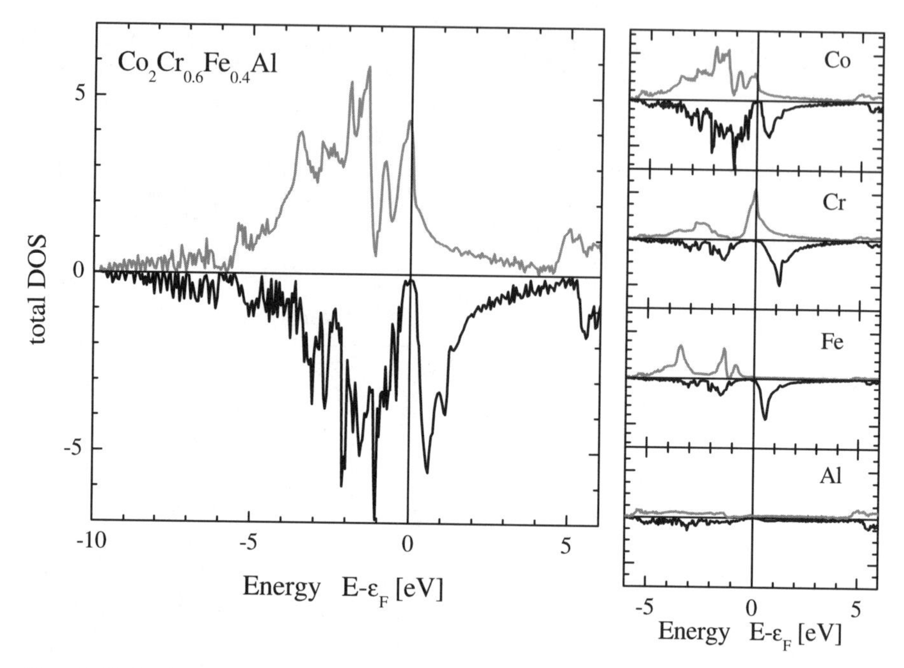

Fig. 7. Full relativistic density of states of $L2_1$ ordered $Co_2Cr_{0.6}Fe_{0.4}Al$. The left part shows the total DOS and the right part the PDOS for the constituting elements

spin polarization at the Fermi-energy from 99% in NiMnSb down to below 80% in MnBi.

More results from full relativistic calculations of spectroscopic properties are discussed below in comparison to the experiment.

3 Bulk Properties

3.1 Preparation and Structure

Polycrystalline samples of $Co_2Cr_{1-x}Fe_xAl$ were usually synthesized by melting the elements in an arc melter in a noble gas atmosphere. Several variations of the temperature variation during the solidification process were reported, in order to improve the crystallographic structure and minimize anti-site defects. In their first experiments with this compound de Groot et al [48] reported a quite long annealing period of several days at 1100 K followed by a slow cooling rate for compounds with $x = 0$ and $x = 1$. In [49] the procedure was refined for $x = 0$ to an annealing period of 1 day at 1200 K followed by 3 days at 1100 K in argon and finally a quenching into H_2O. In [18] re-grinding and remelting of the samples with $x = 0.4$ were performed to improve the homogeneity. Thereafter the samples were also annealed at 1100 K for 1 day followed by a rapid cooling in argon due to a shut down of the heating. Similar samples with $x = 0.4$ that were annealed were compared with samples that were not annealed in [50]. In recent works [20,24] repeated melting and annealing at 1100 K for up to 5 days was reported. In [24] the typical small loss of Al due to oxidizing was re-added during the remelting processes. To our experience, starting with lumps of the pure metals instead of powder, leads to a better ordering of the sublattices. From XPS investigations we know that samples synthesized by powder have a much higher oxygen contamination. Oxygen impurities seem to increase the site disorder of the Heusler alloys. Similarly, ball milling and grinding deteriorates the $L2_1$ structure.

$Co_2Cr_{1-x}Fe_xAl$ crystallizes in a cubic structure ($F\,m\overline{3}m$) with lattice constants of $a = 5.73(1)$ Å. In the case of completely ordered Heusler structure the Al atoms form a fcc close packed structure with the octahedral sites completely filled with Fe and Cr atoms and the tetraedral holes filled with Co atoms. Structural characterization has been performed with X-ray diffraction (XRD) as the standard method. Due to the little differences of the scattering factors between the 3d-metals Cr, Fe and Co, structural information beyond a simple confirmation of a single cubic phase can only be gained on the measurement of the comparatively small (10% of the (220)-peak) (111) and (002) superstructure peaks (see Fig. 8). Thus the XRD measurements are extremely difficult and time consuming. The simulated powder diffraction pattern of Co_2YAl with the Y site occupied by the mean value of Cr and Fe shows the decisive (111) and (200) peaks for different types of disorder. These superlattice peaks both vanish for the case of randomly occupied lattice sites ($A2$ structure). In the case of randomly occupied Y sites with Cr, Fe and Al ($B2$) only the (200) superlattice peak can be seen, while the (111) peak vanishes. For this $B2$ structure the Co atoms still form a simple cubic lattice that may be responsible for the half-metallic electronic structure [21] (see also Sect. 2). The half-metallic character is destroyed when Co atoms are partly exchanged by Fe or Cr (X structure). This type of disorder shows up as a (111) superlattice peak with higher intensity than the (200) peak. For the fully ordered Heusler ($L2_1$) structure the (111) peak is smaller than the (200) peak.

Experimentally reported atomic structures are summarized in Table 7. Although a $L2_1$ structure was reported in several references, a X-ray diffraction pattern was published only in [18,20]. It clearly shows both superlattice peaks however with the (111) peak being larger than the (200) peak indicating some degree of the X or $DO3$ – type disorder. Our own X-ray measurements (see Fig. 9) show only the (200) peak indicating a $B2$-structure.

For a structural characterization one can also apply neutron diffraction in order to exploit the different scattering factors for Fe, Cr and Co, thus enabling a better diagnosis of the different types of disorder. The measured

Table 7. Experimentally obtained structural, magnetic and electronic properties of arc-melted $Co_2Cr_{1-x}Fe_xAl$ compounds. Lattice constant a, Magnetization M, Curie temperature T_C, electrical resistivity ρ and estimated spin polarization at the Fermi edge P as explained in the text

x	Annealing	Structure	a(pm)	M(μ_B/f.u.)	T_C(K)	$\rho(\mu\Omega$cm)	P(%)	Ref.
0	1200 K(24h)	$L2_1$	589	1.5	334	–	–	[49]
0	1100 K(5d)	$L2_1$	574(3)	–	–	57	–	[20]
0	1100 K(24h)	$L2_1$	572.7	2.9	>400	–	–	[51]
0	no	$B2$	572.7	1.6	340	–	–	[52]
0.4	–	$L2_1$	573.6	3.65	665	60	80	[24]
0.4	1100 K(24h)	$L2_1$	573.7	3.3	650	–	40	[18]
0.4	no	$B2$	573.7	3.5	–	90	49	[50, 52, 53]
1.0	no	$B2$	573.0	5.3	–	–	–	[52]

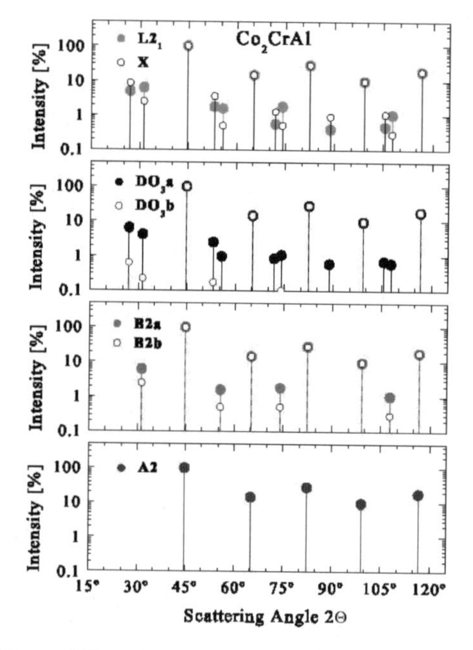

Fig. 8. Simulated X-ray diffraction pattern in Bragg-Brentano geometry for different types of ordered or disordered Co_2CrAl alloys.
Sublattice occupation defined as $Co_{8c}Cr_{4a}Al_{4b}$ for $L2_1$, $Co_{4a}Co_{4b}Cr_{4c}Al_{4d}$ for X, $(Co_2Cr)_{8c,4b}AL_{4a}$ and $(Co_2Al)_{8c,4b}Cr_{4a}$ for $DO3a$ and $DO3b$, $Co_{1a}(Cr_{1/2}Al_{1/2})_{1b}$ and $(Co_{1/2}Cr_{1/2})_{1a}(Co_{1/2}Al_{1/2})_{1b}$ for $B2a$ and $B2b$, $(Co_{1/2}Cr_{1/4}Al_{1/4})_{2b}$ for A2 – like disorder

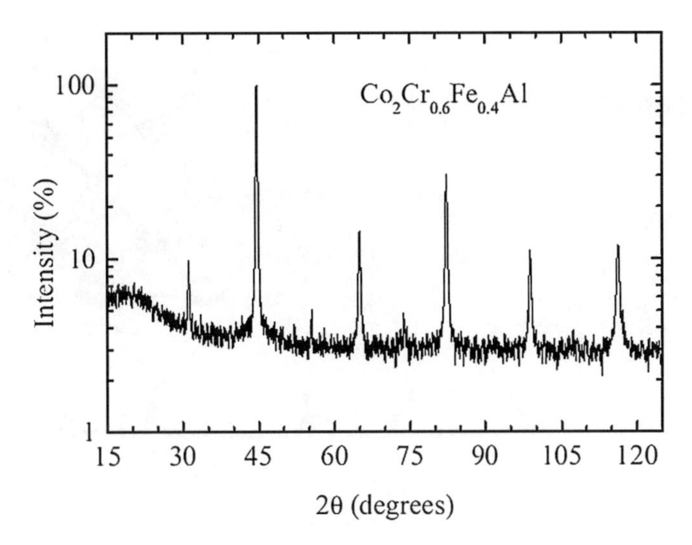

Fig. 9. X-ray powder diffraction pattern in Bragg-Brentano geometry for $Co_2Cr_{0.6}Fe_{0.4}Al$, indicating a $B2$ – type structure

ambient temperature neutron powder diffraction diagram of $Co_2Cr_{1-x}Fe_xAl$ with $x = 0.4$ (Fig. 10) reveals the well-ordered structure of a single phase cubic lattice with a lattice constant of $a = 5.737$ Å. The data set can be refined with a fully ordered Heusler ($L2_1$) phase with Al atoms on an fcc lattice. However, the neutron scattering factors for Cr and Al are almost identical. Therefore, the fully ordered $L2_1$ structure cannot be distinguished from the fully ordered $B2$ structure. The refinement results in an anti-site disorder of 10% in rough agreement with results from a Mößbauer study [18]. Besides the Mößbauer sextett characteristic for Fe in the ferromagnetic state optimized bulk samples of $Co_2Cr_{1-x}Fe_xAl$ with $x = 0.4$ still show a paramagnetic single line with a spectroscopic weight of 7%. The paramagnetic state of the Fe atoms might be due to Fe atoms with only Al atoms as nearest neighbors or small Fe clusters in a simple cubic structure.

There is no clear relation between the reported annealing procedure and structural properties. While longer annealing periods at 1100 K seemed to be successful in forming a $L2_1$ or at least a $B2$ structure in many cases [18,20,24, 48,49], other annealing experiments at the nominally identical temperature of 1100 K result in disorder, i.e. a transformation of the $B2$ structure into a $A2$ structure, combined with a considerable reduction of the magnetic moment and an increase of the contribution of the paramagnetic Mößbauer line [50]. A possible explanation might be a structural martensitic phase transition occurring close to or below the nominal annealing temperature. Thus it seems that a rapid cooling after melting or annealing is crucial for improving the structural quality of the samples.

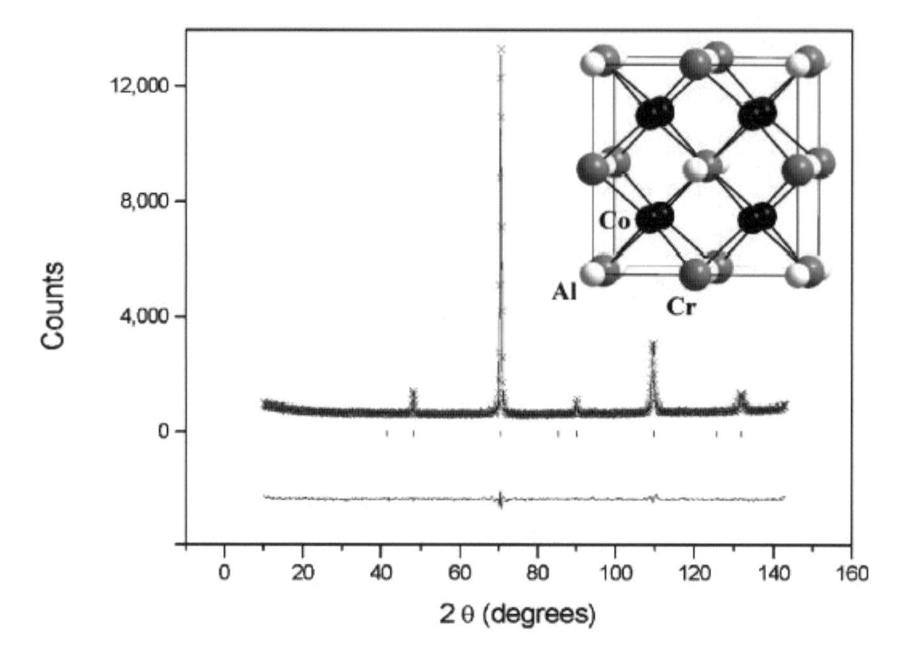

Fig. 10. Neutron diffraction diagram of $Co_2Cr_{1-x}Fe_xAl$ powder and difference between experimental data and refined data. (from [18])

3.2 Magnetic Properties

Magnetization

The easy experimental access to the mean magnetization or the magnetic moment per formula unit (f.u.) provides a quick test for a sample survey. Half-metallic Heusler alloys are expected to follow the Slater-Pauling behavior for the magnetization [19]: $M = (N - 24)\mu_B/\text{f.u.}$, where M is the total spin moment and N is the total number of valence electrons. For $Co_2Cr_{1-x}Fe_xAl$ one therefore expects $M = (3+2x)\mu_B/\text{f.u.}$. Though $Co_2Cr_{1-x}Fe_xAl$ with $x > 0.6$ is not predicted to be half-metallic, according to calculations this system still approximately follows this law [19]. Since all Heusler compounds show a very soft magnetic behavior, i.e. small remanent magnetization and almost no hysteresis, a sufficiently large external magnetic field (ca. 1 Tesla) has to be applied in order to align all moments and to determine the saturation magnetization that can be compared to calculated values. The saturation magnetization increases with Fe content as expected. However, a reduction of experimental values as compared to the calculated moments is observed in many cases (see Table 7). Increasing the Fe content tends to diminish this reduction and the magnetization of Co_2FeAl reaches the full predicted value of 5 μ_B [52]. As an exception, a high magnetic moment of 3 μ_B was reported for Co_2CrAl in [18].

The reduction of the magnetic moment was tentatively attributed to the paramagnetic Fe clusters detected by Mößbauer spectroscopy [18]. However, more insight can be gained by the element specific information obtained by magnetic circular dichroism in X-ray absorption (XMCD) [50]. Results obtained for $Co_2Cr_{0.6}Fe_{0.4}Al$ are shown in Fig. 11.

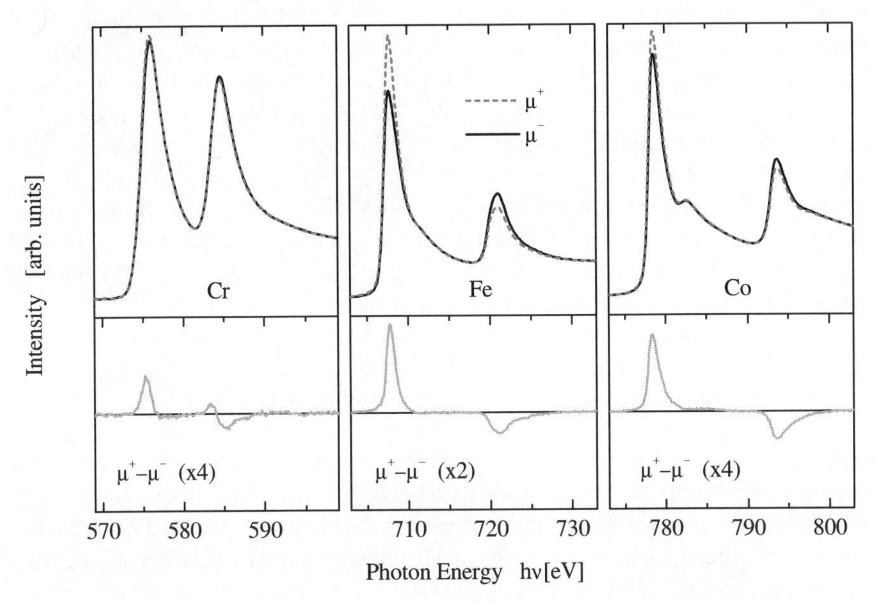

Fig. 11. Cr, Fe and Co 2p $\rightarrow$ 3d XAS-MCD spectra in the quenched $Co_2Cr_{0.6}Fe_{0.4}Al$ compound. Solid ($\mu-$) and dashed ($\mu+$) lines show the XAS spectra measured with external field applied parallel and anti parallel to the surface normal. The bottom panels represent the XMCD spectra. (from [16])

The XAS spectrum was measured by the total electron yield method, measuring directly the sample current while scanning the photon energy. For this method it is assumed that the total yield of photoelectrons I^i is proportional to the absorption μ^i. The helicity of the light was fixed while two spectra with opposite directions of the external field, defined as μ^+ and μ^-, were acquired consecutively. The magnetic field applied to the sample (up to 1 Tesla) during the measurement was aligned with the surface normal and at an angle of 30 degrees with respect to the incident photon direction. Before the MCD measurement the surfaces were scraped in situ in ultrahigh vacuum in order to remove the surface oxide layer. The success of the scraping was confirmed by the vanishing of the O 1s absorption edge. This in-situ cleaning process proves to be crucial because of the high reactivity and because of the selective oxidation of Al and Cr at the surface that destroys the stoichiometric composition while keeping the metallic behavior and magnetism for Fe and Co [50].

In order to quantitatively discuss spin μ_s and orbital μ_l momenta contribution to the magnetic moment, the magnetooptical sum rules may be applied as discussed in [54, 55], neglecting the magnetic dipole operator as proposed in [56]. Absolute magnetic moments cannot be determined because the proportionality factor $N_{\rm h}$, i.e. the number of d-holes. $N_{\rm h}$ is different for the 3 elements Cr, Fe and Co and in metallic systems it differs from the corresponding atomic values. $N_{\rm h}$ cannot be determined quantitatively from the XAS spectra. Even for pure elements varying values have been given, i.e. $N_{\rm h}(\rm Co) = 2.43 - 2.8$ [57–59]. For Co and Fe we have chosen the most likely values as determined for metallic systems [57], $N_{\rm h}(\rm Co) = 2.49$ and $N_{\rm h}(\rm Fe) = 3.39$, by calibration with bulk magnetization. For Cr, we set $N_{\rm h}(\rm Cr) = 6$ to the atomic ground state value. These values are in agreement (within an error of 10%) with the theoretical calculations discussed in Sect. 2.

While the error due to data fluctuations is negligible, an experimental error of 10% for the determination of magnetic moments is due to the error of the polarization and to the background subtraction. An additional error of the same order of magnitude arises from the uncertainty of $N_{\rm h}$, which only affects absolute values, and from the intermixing between the L_{II} and L_{III} edges, which mainly affects the ratio of spin and orbital moments particularly for Cr. Moreover, the magnetic dipole operator cannot be neglected for strongly correlated electronic systems, Heusler alloys being on the borderline. We anticipate that the total error for values of the atomic magnetic moments, when normalized such that the compositionally weighted sum coincides with the mean magnetization, is less than 25%.

Results for the 3 samples are included in Fig. 6. For samples that were quenched (see Sect. 3.1) we obtained spin magnetic moments in agreement with band structure calculations for Co and Fe atoms whereas Cr atoms possess a reduced magnetic moment. It is obvious that the experimentally observed reduced magnetization with respect to the theoretical prediction for the $L2_1$ structure is to a large extent due to a reduction of the Cr moment. The reduced Cr moment may be attributed to the Cr atoms occupying the Co sites in the $L2_1$ structure, as suggested by Miura et al. [21], who predicted the Cr atoms on Co atom sites (X-type disorder) to be antiferromagnetically coupled with the Co atoms. Annealing has led in our case to a further reduction of the small Cr moment in addition to a dramatic decrease of the Fe moment [50]. This means that even the optimum prepared samples do not form the perfect $B2$ structure as concluded from X-ray and neutron diffraction but contain a certain degree of disorder towards the $A2$ structure. In this sense, for the Cr containing compounds the deviation of the average magnetization from the Slater-Pauling behavior is a crucial measure for the unwanted type of disorder that destroys the half-metallicity.

Orbital Moments and Magnetic Anisotropy

For a very long time the Heusler alloys have been considered as prototypes of a material with completely quenched orbital moment moments because of their cubic structure and the observation of a g-factor coinciding with the free electron value [60]. Using the sum rules, XMCD reveals element-specific orbital magnetic moments (see Fig. 6 and Table 6). Without further assumptions on the number of d-holes one can determine the ratio $r = \mu_L/(\mu_S+\mu_T)$. The magnetic dipole term μ_T is usually less than 20% of μ_S and may be neglected. Values for r for Fe and Co being even larger than values obtained for pure elements were found [61]. We note that due to the XMCD mean probing depth of about 1 nm the bulk properties are augmented by surface effects that are due to a loss of symmetry at the surface, as well as to strain induced by the scraping that was applied to clean the surface.

The orbital moment contribution is smallest for the mixed $Co_2Cr_{1-x}Fe_xAl$ compound with $x = 0.4$. Usually the orbital contribution becomes larger when the $3d$ states are more localized. Therefore, the minimum of r values for $x = 0.4$ suggests that the $3d$ states are most localized for the pure ternary compounds [61]. It is interesting to note that a maximum magnetoresistance effect was found for pressed powder pellets with decreasing values when the composition deviates from $x = 0.4$.

The orbital to spin moment ratio r as determined by XMCD reveals a surprising dependence on the applied field [52]. For $Co_2Cr_{0.6}Fe_{0.4}Al$ the Fe and Co orbital to spin moment ratio r shows constant values within error limits as a function of external field: $\Delta r(\mathrm{Fe}) = 0.01 \pm 0.02$ and $\Delta r(\mathrm{Co}) = 0.02 \pm 0.02$. For Co_2FeAl we observed a decrease of $\Delta r(\mathrm{Fe}) = 0.04 \pm 0.02$ and $\Delta r(\mathrm{Co}) = 0.06 \pm 0.02$. The observed field dependence of r can be discussed in terms of a relation to magnetocrystalline anisotropies in the following way:

The magnetization loops show that the remagnetization in the Heusler compounds is mainly a reversible process. Reversible rotation of magnetization and reversible domain wall movements contribute to the reversible magnetization process. The domain walls adjust in moderate external fields without the cost of energy to match the mean demagnetization field and the external field. In a polycrystalline material crystallites are randomly oriented. If the anisotropy is large and the external field small, the magnetization vector will stay along the easy axis in every randomly oriented crystallite, while domain walls are being displaced. This explains the fact that the orbital moment do not depend on external fields for small fields. With increasing external field a reversible rotation of the magnetization vector takes the place of the domain wall displacement. The magnetization vector rotates into the direction of the external field even if this direction deviates from the (local) easy axis. Figure 12 illustrates this effect. The orbital magnetic moment is largest when the magnetization is directed along the easy axis of the magnetocrystalline or magnetoelastic anisotropy, while it is suppressed to some extent upon rotation into a hard axis [62]. This effect would thus explain

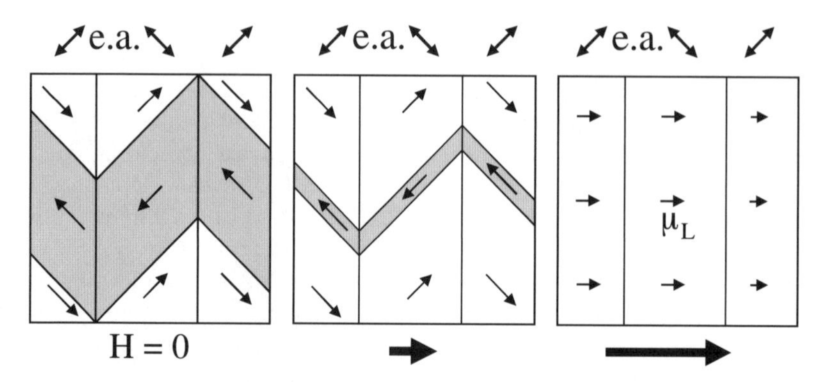

Fig. 12. Example of antiparallel domains in a [110] oriented grain. The easy axis (e.a.) of the magnetic anisotropy is indicated at the top. With increasing field domains parallel to the field increase on the expense of antiparallel domains. If the external field becomes very large, the magnetization will rotate into the hard direction parallel to the field which suppresses the orbital moment to some extent (indicated by the shorter arrows). (from [52])

the observed quenching of the orbital moment in the Co_2FeAl compound for large fields.

According to Bruno [62], there should be a proportionality between the orbital moment difference $\Delta\mu_L$ for easy and hard axis orientation of the magnetization and the magnetocrystalline anisotropy. Using constant factors as discussed in [52, 62] between magnetic and orbital moment anisotropy, the magnetic anisotropy corresponds to anisotropy fields of $1-2$ Tesla in rough agreement with the observed saturation field. Our observation indicates that the anisotropy is considerably smaller for $Co_2Cr_{0.6}Fe_{0.4}Al$ compared to Co_2FeAl. Again one should note that the anisotropy may to some extent be influenced by surface effects and strain anisotropy that may be enhanced due to the sample preparation.

Magnetic Domains

The sketched domain structure shown in Fig. 12 is motivated by domain structures observed for a $Co_2Cr_{0.6}Fe_{0.4}Al$ sample using photoemission electronmicroscopy (PEEM) (see Fig. 13). Magnetic contrast is obtained by subtracting images for opposite photon helicity at the L_3 edge of either Fe or Co. At the Cr edge the XMCD is too small to obtain sufficient contrast. Dark and bright areas in the XMCD image corresponds to magnetic domains with opposite magnetization and parallel to the projection of the photon beam. Domains with a mean gray value are aligned perpendicular to the photon beam. The images show the occurrence of micrometer-sized domains. On the submicrometer scale a more complicated domain pattern (ripple structure) is resolved. The observed domain structure is typical for non-oriented

Fig. 13. XMCD-PEEM images (size 9.3 × 8.3 μm^2) revealing the micromagnetic domain contrast of $Co_2Cr_{0.6}Fe_{0.4}Al$.
(**a**) and (**b**) show the circular dichroism at the L_3 edges. Dark and bright areas correspond to magnetic domains with opposite magnetization and parallel to the projection of the photon beam. Domains with a mean gray value are aligned perpendicular to the photon beam. (**c**) and (**d**) show the magnetic dichroism at fixed helicity of the photons. (from [52])

polycrystalline material [63]. It is caused by magnetostrictive interactions between adjacent grains. If magnetostriction along the easy axis is not zero, elastic interactions between the grains lead to additional conditions for the domain structure. The trend towards a homogeneous distribution of magnetoelastic and stray-field energy causes the folded bands of basic domains [63] as observed in Fig. 13. Magnetostrictive interactions also explain the surprisingly low magnetic remanence reported for all polycrystalline samples of

$Co_2Cr_{1-x}Fe_xAl$. Without magnetostrictive interactions an averaging over independent grains would lead to a remanence beyond 80% for cubic materials.

3.3 Transport Properties

The resistivity ρ of polycrystalline samples of $Co_2Cr_{1-x}Fe_xAl$ as a function of temperature should reveal the metallic character of the compound. The expected decrease of ρ with decreasing temperature has generally been found. The striking feature is that the residual resistivity at low temperatures is quite large (56 $\mu\Omega$cm to 90 $\mu\Omega$cm) [20, 24, 53], while the overall variation with temperature is weak. Residual resistivity ratios $\rho(300\text{ K})/\rho(0\text{ K} \leq 1.1$ were reported [20]. These magnitudes are much smaller compared to values observed for pure metallic elements and even smaller than values reported for other Heusler compounds, as 6.5 for single-crystalline Co_2MnSi and 2.7 for polycrystalline Co_2MnSi [64]. As pointed out in [20, 64] small amounts of defects and impurities can have dramatic effects on the residual resistivity ratio. For Heusler compounds reduced values may result from scattering contributions from impurities as well as anti-site defects. Therefore, it is a reasonable assumption that small values of the residual resistivity ratio in Heusler compounds indicates large amounts of anti-site defects and the temperature dependence of the resistivity is a key measurement for their detection.

The resistivity behavior at low temperatures bares an additional interesting information. For conventional metallic ferromagnets the contribution of the magnetic scattering to the low temperature resistance is governed primarily by the one-magnon scattering processes, that give rise to a T^2 temperature dependence due to absorption and emission of a single magnon [65]. For half-metallic ferromagnets with fully spin-polarized conduction electrons it was shown that one-magnon scattering processes are suppressed and two-magnon processes can lead to a $T^{9/2}$ temperature dependence [66]. It has been predicted that spin fluctuations initiated at finite temperatures may reduce this behavior to a T^3 temperature dependence of $\rho(T)$ due to unconventional one-magnon scattering processes [65]. This T^3 behavior has indeed been observed experimentally for Co_2CrAl [20] and was attributed to the half-metallic band structure of this compound.

Magnetoresistance effects in polycrystalline samples of $Co_2Cr_{1-x}Fe_xAl$ are comparatively small (<0.3%) and can show negative or positive sign. This behavior was interpreted as a competition between positive contributions due to the effect of the Lorentz force on itinerant electrons and a negative contribution due to the extinction of inelastic electron-electron scattering [20].

3.4 Electronic Structure

The direct experimental determination of the spin polarization at the Fermi edge proves to be difficult. Recently, the analysis of the Andreev reflection process in point contact junctions has attracted much interest. Point contacts between a superconducting tip and a polycrystalline $Co_2Cr_{1-x}Fe_xAl$ sample with $x = 0.4$ have been realized [53]. The observed dI/dU anomalies were compared to ballistic and diffusive models for the suppression of Andreev reflection by spin-polarization. In their analysis Auth et al. [53] considered a temperature increase in the contact region due to inelastic processes for the diffusive model. Using a classical model for the temperature increase in a small constriction an effective temperature can be defined that determines the quasiparticle distribution function and the BCS value of the superconducting gap. With these modifications the experimental dI/dU-spectra could be nicely fitted and the anomalies at small gap voltages could be clearly attributed to coherent transport. However, the ballistic and diffusive model yield a comparable description of the data and lead to incompatible values of the spin polarization. For the ballistic model the fit yields a spin polarization of $P = 0.49$, while it is $P = 0.06$ in the diffusive approach. Furthermore, one should note, that the obtained values provide a lower limit. It has been shown that selective oxidation of the surface can destroy the stoichiometric composition while the metallic character is preserved at the surface. A spin polarization of up to 50% is compatible with the existence of a metallic CoFe alloy at the surface.

First evidence for the predicted high spin polarization was found for $Co_2Cr_{1-x}Fe_xAl$ with $x = 0.4$ by measuring a magnetoresistive effect in pressed composite powder compacts [18]. A negative magnetoresistance effect of 30% was found at room temperature in a magnetic field of only 0.1 Tesla and the effect was nearly temperature independent in the range of 4-300 K. In the simplest assumption, one may model the powder compact as a network of spin polarized tunneling devices [67] (powder magnetoresistance effect). Electron transport from grain to grain will depend on the relative spin directions. An external field large enough to align the spins is than expected to cause a negative magnetoresistance effect as observed in the experiment. From the observed magnetoresistance effect a spin polarization of 40% may be inferred [68]. Annealing the powder compacts reduces the observed magnetoresistance effect to about 25%. Deviations from $x = 0.4$ rapidly suppressed the magnetoresistance effect to zero [16].

In order to overcome the difficulty to model a random network of grains, Clifford et al. [24] performed point contact measurements between two selected crystallites of the $x = 0.4$ compound using a piezo system. Significant magnetoresistance effects at small fields (10 mT) were found only when the conductivity was below $G_0 = 2e^2/h$. In some cases the magnetoresistance effect was as high as 80%. Using Julliéres formula ($\mathrm{MR} = 2P^2/(1+P^2)$) this value corresponds to a spin polarization of 81% which is quite high considering that at room temperature a spin polarization of no more than 90% can be expected due to thermal fluctuations. However, the data reported in [24] showed some scatter and the typical value was about MR = 25%,

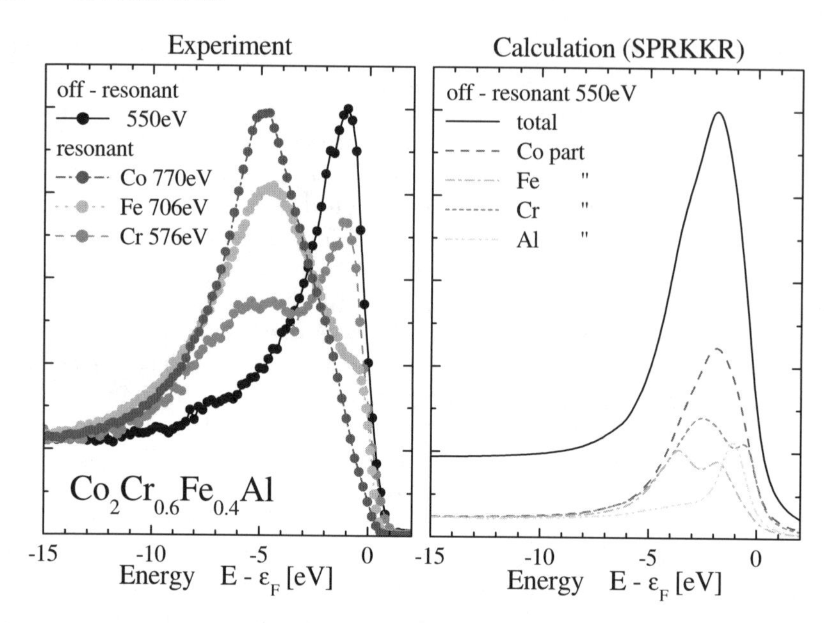

Fig. 14. Resonant photoemission spectra of $Co_2Cr_{0.6}Fe_{0.4}Al$ using synchrotron radiation (*left panel*) and calculated spectra using the KKR method

corresponding to a spin polarization of 38%. Magnetoresistive contributions to the observed effects as discussed previously were excluded in this experiment because independent measurements showed that magnetostriction in $Co_2Cr_{0.6}Fe_{0.4}Al$ is negligible for fields less than 10 mT and still small for larger fields ($\lambda = 5 \times 10^{-6}$ for 100 mT).

Using photoemission spectroscopy one obtains information on the electronic band structure in metals. For the interpretation one has to take into account that usually the kinetic energy of the photoemitted electrons is in a range where the mean free path is only a few Angstroms and the method therefore mainly shows electronic bands of the surface layer. On the other hand, this is just the relevant information for tunneling magnetoresistance devices. Figure 14 shows the measured and calculated photoemission spectra for polycrystalline $Co_2Cr_{0.6}Fe_{0.4}Al$. The surface of the sample was cleaned by mechanical scratching. The resonant excitation of spectra on the absorption edges of the elements in $Co_2Cr_{0.6}Fe_{0.4}Al$ allows an element specific discrimination of the density of states. Thus a direct comparison between experiment and calculation can be made. This comparison indicates that Cr and Al has segregated to the surface to some extent.

The X-ray absorption spectra contains information on the unoccupied density of states above the Fermi energy. For the resonant excitation of the 2*p*-states of the 3*d* transition elements the transition probability is proportional to the density of final states resulting from the 3*d* and 4*s* holes. The spectra, consisting mainly of the $L_3(2p_{3/2})$ and $L_3(2p_{1/2})$ white lines (see Fig. 11), reflect the high density of states resulting from the $3d$ electrons. Comparing spectra from pure Co with those from Co in $Co_2Cr_{1-x}Fe_xAl$ (see Fig. 15) reveals that the two prominent lines are narrower for the Heusler compound indicating a localized character of the $3d$-bands. No splitting of the prominent peak can be detected, indicating the more metallic-like band structure compared to the Co_2YSn compounds [69] with Y=Ti, Zr, and Nb. A prominent shoulder at the L_{III} peak was observed for $Co_2Cr_{1-x}Fe_xAl$ shifted by 4 eV with respect to the maximum to higher photon energy. A similar but less pronounced structure showed up at the L_{II} peak. The same structure occurs in the simulated absorption structure and can be attributed to Al s-states that are hybridized with Co $4s$ states. Thus, the observed shoulder is characteristic for the $Co_2Cr_{1-x}Fe_xAl$ ordered $L2_1$ or $B2$ structure. It vanishes for annealed and presumably disordered samples [50].

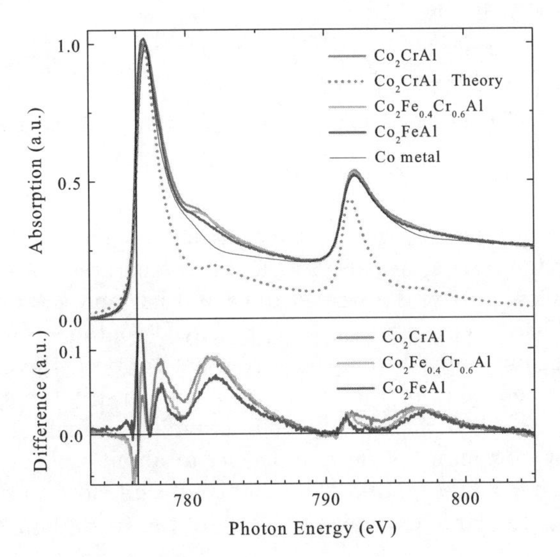

Fig. 15. X-ray absorption spectra (XAS) at the L_3 and L_2 edge of Co for the pure element reference sample and Co in $Co_2Cr_{1-x}Fe_xAl$.
The spectra were normalized at the maximum absorption intensity. The theoretical spectra was simulated using the band structure data calculated by SPRKKR (see Sect. 2.3). The lower panel shows the difference between the XAS of $Co_2Cr_{1-x}Fe_xAl$ and pure Co

Figure 16 shows the spin- and orbital density of states as derived from the Cr XMCD spectra using the sum rules. Some care has to be taken in the interpretation because the spin-orbit splitting of the Cr-$2p$ initial states is comparatively small and the spectra from the two edges overlap to some extent. Therefore we compare the experimental spectra with calculated spec-

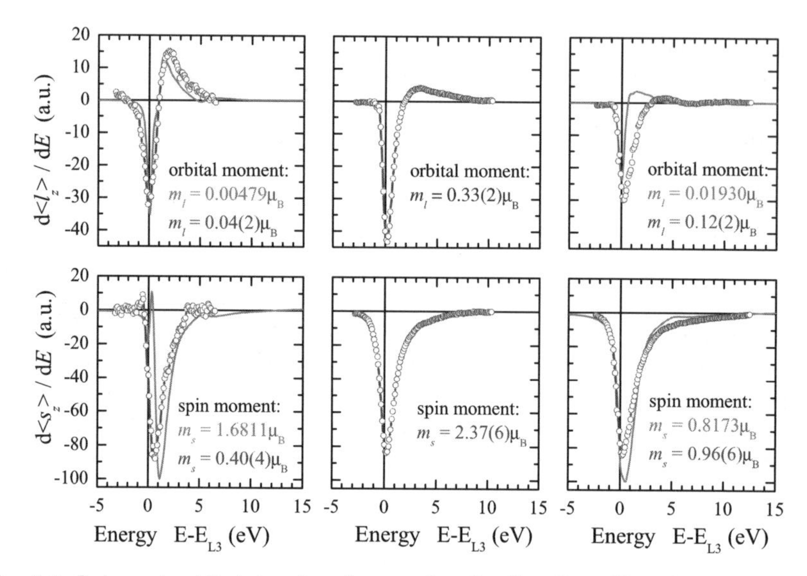

Fig. 16. Spin and orbital density of states for $Co_2Cr_{0.6}Fe_{0.4}Al$ as derived from the experimental XMCD spectra using the sum rules in comparison with calculated values (see Sect. 2.3)

tra using the ab-initio determined band structure. The Cr XMCD spectra is particularly interesting because in the disordered areas of the sample the Cr moments show partly antiparallel to each other and the magnetic signal cancels out. Therefore, the remaining Cr XMCD contains information from the ordered regions, only. While the agreement between theory and experiment is moderately well for the orbital moment density, the spin density of states can be described remarkably well, assuming a small artificial energy. The prominent maximum in the spin density at about 1 eV above the Fermi energy can clearly be recognized. The change of sign close to the Fermi edge is indicative for the large exchange splitting of the Cr 3*d* bands.

4 Thin Film Properties

4.1 Preparation and Structure

Polycrystalline as well as epitaxial $Co_2Cr_{1-x}Fe_xAl$ films were grown by dc magnetron sputtering. Using glass [70] or naturally oxidized Si wafers [71] as substrates, polycrystalline $Co_2Cr_{1-x}Fe_xAl$ films could be grown with a partly disordered *B*2 structure [71]. In this case a trend of increasing anti-site disorder with increasing Cr content was observed [71].

Single crystalline $Co_2Cr_{1-x}Fe_xAl(001)$ films were grown on MgO(001) at elevated temperatures (800 K) [70]. The in-plane $Co_2Cr_{1-x}Fe_xAl(001)[100]$

Table 8. Structural, magnetic and electronic properties of $Co_2Cr_{1-x}Fe_xAl$ films grown on several substrates by magnetron sputtering.
Lattice constant a, Magnetization M, Curie temperature T_C, electrical resistivity ρ and estimated spin polarization at the Fermi edge P as explained in the text

x	Substrate	T_s(K)	Structure	M(μ_B/f.u.)	ρ($\mu\Omega$cm)	P(%)	Ref.
0	SiO_x	300	$B2$	0.9	–	14	[71]
0	MgO(001)	800	$B2$	1.0	180	–	[70]
0.4	SiO_x	300	$B2$'	2.0	–	31	[23]
0.4	MgO(001)	800	$B2$	2.5	195	–	[70]
0.4	$Al_2O_3(11\bar{2}0)$	1000	$B2$	2.0	230	11	[72, 73]
1.0	SiO_x	300	$A2$	5.0	–	–	[23]
1.0	MgO(001)	800	$B2$	5.0	39	–	[70]

direction was oriented parallel to the substrate MgO(001)[110] direction, i.e. the films grow with a 45^0 rotation relative to the substrate. The misfit between films and substrate in this case is -3.8%. The lattice parameters were similar to bulk values and showed no significant variation with the Fe content. A Cr seed layer diminishing the misfit to -0.4% slightly decreased the FWHM of diffraction peaks. The FWHM also decreased for increasing Fe content. Due to the occurrence of the (002) peak the structure could be clearly identified as a $B2$ structure. No discrimination between $B2$ and $L2_1$ was considered.

Single crystalline $Co_2Cr_{0.6}Fe_{0.4}Al(110)$ films can be prepared epitaxially on a-plane $(11\bar{2}0)$ Al_2O_3 substrates [72]. In this case the in-plane $[1\bar{1}0]$ direction of the film is parallel to the [0001] direction of the substrate. X-ray diffraction intensities were observed for all (hkl) peaks where h, k, l are all even, but no intensity could be found for h, k, l all odd, i.e. the films do not possess the fully ordered $L2_1$-structure but rather a $B2$ structure. The structural properties improve with increasing substrate temperature (≤ 1000 K). On the other hand, high substrate temperature led to the growth of islands of typically 200 nm dimension with deep trenches between the islands that prevent electrical conductivity.

4.2 Magnetic Properties

Up to now, the thicknesses of the investigated films were in the range of a few 100 nm. In this thickness regime interface effects can be neglected. The magnetic properties of the films are expected to be similar as in the bulk, except the presence of a pronounced shape anisotropy. In contrast to the bulk behavior, $Co_2Cr_{1-x}Fe_xAl$ thin films showed a square shaped hysteresis loop with the remanent magnetization almost reaching the saturation value

[70,72]. Films containing Cr showed a substantial paramagnetic component of the order of $0.2\mu_B$/f.u. [70]. An in-plane magnetic anisotropy was measured for epitaxial $Co_2Cr_{0.6}Fe_{0.4}Al(110)$ films on *a*-plane-$Al_2O_3(11\bar{2}0)$ [72] with an easy axis along the [001] direction.

The saturation magnetization increases almost linearly with the Fe content starting at $1\mu_B$/f.u. for $x = 0$ to $5\mu_B$/f.u. for $x = 1$, the latter value being in accordance with the theoretical prediction of $4.9\mu_B$/f.u. [19]. As observed for bulk samples, the saturation magnetization is increasingly reduced compared to theoretical values with increasing Cr content. The reduction of the magnetization is slightly more pronounced for thin films with respect to bulk samples comparing films and bulk material with the same Cr content. This might indicate that the chemical order is more difficult to achieve in thin films than in bulk samples, although the structural quality was quite good. For the case of films with $x = 0.4$ we could show by XMCD that the Cr moment almost vanishes. As in the case of bulk material the reduction might therefore be attributed to Cr magnetic moments being partly directed antiparallel to the mean magnetization. Because it is mainly the Cr moment that is missing, the trend of a strong magnetization reduction with increasing Cr content is easy to understand.

4.3 Transport Properties

The resistivity ρ of $Co_2Cr_{1-x}Fe_xAl$ films does not show a pronounced temperature dependence. In general one observes a roughly linear behavior above $T = 100$ K while non-linear deviations occur at lower temperatures [70]. For films with $x = 1$ $\rho(T)$ resembles to a conventional metallic ferromagnet. Below 50 K, a quadratic behavior as expected for conventional spin-wave scattering is observed. Above 100 K $\rho(T)$ increases linearly with temperature. For films containing Cr, however, the resistivity is 5 times higher and the temperature variation $d\rho/dT$ for $T > 100$ K changes with increasing Cr content from a metallic behavior ($d\rho/dT > 0$) for $x > 0.4$ to a semiconducting behavior ($d\rho/dT < 0$) for $x < 0.4$. For $x = 0.4$ the resistivity is nearly temperature independent with an absolute value of 200 $\mu\Omega$cm [72]. Below 50 K the Cr containing films show an upturn in resistivity as the temperature is decreased [70, 72]. This might be attributed to a charge carrier localization at defect sites but a clear explanation for this effect does not exist at present. Only a small negative magnetoresistance effect is observed [70,72] in contrast to results on compressed powder pellets [18].

Measurements of the Hall effect are dominated by a strong anomalous Hall contribution and the normal Hall effect is only visible at high magnetic fields. While the normal Hall effect arises from the Lorentz force on charge carriers and depends on their sign, the anomalous Hall effect is due to the spin-orbit interaction and therefore is proportional to the perpendicular magnetization component. Figure 17 shows the Hall resistivity as a function of the applied field. From the shape anisotropy of a thin film one would expect that the

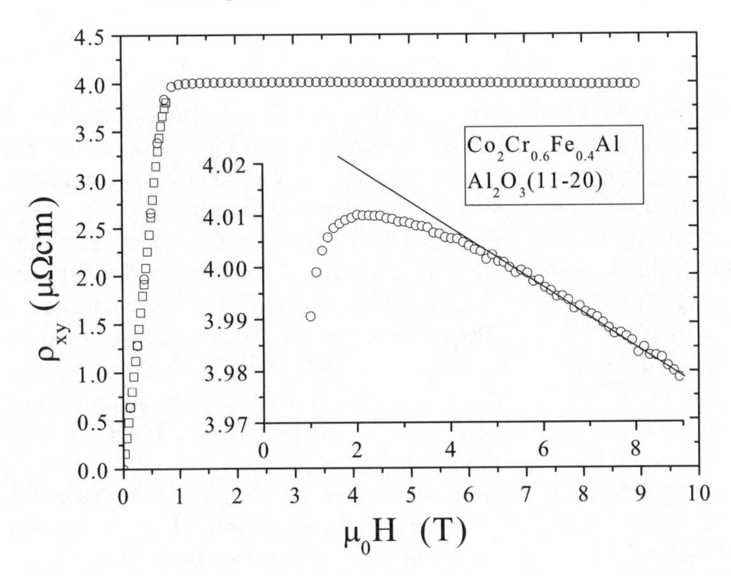

Fig. 17. Hall resistivity vs. magnetic field measured at 65 K of a 240 nm $Co_2Cr_{0.6}Fe_{0.4}Al$ grown on $Al_2O_3(11\bar{2}0)$ at 900 K. from [72]

magnetization component perpendicular to the film should saturate at $H_s = M_s \approx 1$ Tesla. However, a linear behavior of the Hall resistivity is observed only above a surprisingly high field of 5 Tesla [72]. At somewhat smaller fields the Hall coefficient [70] indicate that electrical conduction is dominated by holes for the Cr-containing films ($x \leq 0.5$) with a decreasing slope for decreasing Cr content. For $x = 1$ the electrical conduction is dominated by electrons. From the negative slope observed for high field values for films with $x = 0.4$ a nominal charge carrier density of approximately 5 electrons per unit cell is calculated [72]. These observations indicate that the Fermi surface is partly compensated and that both electron- and hole-like contributions to the carriers do exist, in agreement with calculations (see Sect. 2). The alloy with $x = 0.4$ seems to be close at a compensation point. Finally we should mention that if several types of carriers are present, their contribution to the Hall constant should be weighted with their respective mobilities and thus the scattering rates should be taken into account.

4.4 Spin Polarization

At first sight the obvious method for the measurement of the spin polarization of half-metallic thin films is the preparation of magnetic tunneling devices. One expects a pronounced increase of the magnetoresistance for half-metallic electrodes. Following the Julliére model [74], the tunneling magnetoresistance (TMR) ratio of a junction is related to the spin polarization P of the electrodes:

$$\mathrm{TMR} = \frac{R^{\uparrow\downarrow} - R^{\uparrow\uparrow}}{R^{\uparrow\downarrow}} = \frac{2P_1P_2}{1+P_1P_2} . \tag{2}$$

There are severe objections to this simple interpretation. The Julliére model is not able to explain the temperature and bias voltage dependence of the tunneling magnetoresistance. In equation 2 P_i denotes the polarization of the bulk material. The tunneling conductivity, however, is only sensitive to the local density of states in the vicinity of the barrier which may differ considerably from the one of the bulk material. Additionally, the Julliére model does not take into account the various direct effects of the barrier on the tunneling probability [75–77]. Because the model is commonly used to estimate the spin polarization of the electrodes, we recalculated the observed TMR into a spin polarization using (2), noting that this value defines an effective tunneling spin polarization [78] that is consistent with TMR measurements using a superconducting counter electrode.

Recently, tunneling junctions with $Co_2Cr_{1-x}Fe_xAl$ electrodes could be realized [23, 71, 72]. Spin-valve type tunneling junctions that consist of a series of $Co_2Cr_{1-x}Fe_xAl$ / AlO_x(1.4 nm) / CoFe layers were prepared on thermally oxidized Si substrates without any buffer layers by sputtering at room temperature [71]. In this case, the $Co_2Cr_{1-x}Fe_xAl$ films were polycrystalline with a $B2$-like structure. The maximum TMR was obtained for $x = 0.4$ [71] with values of 19% at room temperature and 27% at 5 K. For $x = 0$ the TMR is only 13% at 5 K. The TMR of the largest value of 27% leads to the spin polarization of 30% for $Co_2Cr_{0.6}Fe_{0.4}Al$ by applying 50% spin polarization of the CoFe alloy into equation 2. Tunnel junctions consisting of $Co_2Cr_{0.6}Fe_{0.4}Al/AlO_x$(1.4 nm)/Co were prepared on $Al_2O_3(11\bar{2}0)$ [73]. In this case, the films were sputtered at 600 K which is far below the optimum growth temperature of 1000 K with the consequence of a smooth interface but a polycrystalline structure [72]. The maximum TMR ratio observed was 10.8% at 4 K and 6% at room temperature, which is somewhat smaller than observed in [71] for similar $Co_2Cr_{0.6}Fe_{0.4}Al$ electrodes. The difference might be explained by the fact that in [73] the Co electrode was not optimized for preventing a switching of both electrodes at the critical field. The observed magnetization loops indicate that at zero field the two electrodes are not fully aligned antiparallel to each other, which of course diminishes the observable TMR effect.

5 Summary and Outlook

For Co_2CrAl a ground state magnetic moment of $1.55\mu_B$/f.u. (formula unit) has been reported by Buschow and van Engen [31] and it was considered that Co atoms mainly carry the magnetic moment in this material whereas the contribution of Cr and Al atoms remains small. According to recent band structure calculation, all constituents of the compound should possess a magnetic moment, i.e. $0.77\mu_B/atom$ for Co, $1.63\mu_B/atom$ for Cr and $-0.10\mu_B/atom$ for Al. For the total moment a value of approximately $3\mu_B$/f.u. is obtained, in agreement with the thumb rule for the magnetic moment, μ of half-metallic Heusler compounds from 3d transition metals $\mu/\mu_B = N - 24$, with N being the cumulated number of valence electrons (4s, 3d for the 3d-transition metals and 3s, 3p for Al). According to several band structure calculations Co_2CrAl reveals a half-metallic character of the density of states, i.e. a gap at the Fermi level in the minority band and a high density of states in the majority band. This property is conserved, when Cr is partly replaced by Fe. Therefore this Fe doping is promising for tuning the electronic properties, i.e. shifting a van Hove singularity to the Fermi edge. However, the band structure calculations reveal that a rigid band model is too simple to correctly predict the result of such a doping.

A comparison of experimental results with calculations shows that site disorder is the main obstacle for preparing half-metallic compounds. The preparation of $Co_2Cr_{1-x}Fe_xAl$ films that are $L2_1$ ordered seems to be even more difficult. While the disorder type with a partly exchange of Cr/Fe with Al atoms (B_2-type) seems to effect the spin polarization only little, an exchange of Co atoms with Cr, Fe or Al destroys the half-metallic character. The measurement of the element specific Cr moment provides an efficient tool for characterizing just this type of disorder in $Co_2Cr_{1-x}Fe_xAl$.

Acknowledgement

This work is supported by the DFG in the Forschergruppe: Neue Materialien mit hoher Spinpolarisation.

The authors are very grateful for help with experiments and calculations by P.-C. Hsu, Y. Hwu, W.-L. Tsai (Academia Sinica, Taipei, Taiwan), F. Kronast, S. Cramm, H.A. Dürr (BESSY, Berlin, Germany), H. Adrian, N. Auth, T. Block, G. Jakob, M. Jourdan, H.C. Kandpal, G. Schönhense, S. Wurmehl (Johannes Gutenberg – Universität, Mainz), H.-J. Lin (NSRRC, Hsinchu, Taiwan).

The authors thank all members of NSRRC (Hsinchu, Taiwan) and BESSY (Berlin, Germany) for their help during the experiments at the synchrotron facilities.

References

1. G.A. Prinz: Science **282**, 1660 (1998)
2. R. Hoffmann. In: *A chemists view of bonding in Extended Structures*, (VCH Verlagsgesellschaft mbH, Weinheim 1980)
3. J.K. Burdett. In: *Chemical bonding in solids*, (Oxford University Press Inc., New York 1995)
4. A. Simon: Angew. Chem. Int. Ed. Engl. **36**, 1789 (1997)

5. C. Felser, R. Seshadri, A. Leist, and W. Tremel: J. Mater. Chem. **8** 787 (1998)
6. J.D. Corbett, R.A. Sallach, and D.A. Lokken: Adv. Chem. Ser. **71**, 56 (1967)
7. C. Michaelis, W. Bauhofer, H. Buchkremer-Hermanns, R.K. Kremer, A. Simon, and G.J. Miller: Z. Anorg. Allg. Chem. **618**, 98 (1992)
8. C. Felser, K. Ahn, R.K. Kremer, R. Seshadri, and A. Simon: J. Solid. State Chem. **147**, 19 (1999)
9. H. Matsuda, K. Endo, K. Ooiwa, M. Iijima, Y. Takano, H. Mitamura, T. Goto, M. Tokiyama, and J. Arai: J. Phys. Soc. Jpn. **69**, 1004 (2000)
10. K. Endo, H. Matsuda, K. Ooiwa, M. Iijima, T. Goto, K. Sato, and I. Umehara: J. Magn. Magn. Mater. **177**, 1437 (1998)
11. K. Endo, M. Tokiyama, H. Matsuda, K. Ooiwa, T. Goto, and J. Arai: J. Phys. Soc. Jpn. **73**, 1944 (2004)
12. J. Pierre, K. Kaczmarska, and J. Tobola: Europ. Phys. J. B **18**, 247 (2000)
13. D. Jung, H.-J. Koo, and M.-H. Whangbo: J. Mol. Struct. (Theochem) **527**, 113 (2000)
14. I. Galanakis, P.H. Dederichs, and N. Papanikolaou: Phys. Rev. B **66**, 174429 (2002)
15. I. Galanakis, P.H. Dederichs, and N. Papanikolaou: Phys. Rev. B **66**, 134428 (2002)
16. C. Felser, B. Heitkamp, F. Kronast, D. Schmitz, S. Cramm, H.A. Dürr, H.-J. Elmers, G.H. Fecher, S. Wurmehl, T. Block, D. Valdaitsev, S.A. Nepijko, A. Gloskovskii, G. Jakob, G. Schonhense, and W. Eberhardt: J. Phys.: Condens. Matter **15**, 7019 (2003)
17. Ph. Mavropoulos, I. Galanakis, and P.H. Dederichs: J. Phys. Cond. Matter **16**, 4261 (2004)
18. T. Block, C. Felser, G. Jakob, J. Ensling, B. Muhling, P. Gutlich, and R.J. Cava: J. Sol. St. Chem. **176**, 646 (2003)
19. I. Galanakis: J. Phys.: Condens. Matter **16**, 3089 (2004)
20. M. Zhang, Z. Liu, H. Hu, G. Liu, Y. Cui, J. Chen, G. Wu, X. Zhang, and G. Xiao: J. Magn. Magn. Mat. **277**, 30 (2004)
21. Y. Miura, K. Nagao, and M. Shirai: Phys. Rev. B **69**, 144113 (2004)
22. S. Ishida, S. Kawakami, and S. Asano: Materials Trans. **45**, 1065 (2004)
23. K. Inomata, S. Okamura, R. Goto, and N. Yezuka: Jpn. J. Appl. Phys. **42**, L419 (2003)
24. E. Clifford, M. Venkatesan, R. Gunning, and J.M.D. Coey: Solid State Comm. **131**, 61 (2004)
25. P. Blaha, K. Schwarz, P. Sorantin, and S.B. Tricky: Comput. Phys. Commun. **59**, 399 (1990)
26. K. Schwarz, P. Blaha, and G.K.H. Madsen: Comp. Phys. Comm. **147**, 71 (2002)
27. J.P. Perdew, J.A. Chevary, S.H. Vosko, K.A. Jackson, M.R. Pederson, D.J. Singh, and C. Fiolhais: Phys. Rev. B **46**, 6671 (1992)
28. O.K. Andersen: Phys. Rev. B **12**, 3060 (1975)
29. S.Y. Savrasov: Phys. Rev. B **54**, 16470 (1996)
30. S. Öĝüt and K.M. Rabe: Phys. Rev. B **51**, 10443 (1995)
31. K.H.J. Buschow and P.G. van Engen: J. Magn. Magn. Mater. **25**, 90 (1981)
32. A. Slebarski, M. Neumann, and B. Schneider: J. Phys.: Condens. Matter **13**, 5515 (2001)
33. G.E. Bacon and J.S. Plant: J. Phys. F: Met. Phys. **1**, 524 (1971)
34. Y. Feng, J.Y. Rhee, T.A. Wiener, D.W. Lynch, B.E. Hubbard, A.J. Sievers, D.L. Schlagel, T.A. Lograsso, and L.L. Miller: Phys. Rev. B **63**, 165109 (2001)

35. H. Akai: Phys. Rev. Lett. **81**, 3002 (1998)
36. T. Kotani and H. Akai: Phys. Rev. B **54**, 16502 (1996)
37. U. von Barth and L. Hedin: J. Phys. C **5**, 1629 (1972)
38. V.L. Moruzzi, J.F. Janak, and A.R. Williams. In: *Calculated Properties of Metals*, (Pergamon Press, New York 1978)
39. S.H. Vosko, L. Wilk, and N. Nusair: Can. J. Phys. **58**, 1200 (1980)
40. S.H. Vosko and L. Wilk: Phys. Rev. B **22**, 3812 (1980)
41. J.P. Perdew and W. Yue. *Phys. Rev. B*, 33:8800, 1986
42. E. Engel and S.H. Vosko. *Phys. Rev. B*, 47:13164, 1993
43. H. Ebert. In: *Electronic Structure and Physical Properties of Solids*, Lecture Notes in Physics, vol 535, ed by H. Dreyssé, (Springer, Berlin 1998) p. 191
44. O. Eriksson, B. Johansson, and M.S.S. Brooks: J. Phys.: Condens. Matter **1**, 4005 (1989)
45. H. Ebert and M. Battocletti: Solid State Comm. **98**, 785 (1996)
46. H. Ebert and G. Schlütz: J. Appl. Phys. **69**, 4627 (1991)
47. Ph. Mavropoulos, K. Sato, R. Zeller, P.H. Dederichs, V. Popescu, and H. Ebert: Phys. Rev. B **69**, 054424 (2004)
48. R.A. de Groot. F.M. Mueller, P.G. van Engen, and K.H.J. Buschow: Phys. Rev. Lett. **50**, 2024 (1983)
49. A.W. Carbonari, R.N. Saxena, Jr. W. Pendl, J. Mestnik Filho, R.N. Attili, M. Olzon-Dionysio, and S.D. de Souza: J. Magn. Magn. Mater. **163**, 313 (1996)
50. H.J. Elmers, G.H. Fecher, D. Valdaitsev, S.A. Nepijko, A. Gloskovskii, G. Jakob, G. Schonhense, S. Wurmehl, T. Block, C. Felser, P.-C. Hsu, W.-L. Tsai, and S. Cramm: Phys. Rev. B **67**, 104412 (2003)
51. T. Block. In: *Neue Materialien für die Magnetoelektronik: Heusler- und Halb-Heusler-Phasen*, PhD Thesis, (Universität Mainz, Germany 2002)
52. H.-J. Elmers, S. Wurmehl, G.H. Fecher, G. Jakob, C. Felser, and G. Schönhense: Appl. Phys. A **79**, 557 (2004)
53. N. Auth, G. Jakob, T. Block T, and C. Felser: Phys. Rev. B **68**, 024403 (2003)
54. B.T. Thole, P. Carra, F. Sette, and G. van der Laan: Phys. Rev. Lett. **68**, 1943 (1992)
55. P. Carra, B.T. Thole, M. Altarelli, and X. Wang: Phys. Rev. Lett. **70**, 694 (1993)
56. S. Imada, T. Muro, T. Shishidou, S. Suga, H. Maruyama, K. Kobayashi, H. Yamazaki, and T. Kanomata: Phys. Rev. B **59**, 8752 (1999)
57. C.T. Chen, Y.U. Idzerda, H.-J. Lin, N.V. Smith, G. Meigs, E. Chaban, G.H. Ho, E. Pellegrin, and F. Settle: Phys. Rev. Lett. **75**, 152 (1995)
58. G.Y. Guo, H. Ebert, W.M. Temmerman, and P.J. Durham: Phys. Rev. B **50**, 3861 (1994)
59. P. Söderlind, O. Eriksson, B. Johansson, R.C. Albers, and A.M. Boring: Phys. Rev. B **45**, 12911 (1992)
60. S. Chikazumi. In: *Physics of Ferromagnetism*, (Oxford Science Publishing, New York 1997)
61. H.-J. Elmers, S. Wurmehl, G.H. Fecher, G. Jakob, C. Felser, and G. Schönhense: J. Magn. Magn. Mater. **272–276**, 758 (2004)
62. P. Bruno: Phys. Rev. B **39**, 865 (1989)
63. A. Hubert and R. Schäfer. In: *Magnetic Domains*, (Springer Verlag, Berlin Heidelberg New York 1998)

64. M.P. Raphael, B. Ravel, Q. Huang, M.A. Willard, S.F. Cheng, B.N. Das, R.M. Stroud, K.M. Bussmann, J.H. Claassen, and V.G. Harris: Phys. Rev. B **66**, 104429 (2002)
65. N. Furukawa: J. Phys. Soc. Jpn. **69**, 1954 (2000)
66. K. Kubo and N. Ohata: J. Phys. Soc. Jpn. **33**, 21 (1972)
67. A. Gupta and J.Z. Sun: J. Magn. Magn. Mater. **200**, 24 (1999)
68. J.M.D Coey, J.J Versluijs, and M. Venkatesan: J. Phys. D: Appl. Phys. **35**, 2457 (2002)
69. A. Yamasaki, S. Imada, R. Arai, H. Utsunomiya, S. Suga, T. Muro, Y. Saito, T. Kanomata, and S. Ishida: Phys. Rev. B **65**, 104410 (2002)
70. R. Kelekar and B.M. Klemens: J. Appl. Phys. **96**, 540 (2004)
71. K. Inomata, N. Tezuka, S. Okamura, H. Kurebayashi, and H. Hirohata: J. Appl. Phys. **95**, 7234 (2004)
72. G. Jakob, F. Casper, V. Beaumont, S. Falk, H.-J. Elmers, C. Felser, and H. Adrian: J. Magn. Magn. Mater., submitted (2004)
73. A. Conca, S. Falk, G. Jakob, M. Jourdan, and H. Adrian: J. Magn. Magn. Mater., submitted (2004)
74. M. Julliére: Phys. Lett. A **54**, 225 (1975)
75. J.M. De Teresa, A. Barthélémy, A. Fert, J.P. Contour, R. Lyonnet, F. Montaigne, P. Seneor, and A. Vaures: Phys. Rev. Lett. **82**, 4288 (1999)
76. Ph. Mavropoulos, N. Papanikolaou, and P.H. Dederichs: Phys. Rev. Lett. **85**, 1088 (2000)
77. H.F. Ding, W. Wulfhekel, U. Schlickum, and J. Kirschner: Eur. Phys. Lett. **63**, 419 (2003)
78. E.Y. Tsymbal, O.N. Mryasov, and P.R. LeClair: J. Phys.: Condens. Matter **15**, R109 (2003)

Epitaxial Growth of NiMnSb on GaAs by Molecular Beam Epitaxy

Willem Van Roy[1] and Marek Wójcik[2]

[1] IMEC, Kapeldreef 75, B-3001 Leuven, Belgium
vanroy@imec.be
[2] Institute of Physics-Polish Academy of Sciences, Al. Lotnikow 32/46, 02-668 Warszawa, Poland
wojci@ifpan.edu.pl

Abstract. The similarity in crystal structures allows for the epitaxial growth of the candidate half-metal NiMnSb on GaAs. We discuss the growth by molecular beam epitaxy using individual sources for Ni, Mn, and Sb on GaAs(001), (111)A and (111)B substrates. We focus the discussion on the aspects that are crucial for obtaining highly polarized films suitable for spin injection into a semiconductor: control of the interface quality and polarity; composition control to within 1% using a special Rutherford backscattering (RBS) calibration procedure; and reduction of point and extended defect density to the 1% level. We pay particular attention to the zero-field nuclear magnetic resonance technique (NMR) that was used to obtain this level of sensitivity to the local disorder.

1 Introduction

The conduction electrons in half-metallic materials are fully spin-polarized, making them very attractive for applications such as the generation (injection) and detection of spin-polarized currents in a (non-magnetic) semiconductor. These are seen as key ingredients for the realization of semiconductor spintronics applications. While this is possible using optical techniques or by the application of a magnetic field, these methods are not suited for large scale integration, therefore electrical injection and detection are preferred. Original experiments using regular ferromagnetic *Ohmic* contacts have not been successful, and it is now well established that this is due to the so-called "conductivity mismatch" problem [1]. The most successful solution so far is the insertion of a tunnel barrier [2–4], which has led to injected spin polarizations up to 40% at low temperature [5], and 16% at room temperature [6]. An alternative solution to the conductivity mismatch problem is the use of "half-metallic" contacts, where this term should not be interpreted in the loose sense of "highly polarized" contacts, but in the strict sense of "100% spin polarization": even for an optimistic ratio of ferromagnet-to-semiconductor conductivities, a spin-polarization of 99.0% in the "half-metal" results in less than 0.1% spin-polarized current in the semiconductor.

Among the proposed half-metallic materials, the Heusler and half-Heusler compounds are attractive since their crystal structure and lattice constants

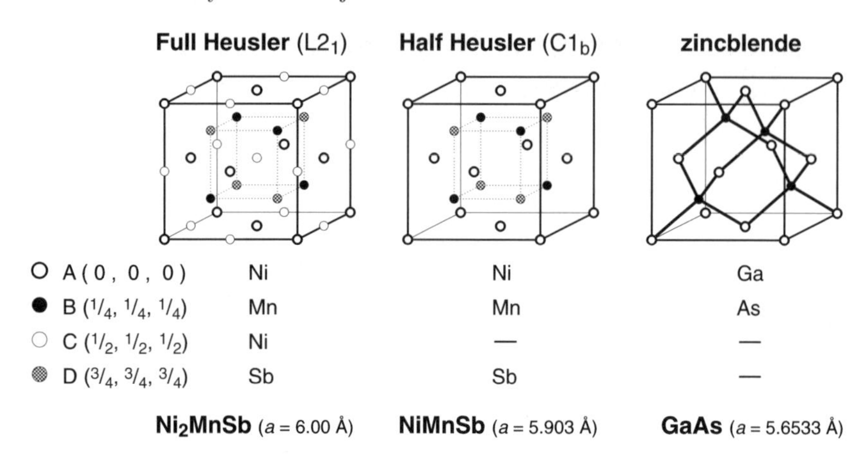

Fig. 1. The Heusler and zincblende crystal structures consist of 4 interpenetrating fcc sublattices A, B, C, and D. In a full-Heusler material such as Ni_2MnSb all sublattices are occupied. This structure can also be viewed as a simple cubic structure with Ni atoms on positions A and C interpenetrating an ordered NaCl structure with Mn and Sb on positions B and D. In a half-Heusler material such as NiMnSb, the C sublattice is unoccupied (or "occupied" by a vacancy □). In a zincblende semiconductor such as GaAs the C and D sublattices are empty. The diamond structure of Si is identical, with Si occupying both sublattices A and B

are closely related to the diamond and zincblende structures of most industrially relevant semiconductors (Fig. 1), therefore allowing for single crystalline integration is the best hope of suppressing interface defects that may reduce the spin polarization.

In this chapter we will discuss the epitaxial growth of NiMnSb on GaAs by molecular beam epitaxy (MBE), and focus especially on the aspects that are important for the realization of spin-selective contacts to a semiconductor. Section 2 gives an overview of the challenges that need to be overcome to realize NiMnSb-based contacts suitable for spin injection into a semiconductor. This is followed by a description of the crystal growth in Sect. 3, and a detailed study of the local structure and disorder by nuclear magnetic resonance (NMR) in Sect. 4. We finish with a brief summary of the spin injection results using NiMnSb-based contacts in Sect. 5.

2 Challenges

The half-metallic properties of Heusler alloys are strongly affected by structural defects such as segregation and non-stoichiometry, point and extended defects, and even the discontinuity of the crystal structure at hetero-interfaces. In this section we give a summary of the various challenges that have to be overcome in order to create a half-metallic contact.

2.1 Surface and Interface Segregation

Earlier work on the spin polarization of free NiMnSb surfaces has shown that the polarization of the inverse photo-emission normal to a (001) surface is dramatically reduced when the surface becomes non-stoichiometric (Sb-rich) after sputter-cleaning [7]. Similarly, it can be expected that non-stoichiometry *of the Heusler HMF* at the semiconductor interface results in a loss of spin-polarization. It has also been shown for Fe/GaAs spin injectors that non-stoichiometry *at the semiconductor side* results in a loss of spin injection: the spin polarization of the electrons injected across a tunneling Schottky barrier was found to decrease with increasing thickness of an interfacial GaAs layer with expanded lattice constant, attributed to Fe indiffusion [8].

For some applications, buffer layers have been used to improve the crystallinity and/or morphology of Heusler films grown on III–V semiconductors, e.g., lattice matched ScErAs buffers between GaAs and Ni_2MnGa [9]. This solution is only possible for spin injection applications if the interlayer does not interfere with the spin injection process: it should be non-metallic and non-magnetic. This may not be the case for ScErAs, which is semimetallic[1] and shows exchange splitting due to the interaction between the quasifree charge carriers and the $4f$ Er spins [10].

The target of our work has therefore been to make direct NiMnSb/GaAs interfaces without interdiffusion and without the formation of interfacial compounds.

2.2 Surface and Interface States; Planar Defects Intersecting the Interface

An even more fundamental problem is the spin polarization at perfect interfaces. The half-metallicity in Heusler alloys is a symmetry effect [11], and is thus very sensitive to any imperfections in the crystal. One such imperfection is the interruption of the periodicity at a free surface or an interface, and it has been shown by de Wijs and de Groot [12] that all stoichiometric, bulk-terminated (001) and (111) free surfaces of NiMnSb have non-halfmetallic surface states. The same goes for NiMnSb/insulator and NiMnSb/semiconductor interfaces in general. Only one particular (111) interface to zincblende semiconductors was shown to preserve the half-metallic character. de Wijs and de Groot considered only CdS and InP as these are nearly lattice matched to NiMnSb, and in both cases only the (111) interface with an Sb-anion (S or P) bond was found to be half-metallic. By extrapolation, this suggests that the NiMnSb/GaAs (111) interface with an Sb–As bond has the highest chance

[1] A semimetal should not be confused with a half-metal: a semimetal is a metal in which the carrier concentration is several orders of magnitude lower than the 10^{22} cm^{-3} typical of ordinary metals. ScErAs has both electron and hole pockets with a total carrier concentration of the order of a few 10^{20} cm^{-3} [10].

to be half-metallic. Other interfaces showed unpolarized interface states, but recovered half-metallic properties within two trilayers (~7 Å) of the interface.

These calculations also suggest that planar defects such as stacking faults or twins (e.g. caused by the 4.4% lattice mismatch between NiMnSb and GaAs) will result in the local loss of half-metallic properties, but the loss should be limited to a width of ~15 Å centered on the defect plane. Stroud et al. [13] have considered the (Zn,Mn)Se/GaAs spin injection interface in the diffusive regime, and shown that planar defects intersecting the interface result in spin-flip scattering for the electrons traveling in the forward direction. This effect can also be expected for stacking faults intersecting the NiMnSb/GaAs interface, even if the stacking fault itself does not destroy half-metallicity. However, because of the much higher electron concentration in NiMnSb compared to (Zn,Mn)Se, screening of the defect potential is much more effective, and spin-flip scattering will be much less effective.

An interface intersected by stacking faults will thus result in a mixture of half-metallic areas injecting 100% polarized electrons with areas injecting both spin orientations. Since the lateral distance between defects is much less than the spin diffusion length in the semiconductor (up to several 100 μm at low temperature in n-GaAs), the conductivity mismatch problem is still expected to suppress the spin injection if the contact is Ohmic. As in a regular (non-halfmetallic) spin-injection contact, this issue can be solved by inserting a tunnel barrier such as a tunneling Schottky barrier (Fig. 3).

2.3 Chemical Disorder

Another imperfection which leads to the loss of crystal symmetry and the appearance of states in the minority spin gap, is atomic disorder as first shown by Ebert and Schütz for PtMnSb [14] and Orgassa et al. for NiMnSb [15], and later confirmed for many other Heuslers [16]. Orgassa considered various kinds of disorder and found as a rough guideline that 1% disorder is acceptable (the Fermi level stayed inside a reduced minority spin gap), while 5% disorder results in the loss of half-metallic properties, either because the Fermi-level intersects one of the bands, or because the gap was completely closed.

These low allowable disorder levels pose a double experimental challenge. The deposition temperatures of thin films are much lower than the temperatures at which bulk crystals are synthesised, hence a larger amount of disorder is to be expected. In addition, the determination of such a low level of disorder is not straightforward. Traditional techniques such as XRD are not very sensitive to the chemical nature of the atoms. Other techniques such as EXAFS that are often used to determine the local structure experience difficulties in distinguishing the small disorder signal superimposed on a large background [17], therefore we have selected nuclear magnetic resonance (NMR), where defects result in additional lines on a zero background, resulting in sensitivities much better than 1%.

Because of the presence of a vacancy site in the half-Heusler $C1_b$ structure, it has been suggested that materials such as NiMnSb are more susceptible to disorder than full Heuslers with an $L2_1$ structure, without vacancy site [18]. We will show that this is not necessarily the case, and that very low disorder levels (~1%) can be obtained in NiMnSb thin films. These results are in agreement with NMR experiments showing very low disorder in bulk NiMnSb [19, 20] and the inability to induce disorder by cold work [21] in contrast to many other half- and full-Heuslers, showing that it is the chemical nature rather than the vacancy site which governs the susceptibility to disorder.

2.4 Finite Temperature and Related Effects

A number of theoretical studies have reanalyzed the possibility of true half-metallicity and argued that small disturbances are nearly always present. Skomski and Dowben have shown that spin-wave excitations at finite temperatures result in a finite density of states in the minority spin channel, except at zero temperature [22,23]. Although the effect is quantitatively small, qualitatively the change from zero to non-zero DOS may be very important for spin injection applications. Chioncel et al. generalized these findings with a more general many-body approach to study the dynamical spin fluctuations and the formation of nonquasiparticle states at finite temperatures [24].

Mavropoulos at al. went one step further, and showed that the spin-orbit coupling, which is a relativistic effect, results in a finite density of states in the minority spin channel, even at zero temperature [25]. The magnitude of the effect depends on the strength of the spin-orbit interaction, and was found to be relatively small in NiMnSb (spin polarization ~99%). This is in agreement with the recent experimental finding that the g factor $g = 2.02 - 2.03$ is very close to the free electron g factor of 2, indicating very weak spin-orbit interaction [26].

Although these effects are generally only of the order of 1%, this may be sufficient to prevent spin injection into a semiconductor using an Ohmic contact configuration, thus creating the need for the inclusion of a tunnel barrier.

2.5 Band Line-Up

A final effect that has to be taken into account for the realization of half-metallic contacts to a semiconductor are the band offsets (or the Schottky barrier) between the half-metal and the semiconductor. In most contacts to III–V semiconductors, surface states are present at the interface. These generally pin the Fermi-level around mid-gap, with the notable exception of InAs where the metal Fermi level is pinned within the conduction band of InAs. Even in an ideal interface without interface states (such as the half-metallic NiMnSb(111) interfaces calculated by de Wijs et al. [12]), the position of the

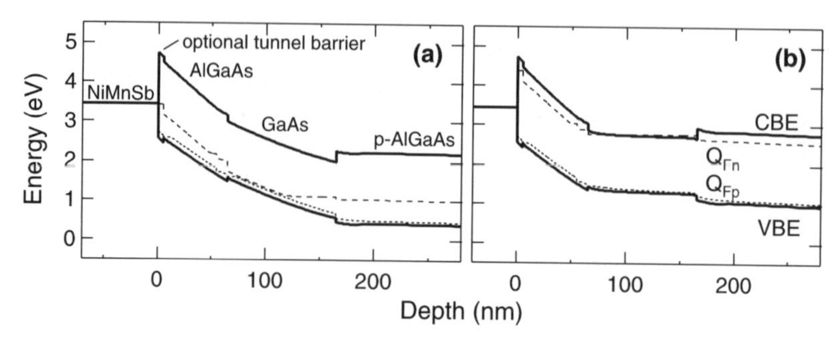

Fig. 2. Simulated band diagrams at 40 K of a NiMnSb-based AlGaAs/GaAs/p-AlGaAs spin-LED with Fermi-level pinning, at a bias of 3.5 V, (**a**) without impact ionization, and (**b**) with impact ionization. Full lines are the conduction and valence band edges (CBE and VBE), dashed lines are the quasi-Fermi levels for electrons and holes (Q_{Fn} and Q_{Fp})

Fermi-level is determined by the workfunctions of both materials and the presence of any interface dipoles.

In the case of NiMnSb/GaAs(001) and (111) the pinning near mid-gap has been confirmed [27]. As a result, it is impossible to inject *any* electrons from the metal into the semiconductor, as is shown by the Medici^{TM} device simulations for a spin-LED (see Sect. 5) in Fig. 2(a). The electron quasi-Fermi level Q_{Fn} is found far below the conduction band edge, indicating that virtually no electrons are present anywhere in the structure, while the hole quasi-Fermi level Q_{Fp} closely traces the valence band edge, indicating that holes originating from the p-type substrate are flowing freely through the device. Surprisingly, the actual device was emitting (unpolarized) light, indicating that electrons *were* present in the structure. Figure 2(b) shows that impact ionization due to the large hole current can act as a source of electrons that can easily be mistaken as a sign of electron injection from the magnetic contact.

These results indicate that a tunnel barrier may be required to accommodate the work function difference, *even if the contact is half-metallic* and the resistivity mismatch doesn't present a problem. Tunnel barriers are not limited to traditional oxides such as Al_2O_3, but can also be a highly doped semiconductor layer as shown in Fig. 3. For a doping level exceeding $\sim 1 \times 10^{19}\,\mathrm{cm}^{-3}$ the depletion region becomes narrow enough to allow electrons to tunnel through. A Schottky tunnel barrier has the additional advantage over amorphous oxide tunnel barriers that epitaxial growth of the half-metal remains possible. This should result, in principle, in films with the highest possible crystal quality and thus the highest possible polarization, combined with a tunnel barrier which circumvents the resistivity mismatch problem in case the contact polarization does not reach the full 100%.

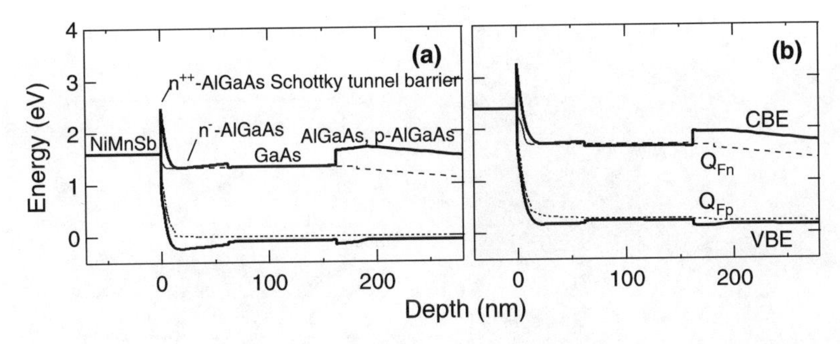

Fig. 3. Simulated band diagrams at room temperature of a NiMnSb-based AlGaAs/GaAs/p-AlGaAs spin-LED with a tunnel Schottky contact, at a bias of (**a**) 1.6 V and (**b**) 2.4 V. Full lines: conduction and valence band edges (CBE and VBE), dashed lines: quasi-Fermi levels for electrons and holes (Q_{Fn} and Q_{Fp})

3 Growth Results

3.1 Experimental Details

All growth experiments were done in a Riber 32P chamber with a base pressure of less than 1×10^{-10} Torr, equipped with solid sources for Ga, Mn, and Ni, a valved cracker source for As and a non-valved cracker source for Sb. Both crackers were always operated with their cracking zones between 900 and 950°C, resulting in As_2 species and a mixture of Sb_2 and Sb_1. The substrates were glued with indium solder on a molybdenum sample holder. The substrate temperatures T_{sub} quoted in this chapter are *best aproximations of the actual substrate temperature*, and were deduced from thermocouple readings T_{TC} by linearly extrapolating between the GaAs desoxidation temperature ($T_{sub} = 585$°C for a typical TC reading around $T_{TC} = 740^o$) and an assumed real temperature $T_{sub} = 0$°C at TC-reading 0°, an arbitrary but reasonably good estimation for our growth chamber.

Most growth experiments started with the growth of a GaAs buffer layer in the same growth chamber, except when NiMnSb was deposited on AlGaAs/GaAs based spin LEDs. These were grown in a different chamber and covered with an Sb-cap that was subsequentially removed by heating under an As flux. Initial experiments showed the importance of a low As background pressure in the chamber for realizing abrupt interfaces [28]. XRD measurements showed the presence of a Mn_2As secondary phase in NiMnSb films grown on GaAs(001) immediately after finishing the GaAs buffer, i.e., with an As background in the chamber around 2×10^{-9} Torr. TEM cross-sections confirmed that this layer was formed at and *below* the interface, indicating that Mn was diffusing *into* the substrate and consuming GaAs during the formation of Mn_2As. This interface reaction can be successfully prevented by allowing sufficient time after the growth of the GaAs buffer for the background pressure to reach the low 10^{-10} Torr range. The TEM cross-section

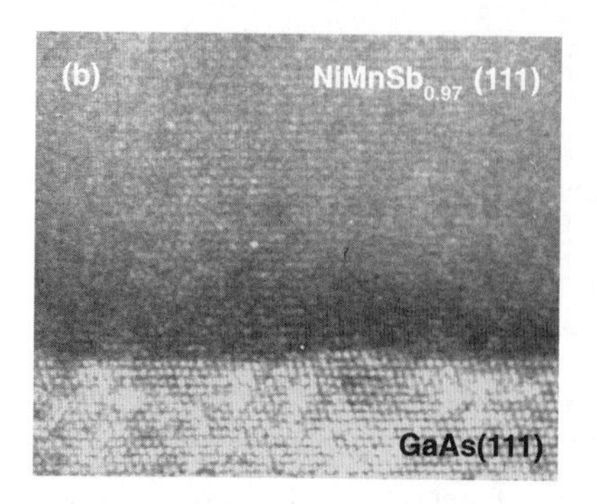

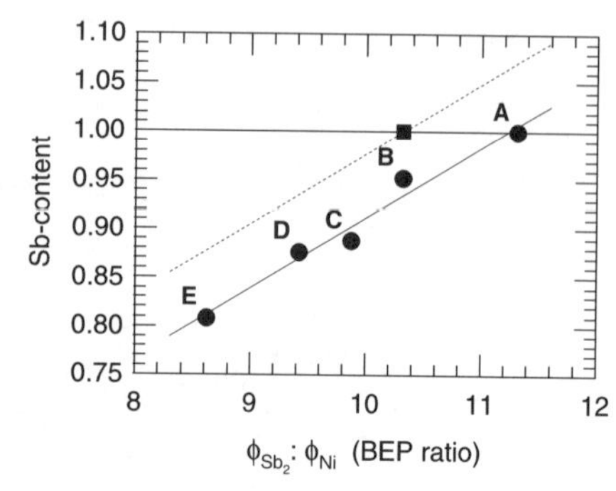

Fig. 4. (**a**) TEM cross-section of a $Ni_{1.00}Mn_{0.99}Sb_{0.97}$ film on GaAs(111)B (film C) grown at $T_{sub} = 240°C$ and 44 nm/h. (**b**) Sb content of the films as determined by RBS. (●) original measurement showing the linear dependence of Sb incorporation on the Sb-flux ϕ_{Sb_2} [30]; (■) recalibration based on sample B, yielding film compositions in the notation 1.00:0.99:x_{Sb} (for full details see [31])

in Fig. 4(a) shows the sharp and abrupt interface of a $Ni_{1.00}Mn_{0.99}Sb_{0.97}$ film grown on GaAs(111)B.

Because of the similarity in crystal symmetry, NiMnSb can be grown epitaxially on GaAs(001) [28, 29] as well as on GaAs(111)A and B [30], the epilayer always assuming the same crystal orientation as the substrate. All growth experiments discussed in the remainder of the chapter started with the simultaneous co-evaporation of Ni, Mn, and Sb at the selected substrate temperature. The films are usually capped by 5 to 10 nm low-temperature GaAs serving as an oxidation barrier.

In atomic layer epitaxy experiments on GaAs(111)B substrates at $T_{sub} = 240°C$, where the elements were supplied sequentially, both the sequences Mn–Ni–Sb and Sb–Ni–Mn resulted in epitaxial growth. Reflection high energy electron diffraction (RHEED) showed very slight roughening at layer thicknesses of 3–7 Å with full recovery of the smooth surface within 15–20 Å, suggesting that both Sb-polar and Mn-polar NiMnSb can be grown on As-polar GaAs(111)B. However, we did not have the means for checking the resulting stacking order after the growth, and it is conceivable that differences in interface energy force the atoms to swap places during the initial stages of the growth.

3.2 Stoichiometry Control

A first important growth parameter is the accurate control of the composition. The incorporation of the group-V element (Sb) is not self-limiting, in contrast to arsenic- and antimony-based III–V semiconductors. As a result, the fluxes of all three elements Ni, Mn, and Sb have to be accurately controlled. This requires also the capability to *measure* the composition of the films with an accuracy of ∼1%. This resolution is beyond the capabilities of most chemical techniques such as Auger, EDX, XPS, SIMS, etc., as these suffer from matrix effects which make an absolute measurement difficult to perform.

We have therefore chosen Rutherford backscattering (RBS), which is a physical technique where the scattering cross-sections of the various elements are well-known. The main drawback of this technique is its poor mass resolution for heavy elements. As a result the signals for the various elements overlap, and a quantitative analysis of the spectra requires curve fitting as shown in Fig. 5(a). This can be circumvented by isolating heavy elements on top of a lighter substrate. When there is sufficient mass difference and at sufficiently high primary energies, the signal from a thin film becomes completely isolated from the substrate. The total dose (at cm^{-2}) can then be determined by integration of the entire peak, which is much more accurate than the curve fitting approach. In the case of NiMnSb, we assumed a unity sticking coefficients for Ni and Mn (atomic masses 58–60–62 and 55), and used calibration films on Si (mass 28–29–30) to accurately determine their deposition rates. Sb (atomic mass 121–123) is heavier than all other elements in our samples including Ga and As (masses 69–71 and 75) and can be determined in the actual NiMnSb films as illustrated by Fig. 5(b).

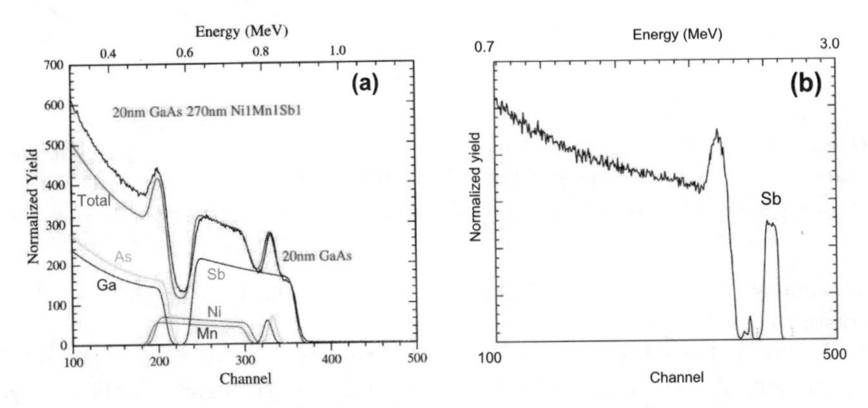

Fig. 5. RBS spectra of NiMnSb. (**a**) 330 nm thick NiMnSb film measured at 1 MeV primary energy with decomposition into the different elements. (**b**) 125 nm thick NiMnSb film (sample B) measured at 3 MeV primary energy: the Sb-peak is isolated from the rest of the spectrum, allowing accurate determination of the total dose by integration

This is particularly important, as the incorporation of Sb depends on the growth conditions. Figure 4(b) shows the Sb content of a series of 125 nm thick NiMnSb films grown on GaAs(111)B at 42 nm/h and $T_{sub} = 240°C$ as function of the Sb beam-equivalent-pressure (BEP) flux ϕ_{Sb_2}. The black circles clearly show a linear dependence between the incorporated amount of Sb and the flux ϕ_{Sb_2}. A discrepancy with XRD and NMR results (Sects. 3.4 and 4.3) forced us to re-examine the RBS calibrations on a different

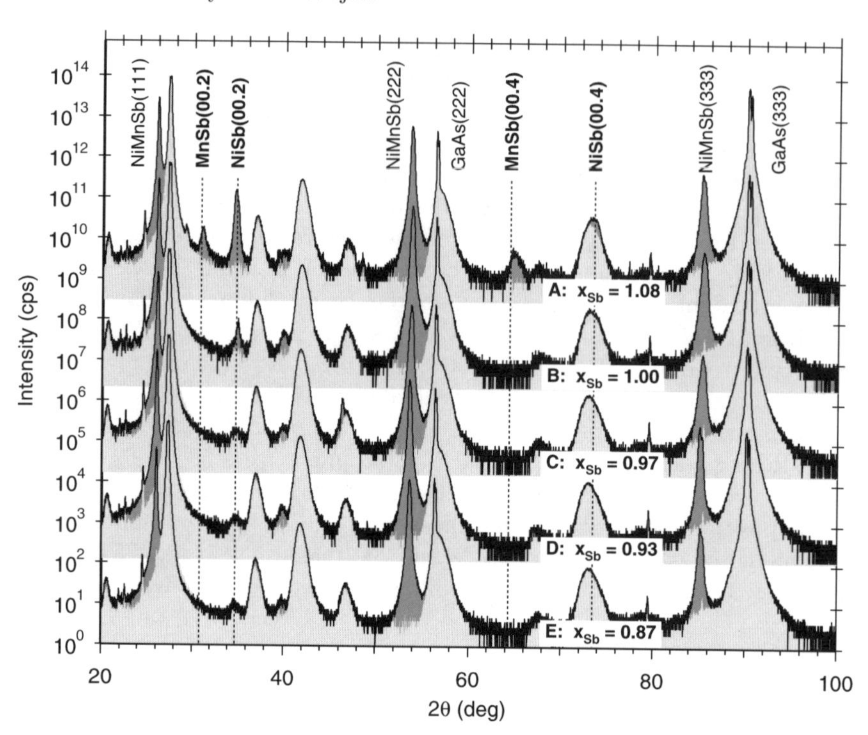

Fig. 6. XRD θ–2θ scans of 125 nm thick $Ni_{1.00}Mn_{0.99}Sb_{x_{Sb}}$ films grown on GaAs(111)B at $T_{sub} = 240°C$. The curves are offset vertically for clarity. Light grey: substrate reference scan, dark grey: NiMnSb film. The films with $x_{Sb} = 1.08$ and 1.00 show NiSb and MnSb additional phases

instrument, resulting in an Ni:Mn:Sb ratio 1.00 : 0.99 : 1.00 for the best film (B) of this series (for full details see [31]).

Figure 6 shows the XRD θ–2θ scans of the same set of films. The spectra are superimposed on a reference scan of a bare substrate to allow easy separation of the substrate and epilayer peaks. (The measurements were taken on a low-resolution, high-intensity instrument, resulting in a large number of spurious peaks). The epilayers are characterized by a well-defined NiMnSb(111) phase in registry with the (111)B substrate. The nearly stoichiometric $Ni_{1.00}Mn_{0.99}Sb_{1.00}$ film shows a small additional peak caused by hexagonal NiSb. The Sb-rich $Ni_{1.00}Mn_{0.99}Sb_{1.08}$ film shows stronger NiSb peaks as well as MnSb. The presence of additional phases shows a very good correlation with the composition determined by RBS. NMR measurements discussed below confirm the formation of MnSb in all Sb-rich films, also on (001) substrates where XRD cannot detect this (NMR, by contrast, is blind to paramagnetic NiSb). *This leads to the conclusion that the formation of these Sb-rich phases (50 at% Sb compared to 33 at% in the $C1_b$ phase) is the preferred way for NiMnSb to accommodate the excess Sb in the films.*

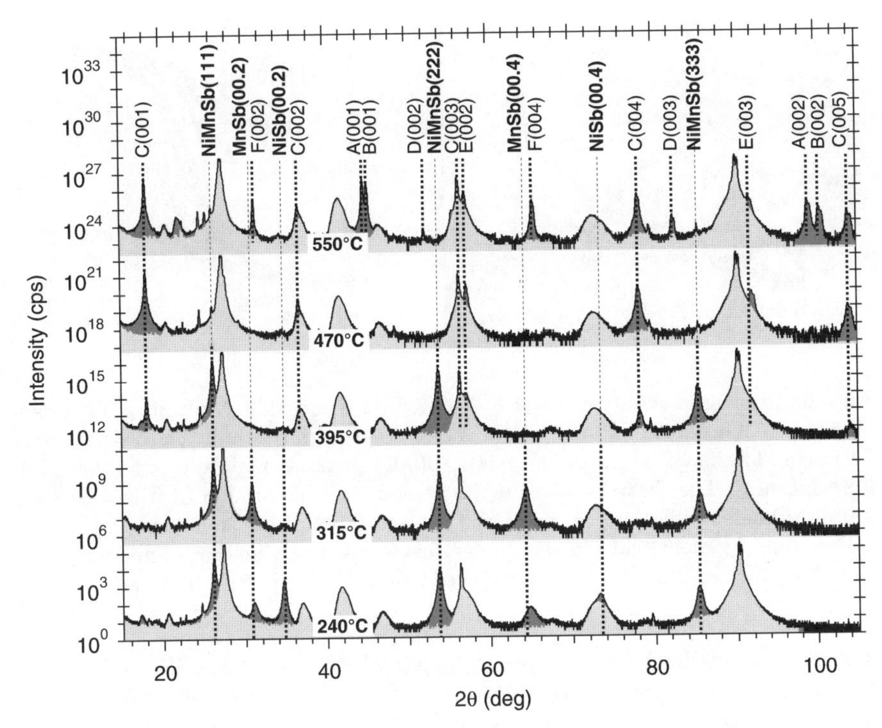

Fig. 7. XRD θ–2θ scans of nominally 125 nm thick NiMnSb films grown on GaAs(111)B with constant Sb-flux $\phi_{Sb_2} = 10 \times \phi_{Ni}$ at substrate temperatures $T_{sub} = 240 \cdots 550°C$. The curves are offset vertically for clarity. Light grey: substrate reference scan, dark grey: NiMnSb film. The films grown at 240°C and 315°C are slightly Sb-rich as shown by the presence of NiSb and MnSb. At 395°C an Sb-poor Heusler phase without a trace of NiSb and MnSb is formed, and an unidentified additional phase C appears. At $T_{sub} \geq 470°C$ the Heusler phase is no longer present, but replaced by up to six unidentified phases labeled A to F. At least two peaks can be indexed to each of the following out-of-plane lattice spacings: 2.01 Å (B), 2.03 Å (A), 3.22 Å (E), 3.51 Å (D), 4.89 Å (C), 5.70 Å (F)

3.3 Substrate Temperature

The Sb incorporation also depends strongly on the substrate temperature as is shown by the XRD scans in Fig. 7. Films grown at $T_{sub} = 240$ and 315°C consist mainly of the Heusler $C1_b$ phase, with some additional hexagonal MnSb and NiSb inclusions (a sign that the films are slightly Sb-rich, see below). At $T_{sub} = 395°C$ the $C1_b$ phase is still present, but another phase (labelled C) appears. At $T_{sub} \geq 470°C$ the $C1_b$ phase is no longer present, but replaced by up to 6 unidentified sets of peaks. Auger depth profiles showed the following approximate compositions for the films grown at elevated temperatures: oxidized GaAs cap – $Ni_{0.42}Mn_{0.17}Sb_{0.28}Ga_{0.13}$ – GaAs substrate at 395°C, oxidized GaAs cap – $Ni_{0.55}Ga_{0.45}$ – $Ni_{0.25}Mn_{0.30}As_{0.45}$ – GaAs substrate at

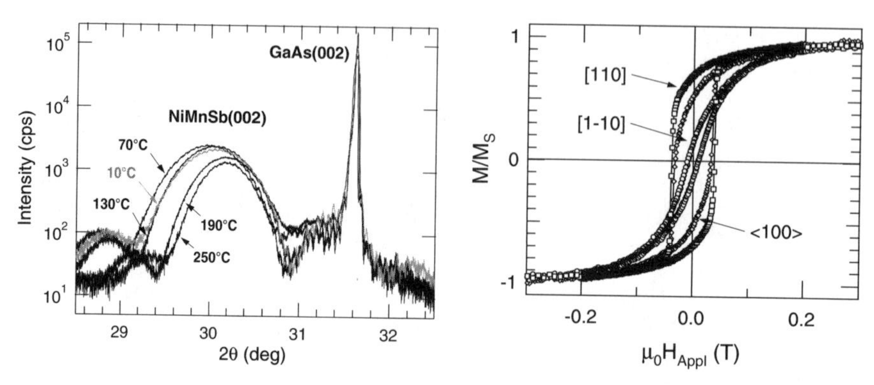

Fig. 8. (**a**) High-resolution XRD θ–2θ scans of 8 nm thin NiMnSb films grown on GaAs/AlGaAs(001) spin LED at $T_{sub} = 10 \cdots 250°C$ and a growth rate of 30 nm/h. The large width of the NiMnSb(002) peaks is mainly due to the small film thickness. The sharp oscillations (most easily visible around 31.0–31.5 deg are finite thickness oscillations in the buried AlGaAs/GaAs LED structure. (**b**) AGFM (alternating gradient feld magnetometer) hysteresis loops at room temperature of the film grown at $T_{sub} = 70°C$, along the $[110]$, $[1\bar{1}0]$, and $\langle 100 \rangle$ directions

470°C, and oxidized GaAs cap – $Ni_{0.43}Ga_{0.43}As_{0.14}$ – $Ni_{0.10}Mn_{0.35}As_{0.55}$ – GaAs substrate at 550°C. At 470°C and above *no Sb at all* was found in the films. The presence of strongly interreacted layers shows that the (growth) interface between NiMnSb and GaAs is not stable at these temperatures.

This behaviour is markedly different from that observed by Turban et al. [32], who studied the MBE growth of NiMnSb on MgO(001) substrates. They obtained films with high structural quality as shown by RHEED, TEM, and EXAFS by growing under a (presumably large) Sb overpressure at $T_{sub} = 630°C$ after an initial seed deposited at $T_{sub} = 350°C$. These high growth temperatures are possible because the MgO substrates are much more chemically inert than GaAs.

MBE growth of NiMnSb is possible down to very low substrate temperatures, as is illustrated by Fig. 8. XRD θ–2θ scans show that the $C1_b$ phase is formed down to $T_{sub} = 10°C$. The measurements are clustered in two groups for $T_{sub} \geq 190°C$ and $T_{sub} \leq 130°C$, which agree with the evolution of the surface roughness during the growth. Films grown at $T_{sub} \geq 190°C$ showed a roughening of the RHEED patterns after ≤ 1 nm, indicative of strain relaxation by the nucleation of misfit dislocations. For $T_{sub} \leq 130°C$ the RHEED patterns remained streaky throughout the growth, showing that the formation of misfit dislocations was delayed or suppressed.

Also the magnetic properties of the low-temperature films are surprisingly good, Fig. 8(b). The saturation magnetization is of the same order as that of high-temperature films, but the coercive field increases especially for the lowest growth temperatures. This may indicate domain wall pinning caused by an increased number of inhomogeneities in the films, which can be caused by the larger (and possibly inhomogeneous) strain-fields, but may also indicate a larger concentration of point and extended defects. These films have not yet been analyzed by NMR.

3.4 Growth Rate

Another parameter which is very important for the amount of defects in the film is the growth rate. Initial growth experiments done at a growth rate of 330 nm/h resulted in films with good XRD properties, consisting of single-phase NiMnSb(001) or (111), following the substrate orientation, and, in Sb-rich films grown on GaAs(111), additional NiSb and MnSb(00.l) peaks visible. When we analysed the films by NMR, we found a relatively large amount of disorder in the films (either point or extended defects). A film with approximately stoichiometric composition showed that about 20% of all Sb atoms in the Heusler phase have a disturbed environment, and that additionally 13–14% of all Mn and Sb atoms are involved in hexagonal MnSb inclusions (this actually suggests that the film was slightly Sb-rich). More Sb-rich films showed roughly the same amount of local disorder in the Heusler phase and a strongly increased MnSb content, while Sb-poor films showed massive disorder of a diferrent type (point defects) in the main Heusler phase and no formation of MnSb (See [33] and Sect. 4.2).

A reduction in growth rate to 42 nm/h results in a remarkable reduction of the local disorder, with defect densities down to 1.1% of all Mn atoms involved in planar defects, ~0.2% of all Sb atoms involved in a defect, and ~0.5% of all Sb-sites occupied by As impurities (As_{Sb} antisites) (see [31] and Sect. 4.3). The reduction in growth rate has the same positive effect on the local order as an increase in temperature, but without the associated interfacial mixing (Fig. 4(a)), and is a crucial parameter for the growth of highly ordered films on III–V semiconductors.

3.5 Lattice Mismatch and Extended Defects

Stoichiometric and Sb-Rich Films

The lattice mismatch of 4.4% between NiMnSb (a = 5.903 Å) and GaAs (a = 5.6533 Å) is expected to result in a small critical thickness and the generation of misfit dislocations. Indeed, in stoichiometric and Sb-rich films we typically observe a transition from streaky to spotty RHEED patterns at a film thickness of ~$0.9 \cdots 1.1$ nm for both (001) and (111) oriented substrates, indicating surface roughening, followed by a gradual recovery of the RHEED patterns as the growth continues. XRD reciprocal space maps of the NiMnSb(111) samples grown at 42 nm/h and T_{sub} = 240°C (Figs. 4(b) and 6) confirmed that all films were fully relaxed (with actually a small tetragonal distortion of $0.2 \cdots 0.6$% *opposite* to the original strain).

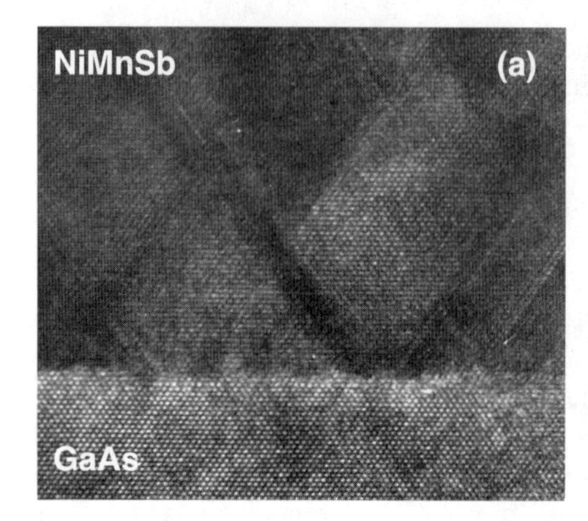

Fig. 9. High-resolution ⟨110⟩ cross-section TEM images of an approximately stoichiometric NiMnSb/GaAs(001) film grown at $T_{sub} = 240°\text{C}$ and 330 nm/h. (**a**) Interface region with interfacial defects, (**b**) region with MnSb or NiSb-like inclsion

Cross-sectional TEM images (Fig. 9) show that the defects run parallel to the $\{111\}$ planes, as expected in an fcc lattice. For films grown on (001) substrates these planes are tilted at an angle of 54.7^o with the surface, and are thus invisible in XRD θ–2θ scans. The defects nucleate at the interface, but they are no simple misfit dislocations. Instead we observed stacking faults or microtwins that penetrate into the growing film, with, in most cases, no clear or well-defined atomic structure. The stacking faults interact with each other, sometimes resulting in a lattice image (Fig. 9(b)) that is consistent with the hexagonal NiSb or MnSb structure, supporting the model that is developed in Sect. 4.2.

TEM images of (111) films showed a slightly different behaviour, with only stacking faults running parallel to the surface and limited to the region at the interface, i.e., there were no stacking faults along any of the $\{111\}$ planes tilted with respect to the surface. Again, the stacking faults were limited to Sb-rich films, and resulted more often in a MnSb- or NiSb-like pattern, in agreement with the XRD characterization (Fig. 6).

As mentioned in Sect. 3.3, the roughening transition was suppressed in (001) films grown at $T_{sub} \leq 130°\text{C}$, suggesting that strain relaxation and the formation of misfit dislocations was suppressed or altered.

Sb-Poor Films

Strain relaxation occurs in a totally different and not clearly identified way in Sb-deficient films. In these films we only notice very little or no surface roughening at all during the growth on both (001) and (111) substrates, and TEM images show a seemingly defect-free crystal lattice (see e.g. Fig. 4(a)). Nevertheless XRD measurements show that the films are completely relaxed. NMR measurements reveal an increased amount of point defects, certainly in films deposited at a high growth rate. Disorder halves the effective lattice constant (when sites A and C, and B and D are fully intermixed the crystal structure becomes the cubic CsCl structure, and when all sites intermix the structure becomes bcc, in both cases with half the lattice constant of the $C1_b$ or $L2_1$ structures), suggesting that it may be easier to form simple misfit dislocations.

Magnetic Anisotropy

The defect structure of the films also influences magnetic properties such as the anisotropy and the coercive field. In our highest quality films (the set of (111)B films grown at 42 nm/h and $T_{sub} = 240°C$), the coercive field could be as small as 0.72 mT [30], while in lower-quality (001) films deposited at high growth rate we obtained 2.0–3.5 mT [28, 29], while films deposited at low temperatures had coercivities up to 20–40 mT (Sect. 3.3 and Fig. 8(b)). We did not observe any systematic difference between Sb-rich and Sb-poor films, or between (001) and (111) films.

The substrate orientation does play a role, however, for the observed in-plane anisotropies. While (111) films are approximately isotropic in the plane, (001) films often show a clear 2-fold in-plane anisotropy (see [28, 29] and Fig. 8(b)) which is not expected for a cubic system. Although an interface contribution cannot be ruled out (the zincblende (001) surface has only 2-fold symmetry), it seems unlikely that this can have such a strong effect on a 330 nm thick film. The most likely origin of the anisotropy is an anisotropic strain relaxation or anisotropic generation of stacking faults with more fault planes intersecting the surface along say [110] than along $[1\bar{1}0]$ as a result of the anisotropic interface bonding at the zincblende surface. (It should be noted that the fault planes form the easy {00.1} basal plane of any hexagonal MnSb inclusions).

The fact that easy and hard axis can be interchanged depending on growth conditions and buffer layers [29] indicates that the anisotropic generation of stacking faults depends on the details of interface nucleation. Surprisingly the values of the coercive fields and the presence of 2-fold anisotropy in Sb-poor films (without visible extended defects) are similar to those in Sb-rich films, indicating that the strain relaxation is anisotropic, even without the formation of extended defects.

Nearly Lattice Matched Growth

Bach et al. have grown NiMnSb by MBE on $In_{0.53}Ga_{0.47}As$ buffer layers lattice matched to InP substrates ($a = 5.869$ Å), with only 0.6% mismatch to NiMnSb [34, 35]. Using XRD reciprocal space maps they could show that the films were pseudomorphic (fully strained) up to a critical thickness of $\sim$80 nm, in good agreement with the thickness where the streaky RHEED patterns turned spotty. The coercive fields of their films are as small as 0.2 mT, in agreement with a lower density of (planar) defects that could cause domain wall pinning. However, they do also observe a strong uniaxial in-plane anisotropy which could only be partially ascribed to an interface effect, and was therefore attributed to growth-induced lattice defects [26]. Indeed, RBS channeling experiments showed that $14 \cdots 25\%$ of the Sb atoms and $25 \cdots 29\%$ of the Mn and Ni atoms were not located on the fcc Heusler lattice, but the measurements did not allow to distinguish between various disorder models [36].

Although the exact growth conditions of the films used for RBS channeling were not given, they reported optimum crystal quality for BEP flux ratios $\phi_{\mathrm{Mn}}:\phi_{\mathrm{Ni}} = 2.4$ and $\phi_{\mathrm{Sb}_{(4)}}:\phi_{\mathrm{Ni}} = 14.2$, and a substrate temperature of 300°C [34] as measured by a thermocouple [35]. The Mn/Ni flux ratio and substrate temperature are identical to ours (we observe $T_{TC} = 300^o$ for $T_{sub} = 240$°C), but the Sb/Ni flux ratio is much larger than ours (~10.0). This might be due to the use of a different Sb species (Sb_4 vs Sb_1/Sb_2), or might indicate Sb-rich growth conditions which are expected to result in MnSb/NiSb inclusions, one of the "disorder" models that were considered [36].

4 Nuclear Magnetic Resonance Study of the Local Structure

4.1 General Remarks

Hyperfine Fields as a Probe of Short Range Order

A particularly useful structural information can be provided by the Nuclear Magnetic Resonance (NMR) experiment, which probes the distribution of hyperfine fields at the nuclei. The hyperfine field is created by the local spatial distribution of electron spin polarization close to the observed nucleus and thus depends sensitively on the topological and chemical environment of this nucleus. In a perfect crystal structure, the nuclei of atoms occupying regular positions in a crystal structure give rise to the well defined resonance lines in the NMR spectrum, which can be regarded as fingerprints of particular lattice order. Any inhomogeneity of the local structure due to the variation of the interatomic distances, the number and nature of the neighbors etc. will result in a modification of the hyperfine field value. Consequently, the NMR lines will be shifted, or broadened depending on the type of defect [37]. In addition, new distinct NMR lines can appear, due to the wrong site occupancy or inclusion of an alien crystal phase [38]. It is therefore possible to obtain structural information from the NMR spectra using an established quantitative relationship between the hyperfine field and the local structure. Since it is difficult for ab initio hyperfine field calculations [39–41] to predict exactly the fine features observed in experiments, the assignment of the resonance frequency to a specific local atomic configuration is usually obtained from the studies of reference materials. These include materials with a perfect structure as well as imperfect reference materials where a certain imperfection such as disorder or off-stoichiometry has been introduced intentionally in a controlled way. The reference frequencies are crucial elements in defect analysis.

Reference Data on ^{123,121}Sb, ^{55}Mn and ^{61}Ni Hyperfine Fields in C1$_b$ NiMnSb

Table 1 collects the reference data from the literature on ^{123,121}Sb, ^{55}Mn and ^{61}Ni NMR frequencies recorded from the well characterized, defect free half Heusler NiMnSb bulk samples at room temperature (RT) and 4.2 K [21, 42–45]. The corresponding ^{123,121}Sb, ^{55}Mn and ^{61}Ni hyperfine fields in NiMnSb, calculated from the relationship between the NMR frequency ν and the hyperfine field HF: $\nu = \gamma$ HF, where γ is the gyromagnetic ratio, are included in Table 1 along with the hyperfine fields reported for bulk full Heusler Ni_2MnSb samples [43–45]. Due to the cubic symmetry all crystallographic positions of Sb, Mn and Ni are equivalent both in case of half Heusler and full Heusler alloys and for each atomic species one isotropic hyperfine field has been observed. In the case of Sb, the two isotopes ^{121}Sb and ^{123}Sb have a different gyromagnetic ratio ($\gamma(^{121}\mathrm{Sb})$ = 10.19 MHz/T, $\gamma(^{123}\mathrm{Sb})$ = 5.518 MHz/T) and give rise to two NMR resonance lines at distinctly different frequencies corresponding to the same hyperfine field: ^{123}Sb line at 164 MHz (152 MHz) and ^{121}Sb line at 300 MHz (281 MHz) at 4.2 K (RT). Incidentally, the ^{55}Mn NMR line ($\gamma(^{55}\mathrm{Mn})$ = 10.553 MHz/T) overlaps with the ^{121}Sb NMR line at the frequency of 300 MHz (281 MHz). As has been shown by Hihara [42], in NMR experiment at RT, these two contributions to the 300 MHz (281 MHz) line can be separated by the application of an additional external DC magnetic field due to the opposite sign of the hyperfine field experienced by ^{55}Mn and ^{121}Sb nuclei.

Table 1. ^{123,121}Sb, ^{55}Mn and ^{61}Ni NMR frequencies and corresponding hyperfine fields in bulk C1$_b$ NiMnSb and L2$_1$ Ni_2MnSb at RT and 4.2 K [21, 42–45]

		NiMnSb		Ni_2MnSb
	T(K)	f(MHz)	B(T)	B(T)
^{123}Sb	300	152	+27.55 [42]	
	4.2	163.7	+29.67 [21]	29.3 [43]
^{121}Sb	300	281	+27.55 [42]	
	4.2	300	+29.67 [21]	29.3 [43]
^{55}Mn	300	281	−26.62 [42]	
	4.2	300	−28.43 [21]	28.5 [44]
^{61}Ni	4.2	?	?	−6.0 [45]

In the experimental spectra obtained in zero field NMR experiments these two lines (^{55}Mn, ^{121}Sb) are not resolved. However, the ^{121}Sb contribution at 300 MHz can be deduced from the ^{123}Sb line by scaling with the isotopic abundance and difference in gyromagnetic ratio. In this way the ^{55}Mn contribution could be resolved. Such an ideal NMR spectrum consisting of two lines has been in fact observed also in case of defect free NiMnSb film [31] discussed in this chapter. An example is given in Fig. 10 where we show the NMR spectrum recorded at RT from a 125 nm thick $Ni_{1.00}Mn_{0.99}Sb_{1.00}$ film grown at $T_{sub} = 240°C$ on GaAs(111)B at a growth rate of 42 nm/h.

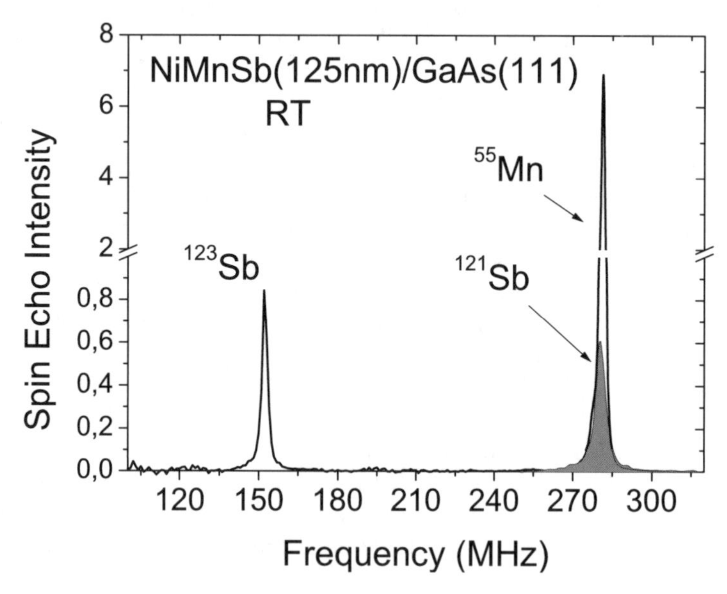

Fig. 10. NMR spectrum recorded at RT from a 125 nm thick $Ni_{1.00}Mn_{0.99}Sb_{1.00}$ film grown at $T_{sub} = 240°C$ on GaAs(111)B at a growth rate of 42 nm/h. The ^{121}Sb contribution to the 281 MHz line is shown by a line with a shaded area

The ^{61}Ni hyperfine field has only been reported in the case of Ni_2MnSb bulk material [45]. Taking into account the very close values of the ^{55}Mn, ^{121}Sb, and ^{123}Sb hyperfine fields for the $C1_b$ (NiMnSb) and $L2_1$ (Ni_2MnSb) structures, it is expected that the ^{61}Ni hyperfine field in NiMnSb will be equal to $\sim$6.0 T as reported for Ni_2MnSb and the ^{61}Ni frequency is expected to be around 23 MHz.

Unfortunately the reference data concerning the NMR studies of local disorder in NiMnSb is very scarce [21, 46]. The only NMR reports directly related to half Heusler NiMnSb phase did not show any influence of the atomic disorder on the NMR spectra. On the other hand, a number of full Heusler X_2YZ, where X = Pd, Ni, Co, Cu; Y = Mn; and Z = Sn, Sb, Ga, Al, have been studied quite intensively and revealed a measurable effect of the atomic disorder on the hyperfine fields in these alloys [46]. Although a direct comparison between the NiMnSb and full Heusler alloys is not possible due to the differences in the character of conductivity, the lack of effect in the case of NiMnSb deserves a closer inspection in view of the reported total indifference to the atomic disorder.

4.2 NiMnSb Films Grown at High Growth Rate

Our NMR study has revealed that the defects developed in NiMnSb thin films depend dramatically on the growth conditions and that two of the most important parameters are the Sb flux [31, 33, 47] and the growth rate. To illustrate this let us first consider a series of samples grown at 330 nm/h with a fixed substrate temperature $T_{sub} = 240°\mathrm{C}$ and large increments of the Sb flux [28]. The NMR spectra recorded from these samples reveal the well pronounced features of all structural defects which have been recognized in NiMnSb films epitaxially grown on GaAs(001) and GaAs(111) substrates. Three different types of structural defects have been recognized in thin films of NiMnSb. These are: inclusions of hexagonal MnSb phase, planar defects such as stacking faults in the fcc structure or antiphase boundaries, and point defects. Each defect gives rise to a characteristic group of lines in the NMR spectrum.

Figure 11 presents NMR spectra recorded at 4.2 K from three $NiMnSb_x$/ GaAs films with different Sb content x. One can readily notice that the NMR spectra recorded from the Sb-rich and Sb-deficient films show a completely different intensity distribution, evidencing that the quality of the crystal structure, the type of structural defects and their population in these films strongly depend on Sb stoichiometry. The spectrum of the nominally stoichiometric film is very similar to that of the Sb-rich film, suggesting that also this film was slightly Sb-rich.

Defects in Sb-Rich Films

NMR spectrum for the nominally stoichiometric film, Fig. 11(b), shows the narrow lines at the frequencies which are expected from the studies of the reference half Heusler NiMnSb bulk samples (see the Table 1 and the discussion above). The ^{123}Sb line is at 164 MHz, and the overlapping lines of ^{121}Sb and ^{55}Mn are at 300 MHz. The ^{61}Ni resonance has not been detected in the available frequency range down to 20 MHz. The exact matching of the observed ^{121,123}Sb and ^{55}Mn resonance frequencies with the corresponding reference frequencies as well as the narrow line-width indicate a very good crystal quality in this sample. However, the structure of this nominally stoichiometric film is not free from defects since the NMR spectrum contains several low intensity lines which are not expected for a defect free structure.

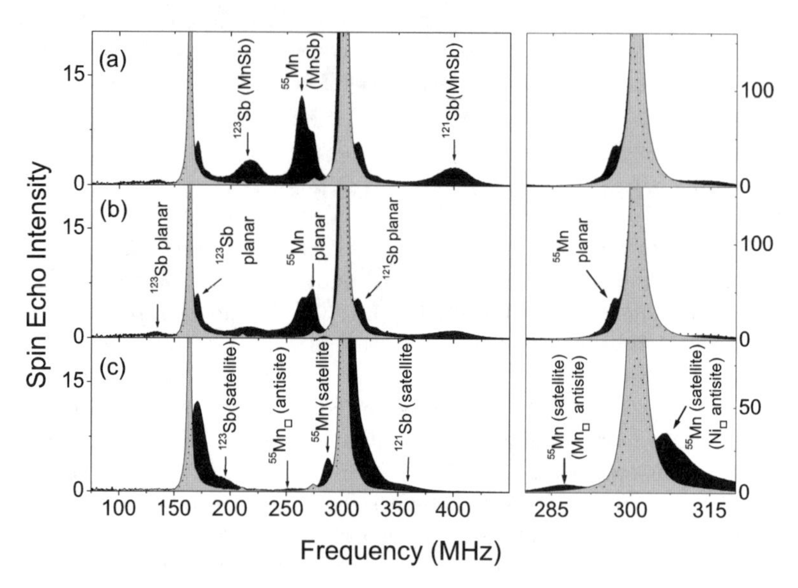

Fig. 11. Black filled curves: NMR spectra recorded at 4.2 K from three 330 nm thick $NiMnSb_x$ films with different Sb content x, deposited at $T_{sub} = 240°C$ and a growth rate of 330 nm/h: (**a**) Sb-rich film on GaAs(001); (**b**) nominally stoichiometric film on GaAs(001); (**c**) Sb-deficient film on GaAs(111). Grey filled curves: NMR reference spectrum from the nearly defect free sample obtained under optimum growth conditions (125 nm thick $Ni_{1.00}Mn_{0.99}Sb_{1.00}$ film B grown at $T_{sub} = 240°C$ on GaAs(111)B at 42 nm/h [31]

A similar low intensity structure has also been observed in the NMR spectra recorded from the other studied samples which were obtained for similar growth parameters and for both GaAs (001) and (111) substrate orientations.

The low intensity structure in the NMR spectrum corresponding to the nominally stoichiometric film can be separated into two groups of lines differing in their character, which reveal two types of defects present in the studied film. One group is related to the formation of an additional hexagonal MnSb phase and the second group revealed the planar defects (stacking faults and/or antiphase boundaries) inside the main NiMnSb half Heusler phase.

In order to facilitate the discussion, the local environment of all lattice sites in NiMnsb are presented up to the third nearest neighbor shell in Table 2.

Hexagonal MnSb Inclusions

The broad lines at 217 MHz (^{123}Sb), 263 MHz (^{55}Mn) and 399 MHz (^{121}Sb) evidence the inclusions of hexagonal MnSb phase [48], see Fig. 11(a,b). This assignment has been confirmed in several NMR experiments going beyond a

Table 2. Number and type of neighbors in NiMnSb structure (□ represents the vacancy site)

Site	1st nn	2nd nn	3rd nn
Ni(A)	4×Mn(B) + 4×Sb(D)	6 × □(C)	12×Ni(A)
Mn(B)	4×Ni(A) + 4×□(C)	6×Sb(D)	12×Mn(B)
□(C)	4×Mn(B) + 4×Sb(D)	6×Ni(A)	12 × □(C)
Sb(D)	4×Ni(A) + 4×□(C)	6×Mn(B)	12×Sb(D)

simple frequency matching between the observed set of ^{123,121}Sb, ^{55}Mn frequencies at 4.2 K and the corresponding NMR frequencies of the bulk MnSb samples reported in the literature [48]. It is important to note that while NMR shows the presence of MnSb in all films with Sb stoichiometry similar to that shown in Fig. 11(b) independently of the GaAs substrate orientation, XRD measurements have revealed the presence of MnSb (00.1) oriented inclusions only in case of (111) orientation of GaAs substrate. Since NiMnSb film grows epitaxially and takes on the same orientation as the substrate, it has been suggested that the growth of MnSb takes place with the hexagonal c (0001) plane parallel to (111) plane of NiMnSb lattice (fcc lattice). Consequently, the MnSb inclusion in NiMnSb half Heusler can be regarded as an extended hexagonal stacking fault in the fcc structure of NiMnSb. A geometrical model based on this idea has been constructed to show a mechanism of such a growth, starting from a single micro-twin defect in NiMnSb [33]. Presence of such micro-twin defects has been revealed by the TEM images taken for Sb-stoichometric and Sb-rich NiMnSb films grown on GaAs substrates of both orientations (001) and (111).

The geometrical model of MnSb incorporation into NiMnSb can be sketched as follows: each of the individual fcc lattices can be regarded as the *abc* sequence of hexagonal planes stacked along ⟨111⟩ and equally spaced by a distance of $d = a_0\sqrt{3}/3 = 3.40$ Å. Taking into account that the entire lattice consists of four interpenetrating fcc lattices equally offset along the [111] direction, it is therefore natural to consider the NiMnSb structure as a layered sequence of (111) planes spaced by $d_1 = a_0\sqrt{3}/12) = 0.85$ Å along the [111] direction (Fig. 12a). In this stack the general *abc* sequence of the fcc lattice is recovered, but it consists of alternating atomic planes belonging to the A, B, C, D sublattices as presented in Fig. 13. It can also be viewed as a sequence of Mn–Ni–Sb–□ planes along the [111] direction stacked in the *abc* sequence. In such a structure a twin fault creates a region that is mirrored around a (111) plane resulting in the change of stacking order from *abcabc* to *cbacba*. The presence of such a twin fault around the Sb plane in NiMnSb structure changes the sequence of planes Mn_c–Ni_a–Sb_b–$\square_c$–Mn_a–Ni_b–Sb_c–$\square_a$ into the sequence $\square_a$–Sb_c–Ni_b–Mn_a–$\square_c$–Sb_b–$\square_c$–Mn_a–Ni_b–Sb_c–$\square_a$. This fault creates a new stacking sequence Mn_a–□–Sb_b–□–Mn_a which involves only Sb and Mn planes with a spacing which is double of the orig-

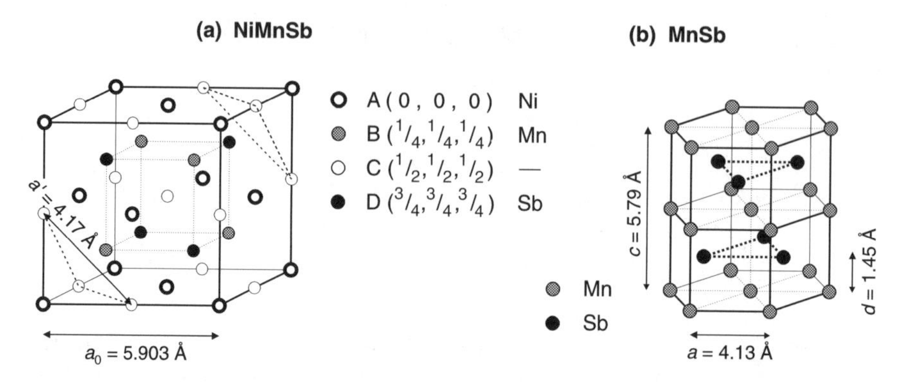

Fig. 12. (**a**) Crystal structure ($C1_b$) of the half Heusler alloy NiMnSb. (**b**) Hexagonal NiAs-type crystal structure of MnSb with *abac* stacking order

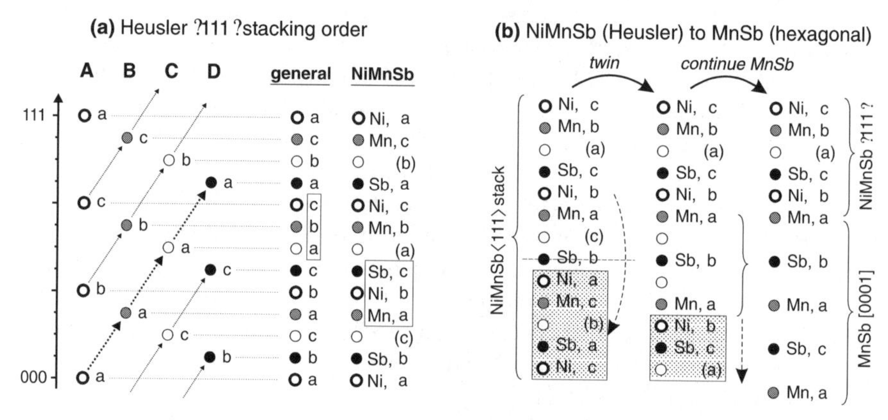

Fig. 13. (**a**) Stacking sequence along [111] direction in the Heusler structure. Left: individual sublattices A, B, C, D, each with the *abc* sequence and offset over one quarter of the body diagonal. Right: overall stacking for the general Heusler structure and for the particular case of NiMnSb. The *abc* stacking sequence is recovered but with alternating atoms on the successive planes. (**b**) Modification of NiMnSb (111) stacking sequence leading to hexagonal MnSb compound. Micro-twin fault around the Sb plane creates a seed consisting of Mn_a–□–Sb_b–□–Mn_a–□–Sb_c–□–Mn_a sequence

inal interplanar distance $2d_1 = 1.70$ Å. This sequence is actually half of a full $Mn_aSb_bMn_aSb_cMn_a$ sequence characteristic for hexagonal MnSb structure presented in Fig. 12(b). In fact, the lattice matching between MnSb and NiMnSb in the hexagonally packed planes is very good (only −1.0% mismatch). Growth of such hexagonal MnSb layer with a thickness extending over many lattice constants in NiMnSb thin film as documented by NMR can be initiated by micro-twin fault presence described above (Fig. 13).

Hexagonal inclusions in NiMnSb are not limited to MnSb: in most cases XRD shows the simultaneous formation of MnSb and NiSb [30,31]. This last compound is paramagnetic and cannot be detected in zero field NMR experiment. However, we expect that NiSb can be formed by the same mechanism as for MnSb although the NiSb case must involve an additional chemical stacking fault interchanging the Mn and Ni layers in a sequence of planes generated by twinning around the Sb plane. Alternatively, NiSb can be created simply by replacement of Mn by Ni in the hexagonal lattice. Formation of these two Sb-rich compounds within the NiMnSb structure suggests that this is the way to accommodate the excess of Sb. The precipitation of the hexagonal phase inclusions must be energetically more favorable than distributing the surplus Sb atoms over other (Mn, Ni, vacancy) sublattices. This is fully confirmed by the NMR spectrum recorded from Sb-rich film nominally described as $NiMnSb_{1.15}$ as shown in Fig. 11(a). An increase of Sb content on the film leads to an increase of the spectrum intensity only at the frequencies corresponding to ^{123}Sb, ^{121}Sb and ^{55}Mn lines from MnSb leaving the intensity of the lines from NiMnSb phase intact. Clearly, the excess of Sb is accommodated by increasing the volume fraction of the secondary phases in the following way: $NiMnSb_{1+x} = (1-x)\times$ NiMnSb + $x\times$ NiSb + $x\times$ MnSb.

Defects within the Main NiMnSb Half-Heusler Phase

The second group of extra lines in the NMR spectrum from the nominally stoichiometric NiMnSb film consists of a rather complex set of ^{121,123}Sb and ^{55}Mn lines [47]. Based on their frequency position these lines can be classified as two ^{55}Mn resonance lines and two pairs of Sb lines corresponding to ^{123}Sb and ^{121}Sb isotopes. Both ^{55}Mn lines have a lower resonance frequency compared to the ^{55}Mn main line. The first ^{55}Mn line can be recognized at 274 MHz (25.96 T) partly overlapping with the ^{55}Mn line (264 MHz) from MnSb. These lines become fully resolved in the RT spectrum due to the different temperature dependence. The second line is clearly visible as a low frequency satellite at 297 MHz (28.12 T) with -3 MHz (0.3 T) frequency spacing to the main ^{55}Mn line at 300 MHz (28.43 T). In the case of Sb, two distinct satellite structures are observed: one on the low frequency side and one on the high frequency side of the main line. They are clearly resolved in the case of the ^{123}Sb isotope, where the resonance frequency does not interfere with any other signal, as shown in Fig. 11. On the high frequency side of the ^{123}Sb line a satellite can be distinguished, spaced by 7.4 MHz. Its counterpart on the ^{121}Sb isotope is observed at 314 MHz. The frequency spacing of 7.4 MHz and 14 MHz observed for the ^{123}Sb and ^{121}Sb isotopes scales with their respective nuclear gyromagnetic ratio γ. On the low frequency side a weak satellite is observed around 133 MHz on the ^{123}Sb isotope. Its counterpart on the ^{121}Sb isotope cannot be directly seen due to the overlapping MnSb signal in the frequency range where they are expected.

The complex crystal structure requires considering two categories of defects related to the above lines. One category involves defects of a simple chemical character due to the antisite disorder where some amount of atoms occupy wrong crystallographic positions. This disorder modifies the chemical short-range order without changing the symmetry. The other category goes beyond the simple point defects and implies an extended defect involving stacking faults, twinning, dislocations, etc. which result in modification of local symmetry.

In order to determine the origin of the low intensity lines, an additional series of $NiMn_{1+x}Sb_{1+y}$, (x, $y = 0.00, 0.05, 0.10$) have been prepared with varying Sb:Ni and Mn:Ni flux ratios [49]. Three spectra recorded from these series for GaAs (111) growth are shown in Fig. 14. The invariance of the low intensity structure to the Mn content provides a sound experimental proof that the observed structure is not related to Mn and Sb atoms placed on a wrong sublattice. While a complete geometrical model has not yet been constructed, these lines are tentatively assigned to planar defects such as

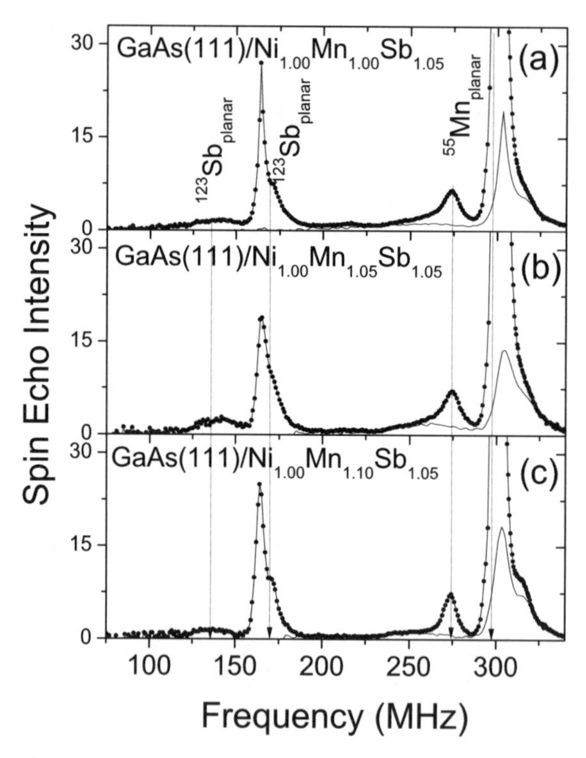

Fig. 14. NMR spectra recorded at 4.2 K from Mn off-stoichiometric $Ni_{1.00}Mn_xSb_{1.05}$/GaAs(111), $x = 1.00, 1.05, 1.10$ thin films. Solid circles: $^{123,121}Sb$ + ^{55}Mn total intensity, solid line: ^{121}Sb contribution showing the invariance of the low intensity structure to the Mn content, as marked by the vertical lines

hexagonal stacking faults inside the fcc structure of NiMnSb similar to those giving rise to micro-twins or antiphase boundaries. Let us observe that the atomic disorder introduced by these defects must be very regular and confined in space as evidenced by the frequency position and the narrow line-width of the main lines. For the sake of further discussion these lines have been labeled as planar defects.

Defects in Sb-Deficient Films

The NMR spectrum from $NiMnSb_{0.85}$/GaAs(111) is presented in Fig. 11(c). The structure of this NiMnSb film is different from the structure of the previously discussed films. This is clearly evidenced by the difference between the NMR spectrum of this film and the spectra discussed above (Fig. 11(a,b)). First of all, the two groups of lines observed in the films with Sb content close to and above the stoichiometric compositions are now missing, meaning that the corresponding structural defects are not present in the film of such composition. This conclusion is fully consistent with the fact that the micro twins are not present in TEM images from this sample and XRD does not show any hexagonal MnSb (NiSb) phase. Instead, a significant line broadening of the ^{123,121}Sb and ^{55}Mn resonances, the presence of broad satellites and the frequency shift of the ^{123,121}Sb main lines clearly indicate that the disorder induced by Sb deficiency extends over the entire sample volume. This implies a chemical disorder where the atomic environments of almost all Sb and Mn atoms are disturbed due to statistically distributed point defects.

We do not observe Mn_{Sb} antisites (these would give rise to an additional ^{55}Mn line located at 330 MHz as will be shown below). On the other hand the point defects can be classified as typical $Ni_{\square}$ antisites on the vacancy sublattice or vacancies on the Sb sublattice. Analysis of NMR spectrum suggests also a presence of some $Mn_{\square}$ antisites on the vacancy sublattice.

The presence of a limited number of isolated $Mn_{\square}$ antisites on the vacancy sublattice can give rise to three new spectrum features. Mn atom misplaced on the vacancy sublattice disturbs the first nearest neighbor shell of Sb and Mn atoms which consists of $4\times$Ni and $4\times\square$ in the ideal case, see Table 1. Such $Mn_{\square}$ antisite as a first nearest neighbor increases the transferred hyperfine field at Sb nucleus and gives rise to a small ^{123}Sb satellite at 195 MHz (3.53 T) shifted by +23 MHz (4.2 T) with respect to the main ^{123}Sb line at 171 MHz (3.10 T) (see Fig. 11(c)). The corresponding ^{121}Sb satellite can be recognized at 360 MHz at the background of ^{55}Mn line tail. The same Mn atom at the vacancy antisite position becomes a first nearest neighbor to Mn in the regular positions giving rise to ^{55}Mn satellite at 287 MHz spaced by −13 MHz (1.23 T). A new ^{55}Mn line due to the $Mn_{\square}$ antisite at the vacancy position can be recognized at 250 MHz (2.37 T), a frequency expected for Mn with 4 Mn and 4 Sb as the first neighbors. Quantitative analysis of the relative intensities of the respective lines in the spectrum gives an estimate of the concentration of Mn misplaced on the vacancy sublattice equal to ~2% of the total Mn content.

The remaining modifications observed in the spectrum from the Sb-deficient film consist of ^{123}Sb(^{121}Sb) line broadening and the frequency up-shift of this line to 171 (316) MHz as well as the presence of additional ^{55}Mn poorly resolved structure with +5 MHz (−0.47 T) spaced satellites on the high frequency side of ^{55}Mn main line. These spectrum features are not related to additional $Mn_{\square}$ antisites, on Sb sublattice for example, which would place Mn as further neighbors of Sb and Mn in regular positions and consequently modifying the hyperfine field to less extent than in the case of $Mn_{\square}$ antisites. It is expected that such Mn antisite atoms on Sb sublattice for example would give rise to an additional ^{55}Mn line located at 330 MHz to account for the observed 5 MHz shift due to each from 6 Mn as a second nearest neighbor (Table 1), but such line is not present in the spectrum.

One possibility is that these changes can be related to a small increase of magnetic moments of Mn atoms on the regular positions. Such an increase of Mn magnetic moment has been reported for $Ni_{1+x}MnSb$ bulk samples with $C1_b$ structure for $0 < x < 0.6$ [19], where Ni surplus atoms were randomly distributed over the vacancy sublattice (C), ie. $Ni_{\square}$ antisites were created and there were no vacancies introduced on Ni (A) sublattice. In order to apply this observation to the present case one has to allow for a small modification of Mn magnetic moment in sublattice B depending on the number of $Ni_{\square}$ antisites as first nearest neighbors. This would require about 30% of all Mn population to have slightly modified magnetic moment in the present case. Since 1 $Ni_{\square}$ antiste affects four Mn neighbours a very crude estimate gives 7.5% of $Ni_{\square}$ antisites. Such a scenario would suggest that Sb deficiency in NiMnSb film could be partly compensated by placing certain amount of Ni and Mn atoms on the vacancy sublattice to provide a better balance between the amount of Ni, Mn and Sb on the regular (A, B and D) sublattices. Such occupancy of 4 sublattices (A, B, C, D) would probably help to preserve $C1_b$ structure of the NiMnSb film [19].

The other possibility is that the observed changes are simply caused by the Sb vacancies on the Sb sublattice (D). Electron band structure calculations by Galanakis et al. [50] reveal a strong hybridisation between Sb 5p states and transition metal 3d4s states. Actually is has been claimed that Sb 5p states can accommodate 3 transition metal valence electrons leaving in the conduction band only d electrons. It can be expected that due to the missing Sb the band structure will be disturbed locally and these changes will modify also the hyperfine fields on the nuclei of Mn and Sb atoms in the nearest surrounding. This last scenario of the disorder due Sb deficiency is consistent with the already discussed reluctance of the NiMnSb structure to fill the vacancy sublattice (C) with a surplus of Sb for the Sb rich stoichiometry. However, the available NMR data cannot determine whether the

Sb off-stoichiometry leads to $Ni_{\square}$ antisites on the vacancy sublattice (C) or to the Sb vacancies on the Sb sublattice (D).

4.3 NiMnSb Films Grown at Slow Growth Rate

The critical role of Sb content for the crystallographic quality of NiMnSb films has been confirmed by an additional NMR study which has been carried out on a set of NiMnSb films grown in a narrow composition region around the ideal 1:1:1 stoichiometry with Sb content as the only parameter which was varied [31]. These studies have shown that an almost defect free NiMnSb/GaAs(111) film can be grown by MBE. The growth was carried at $T_{sub} = 240°C$ on GaAs(111)B and a growth rate of 42 nm/h. The RBS analysis has shown that the Sb content varies linearly with Sb flux with a step of about $0.03 \cdots 0.6$ in the range from 1.08 for the sample A down to 0.87 for the sample E.

The NMR spectra for this series of samples are shown in Fig. 15. Two basic observations can be made immediately. Two main lines (164 MHz, 300 MHz) from Heusler phase are present in all the samples at the frequencies expected for a well ordered phase, although their line-width varies depending on the Sb content with a clear minimum for sample B. The small steps in Sb content make it possible to follow the smooth transition between the defect patterns for Sb-rich and Sb-poor films. Starting from the Sb-rich composition (sample A) and reducing the Sb content, a reduction of intensity of the lines due to MnSb inclusions and planar defects can be observed. Their complete disappearance can be noted in spectra from the two last films (Sb-deficient films D and E), which show the effects due to the point defects as described above. Features of point defects characteristic for Sb deficiency can be already recognized in the NMR spectrum from the sample C where they coexist with the lines indicating traces of the planar defects.

A detailed analysis of the NMR spectra identifies the film B as the film with the highest order. No hexagonal MnSb is found while the point defects typical for the Sb-deficient composition are absent and the amount of planar defects is the smallest. The best estimate of the amount of defects present in this sample can be made based on the intensity of ^{55}Mn line at 274 MHz belonging to the group of lines related to planar defects. Comparing the intensity of this line with the main ^{55}Mn line at 300 MHz, after extraction of ^{121}Sb contribution, one concludes that 1.1% of all Mn atoms belongs to planar defects. This very low level of defects is not affected by the presence of an additional defect which is present in almost all samples from this series. This defect is visible from a small line at 211 MHz and is attributed to As_{Sb} substitutional defects. From the line intensity we deduce that about 0.5% of all Sb sites is occupied by As. The presence of some As in the films is not surprising, since the growth of the GaAs buffer layer in the chamber leaves some As background. But since this line has not been observed in NMR

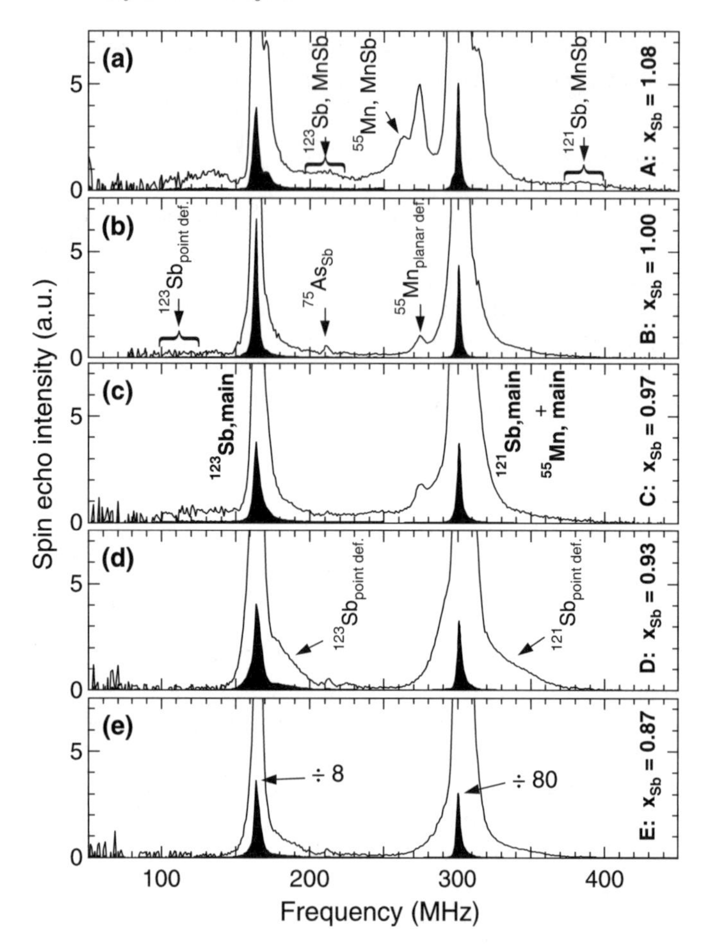

Fig. 15. NMR spectra recorded at 4.2 K from the set of 5 $Ni_{1.00}Mn_{0.99}Sb_{x_{Sb}}$/ GaAs(111)B films shown in Figs. 4(**b**) and 6. Solid lines: low intensity part of the spectra highlighting the defect structure. Filled to zero: full-scale version of the same spectra, with different reduction factors for the sections below and above 250 MHz as indicated in panel (**e**). The features corresponding to the unfaulted Heusler phase are labeled in panel (**c**), defects lines in (**b**) and (**d**), and hexagonal MnSb inclusions in (**a**)

spectra from other samples, the presence of As impurities on the level of 0.5% seems to be accidental rather than systematic.

Concluding, the analysis of the NMR results has shown the lowest level of atomic disorder (planar defects involving Mn atoms at the level of 1% of total Mn atoms) and the best quality of crystal structure (the smallest line-width of the main lines from Heusler phase) for the sample B. This is consistent with an almost perfect 1:1:1 stoichiometry of this film confirmed by additional RBS structural analysis and the first traces of NiSb formation as revealed by XRD [31].

5 Spin Injection Results

The ultimus test for the growth of half-metallic NiMnSb films on GaAs is of course the spin injection experiment itself. For this we have fabricated spin-light emitting diodes (spin-LEDs) as shown in Figs. 2 and 3. Spin-polarized electrons are injected into a 100 nm thick active GaAs region that is sandwiched between two AlGaAs confinement layers. There they recombine optically with unpolarized holes provided by the p-type substrate. The circular polarization of the light emitted along the surface normal provides information about the spin polarization of the injected electrons [6,51].

Various injection configurations have been examined. In an initial experiment we attempted to make "Ohmic" contacts on (111)B GaAs substrates to preserve half-metallic interfaces [12], but as discussed in Sect. 2.5 the band line-up between NiMnSb and GaAs resulted in a Schottky barrier of ~0.8 eV and no electron injection was possible [27,52].

This made it clear that a tunnel barrier is required to accommodate the potential difference between NiMnSb and GaAs. A tunneling Schottky barrier (Fig. 3) seems to offer the ideal compromise: it allows the growth of epitaxial NiMnSb films with the same high quality as on any GaAs substrate, and it has the additional advantage of accommodating the conductivity mismatch problem if the contact would not be truly half-metallic (as is expected after considering all the issues discussed in Sect. 2). We have fabricated tunneling Schottky barriers both on GaAs(001), using a thin 1.5×10^{19} cm^{-3} doped layer, and on GaAs(111)B, using a δ-doped layer. The results were disappointing, with injected spin polarizations $\leq 2\%$ on GaAs(001) and $< 1\%$ on GaAs (111)B at 80 K. These values are much less than the results obtained with CoFe/AlO_x injectors (24% at 80 K [53] and 16% at RT [6]), Fe/GaAs Schottky injectors (32% at 4.2 K [54]), or $Ga_{1-x}Mn_xAs$ Zener injectors (82% at 80 K and 25% at 80 K [55]). We therefore also prepared a MnSb reference injector on the same LEDs, and these resulted in both cases in a 3× larger spin injection than NiMnSb, showing that the tunnel barrier and the LED functioned correctly.

In the case of Fe/GaAs spin injectors it was shown that an expansion of the GaAs lattice constant close to the interface resulted in a strong reduction in injected spin polarization [8]. This was attributed to larger intermixing or possible Fe indiffusion into GaAs for deposition temperatures above room temperature. We have therefore also attempted to grow epitaxial NiMnSb spin injectors on GaAs(001) spin LEDs at temperatures as low as 10°C. Although the structural and magnetic properties of the films were surprisingly good (see Sect. 3.3), the injected spin polarization remained lower than ~2%.

All these values remain smaller than the spin injection obtained from polycrystalline NiMnSb films deposited on AlO_x tunnel barriers, ~6.5% at 80 K [27, 52]. This suggests that all the interfaces between NiMnSb and GaAs that we have been able to prepare have a low interface polarization, smaller even than the interface between polycrystalline NiMnSb and amorphous AlO_x. At present it is not clear whether this is due the chemical instability of the interface, or to structurally sound interfaces that suffer from electronic effects such as discussed by de Wijs and de Groot [12]. One route that has not yet been explored in detail is the control of the plarity of NiMnSb/GaAs(111) interfaces. Although we have shown epitaxy on both As-terminated GaAs(111)B and Ga-terminated GaAs(111)A substrates, and demonstrated atomic layer deposition using both the sequences Mn–Ni–Sb and Sb–Ni–Mn deposition orders on GaAs(111)B, we did not have the means for checking the resulting stacking order after the growth. Therefore it is conceivable that differences in interface energy forced the atoms to swap places during the initial stages of the growth. This aspect merits further experimental and theoretical attention.

6 Conclusions

In order to realize truly half-metallic contacts to a semiconductor based on (half-)Heusler alloys, nearly perfect films have to be grown, with perfect control over the composition and over the interfaces, and without chemical disorder or extended defects. Even if these can be obtained, finite temperature effects may prevent the injection of spins, and band offsets may prevent the injection of any electrons.

These issues can be circumvented by the incorporation of a tunneling Schottky barrier, which combines the best of two worlds: the tunnel barrier accommodates the band offset and removes the requirement for 100% spin polarization, allowing for some defects to be present, while the epitaxial growth remains possible, which should allow *highly polarized* NiMnSb films to be obtained.

Our experiments show that it is possible to obtain epitaxial NiMnSb films on GaAs(001) and (111)A/B that fulfill nearly al requirements, by combining very slow growth rates with an accurate control over the composition by careful RBS calibration.

NMR was used to analyze in detail the concentration and type of local defects, with sensitivities much better than 1 at% that can not be obtained by any other technique.

Our best films are within 1% of the ideal stoichiometry and have a disorder level of ~1%, without indications of any interface reactions. Nevertheless, spin injection experiments on both (001) and (111)B oriented spin LEDs have only shown polarizations up to 2.2%. This is smaller than MnSb reference injectors on the same LEDs, and also smaller than the injection from polycrystalline NiMnSb films on amorphous AlO_x tunnel barriers, suggesting that the interface polarization in all studied configurations was very low. One parameter which remains to be verified is the polarity (stacking order) of NiMnSb(111) films on GaAs(111)B LEDs.

Acknowledgements

It is a pleasure to acknowledge the contributions from, and discussions with, P. Van Dorpe (device simulations and spin injection measurements), B. Brijs and D. Jalabert (RBS), H. Bender (Auger), S. Degroote (high-resolution XRD), O. Richard (TEM), W. van de Graaf (LED growth), G. Borghs, J. De Boeck, and E. Jędryka (NMR). Financial support was given by the Bilateral Agreements between Flanders and Poland (BIL99/22 and BIL03/PL02a) and the EC project G5RD-CT-2001-00535 FENIKS. WVR acknowledges financial support as a Postdoctoral Fellow of the Research Foundation – Flanders (FWO – Vlaanderen).

References

1. G. Schmidt, D. Ferrand, L.W. Molenkamp, A.T. Filip, and B.J. van Wees: Phys. Rev. B **62**, R4790 (2000)
2. E.I. Rashba: Phys. Rev. B **62**, R16267 (2000)
3. A. Fert and H. Jaffrés: Phys. Rev. B **64**, 184420 (2001)
4. H. Jaffrés and A. Fert: J. Appl. Phys. **91**, 8111 (2002)
5. O.M.J. van 't Erve, G. Kioseoglou, A.T. Hanbicki, C.H. Li, and B.T. Jonker, R. Mallory, M. Yasar, and A. Petrou: Appl. Phys. Lett. **84**, 4334 (2004)
6. V.F. Motsnyi, P. Van Dorpe, W. Van Roy, E. Goovaerts, V.I. Safarov, G. Borghs, and J. De Boeck: Phys. Rev. B **68**, 245319 (2003)
7. D. Ristoiu, J.P. Nozières, C.N. Borca, T. Komesu, H.-K. Jeong, and P.A. Dowben: Europhys. Lett. **49**, 624 (2000)
8. R.M. Stroud, A.T. Hanbicki, G. Kioseoglou, O.M.J. van 't Erve, C.H. Li, B.T. Jonker, G. Itskos, R. Mallory, M. Yasar, and A. Petrou: Spintech II conference, Brugge (Belgium), 4–8 August 2003
9. C. Palmstrøm: MRS Bulletin **28**, 725 (2003) and references therein
10. R. Bogaerts, F. Herlach, A. De Keyser, F.M. Peeters, F. DeRosa, C.J. Palmstrom, D. Brehmer, and S.J. Allen Jr.: Phys. Rev. B **53**, 15951 (1996)
11. C.M. Fang, G.A. de Wijs, and R.A. de Groot: J. Appl. Phys. **91**, 8340 (2002)
12. G.A. de Wijs and R.A. de Groot: Phys. Rev. B **64**, 020402 (2001)
13. R.M. Stroud, A.T. Hanbicki, Y.D. Park, A.G. Petukhov, B.T. Jonker, G. Itskos, G. Kioseoglou, M. Furis, and A. Petrou: Phys. Rev. Lett. **89**, 166602 (2002)
14. H. Ebert and G. Schütz: J. Appl. Phys. **69**, 4627 (1991)
15. D. Orgassa, H. Fujiwara, T.C. Schulthess, and W.H. Butler: Phys. Rev. B **60**, 13237 (1999)
16. S. Picozzi, A. Continenza, and A.J. Freeman: Phys. Rev. B **69**, 094423 (2004)

17. B. Ravel, M.P. Raphael, V.G. Harris, and Q. Huang: Phys. Rev. B **65**, 184431 (2002)
18. M.C. Kautzky, F.B. Mancoff, J.-F. Bobo, P.R. Johnson, R.L. White, and B.M. Clemens: J. Appl. Phys. **81**, 4026 (1997)
19. P.J. Webster and R.M. Mankikar: J. Magn. Magn. Mat. **42**, 300 (1984)
20. R.B. Helmholdt, R.A. de Groot, F.M. Mueller, P.G. van Engen, and K.H.J. Buschow: J. Magn. Magn. Mat. **43**, 249 (1984)
21. J. Schaf, K. Le Dang, P. Veillet, and I.A. Campbell: J. Phys. F **13**, 1311 (1983)
22. R. Skomski and P.A. Dowben: Europhys. Lett. **58**, 544 (2002)
23. P.A. Dowben and R. Skomski: J. Appl. Phys. **93**, 7948 (2003)
24. L. Chioncel, M.I. Katsnelson, R.A. de Groot, and A.I. Lichtenstein: Phys. Rev. B **68**, 144425 (2003)
25. Ph. Mavropoulos, K. Sato, R. Zeller, P.H. Dederichs, V. Popescu, and H. Ebert, Phys. Rev. B **69**, 054424 (2004)
26. B. Heinrich, G. Woltersdorf, R. Urban, O. Mosendz, G. Schmidt, P. Bach, L. Molenkamp, and E. Rozenberg: J. Appl. Phys. **95**, 7462 (2004)
27. W. Van Roy, P. Van Dorpe, V. Motsnyi, Z.L.Z.G. Borghs, and J. De Boeck: phys. stat. sol. (b) **241**, 1470 (2004)
28. W. Van Roy, J. De Boeck, B. Brijs, and G. Borghs: Appl. Phys. Lett. **77**, 4190 (2000)
29. W. Van Roy, J. De Boeck, and G. Borghs: J. Cryst. Growth **227–228**, 862 (2001)
30. W. Van Roy, G. Borghs, and J. De Boeck: J. Magn. Magn. Mat. **242–245**, 489 (2002)
31. W. Van Roy, M. Wójcik, E. Jędryka, S. Nadolski, D. Jalabert, B. Brijs, G. Borghs, and J. De Boeck: Appl. Phys. Lett. **83**, 4214 (2003)
32. P. Turban, S. Andrieu, B. Kierren, E. Snoeck, C. Teodorescu, and A. Traverse: Phys. Rev. B **65**, 134417 (2002)
33. M. Wójcik, W. Van Roy, E. Jędryka, S. Nadolski, G. Borghs, and J. De Boeck: J. Magn. Magn. Mat. **240**, 414 (2002)
34. P. Bach, C. Rüster, C. Gould, C.R. Becker, G. Schmidt, and L.W. Molenkamp: J. Cryst. Growth **251**, 323 (2003)
35. P. Bach, A.S. Bader, C. Rüster, C. Gould, C.R. Becker, G. Schmidt, L.W. Molenkamp, W. Weigand, C. Kumpf, E. Umbach, R. Urban, G. Woltersdorf, and B. Heinrich: Appl. Phys. Lett. **83**, 521 (2003)
36. L. Nowicki, A.M. Abdul-Kader, P. Bach, G. Schmidt, L.W. Molenkamp, A. Turos, and G. Karczewski, Nucl. Instr. and Meth. in Phys. Res. B **219-220**, 666 (2004)
37. C. Meny, E. Jedryka, and P. Panissod: J. Phys.: Condens. Matter **5**, 11547 (1993)
38. M. Wojcik, J.P. Jay, P. Panissod, E. Jedryka, J. Dekoster, and G. Langouche: Z. Phys. B **103**, 5 (1997)
39. S. Picozzi, A. Continenza, and A.J. Freeman: Phys. Rev. B **66**, 094421 (2002)
40. A. Deb, N. Hiraoka, M. Itou and Y. Sakurai, M. Onodera, N. Sakai: Phys. Rev. B **63**, 205115 (2001)
41. A. Deb, and Y. Sakurai: J. Phys.: Condens. Matter **12**, 2997 (2000)
42. T. Hihara, M. Kawakami, M. Kasaya, and H. Enokiya: J. Phys. Soc. Japan **26**, 1061 (1969)
43. C.C.M. Campbell: J. Phys. F: Met. Phys **5**, 1931 (1975)

44. T. Shinohara: J. Phys. Soc. Japan **28**, 313 (1970)
45. J.C. Love: Hyperfine Int. Conf. Proc. (L'Aquila) p. 953 (1972)
46. Le Dang Khoi, P. Veillet, and I.A. Campbell: J. Phys. F **8**, 1811 (1983)
47. M. Wójcik, E. Jędryka, W. Van Roy, G. Borghs, and J. De Boeck: Phys. Rev. B, in preparation (2004)
48. J. Boumwa and C. Hass: Phys. Stat. Sol. (b) **56**, 299 (1973)
49. M. Wójcik, E. Jędryka, W. Van Roy, and G. Borghs: unpublished
50. I. Galanakis, P.H. Dederichs and N. Papanikolaou, Phys. Rev. B **66**, 134428 (2002)
51. V.F. Motsnyi, J. De Boeck, J. Das, W. Van Roy, G. Borghs, E. Goovaerts, and V.I. Safarov: Appl. Phys. Lett. **81**, 265 (2002)
52. W. Van Roy, P. Van Dorpe, V.F. Motsnyi, G. Borghs, and J. De Boeck, AVS 49th International Symposium, Denver, CO, November 4–8, 2002
53. P. Van Dorpe, V.F. Motsnyi, M. Nijboer, E. Goovaerts, V.I. Safarov, J. Das, W. Van Roy, G. Borghs, and J. De Boeck: Jpn. J. Appl. Phys. **42**, L502 (2003)
54. A.T. Hanbicki, O.M.J. van 't Erve, R. Magno, G. Kioseoglou, C.H. Li, B.T. Jonker, G. Itskos, R. Mallory, M. Yasar, and A. Petrou: Appl. Phys. Lett. **82**, 4092 (2003)
55. P. Van Dorpe, Z. Liu, W. Van Roy, V.F. Motsnyi, M. Sawicki, G. Borghs, and J. De Boeck: Appl. Phys. Lett. **84**, 3495 (2004)

Growth and Magnetotransport Properties of Thin Co_2MnGe Layered Structures

Thomas Ambrose and Oleg Mryasov

Seagate Research, 1251 Waterfront Place, Pittsburgh, PA, USA
thomas.f.ambrose@seagate.com and oleg.mryasov@seagate.com

Abstract. Co_2MnGe and other related Heusler alloys are investigated in the context of magnetotransport properties in multilayer (ML) structures. The most important factors relevant to the problem of magnetotransport in these MLs are reviewed including (i) the growth of thin film Hesuler alloy structures by Molecular Beam Epitaxy and Magnetron Sputtering techniques; (ii) the effects of finite temperature on the half-metallicity, interface states and band structure matching (iii) the type of measured carrier transport and its relationship with the underlying band structure; and finally (iv) the magnetic interactions in an all Heusler ML structure. We have measured a room temperature current-in-plane (CIP) GMR effect in MLs and spin valves and attribute their small magnetoresistance to the materials high resistivity and low carrier mobility that is an inherent property of the Heusler alloys due to their complicated band structure for these p-type conductors. The main challenges for achieving an enhanced GMR effect in Heusler ML structures are discussed on the basis of our room temperature magneto-transport measurements and theoretical modelling results.

1 Introduction

Today the advancement of computer and electronics technologies has placed stringent demands upon memory capacity, logic speed and performance. Such common state of the art technologies include hard disc drive storage with high areal density, affordable non-volatile solid-state memory and ultra fast logic. These technologies continually improve, pushing the limits of electrical transport and device physics while creating a multi-billion dollar per year industry. The development of advanced materials and thin film processing play a key role in this progression. However as form factors shrink, novel electronic devices will also need to be realized. Remarkably these devices may be heavily dominated on the thin film level by the use of magnetic materials. Obviously the greatest impact thus far is the use of ferromagnets in the disc drive storage industry where areal densities are projected to reach 1 $Tbit/in^2$ and beyond [1]. Equally forthcoming, magnetic random access memory (MRAM) develops rapidly [2] and challenge other solid-state memories as well as providing the groundwork for re-programmable logic devices based upon ferromagnets [3]. Magnetic logic devices will have similar attributes compared to semiconductor based devices but will also include additional features of faster switching speeds and non-volatility. Furthermore the possibilities of using spin-polarized currents provide a new facet to digital logic. For in a ferromagnet two spin currents occur naturally due to exchange split band structure of these materials. Such spin currents automatically provide two states (spin up and spin down) that can be easily gated by simply switching the direction of the magnetization of the ferromagnetic contact. Consequently incorporating these materials into logic devices will have a significant impact on the current technology.

Ferromagnets possess a wide variety of different features and salient qualities. These materials may have giant magnetic moments [4], high anisotropy [5] or soft ferromagnetic properties [6] each of which can be tailored for use in a particular application. Moreover, ferromagnetic materials are the heart of today's field sensing devices due to the discoveries of giant magnetoresistance (GMR) [7] and the subsequent room temperature tunneling magnetoresistance (TMR) [8]. These two effects have revolutionized sensing technology and can be credited with the innovation and foundation of the advanced research program known as Spintronics [9].

Spintronics refers to the benefit of using both magnetic and electrical properties of a single charge-spin entity where the conduction electron not only mediates the current but also carries spin, providing an additional degree of freedom to the transport. At present semiconductor based devices only utilize current transport without exploiting the spin resource part. The usefulness of a logic device based upon the Spintronics ideologies will be far superior to its predecessors being of simpler design and having a higher level of functionality. However a major task will involve incorporating ferromagnets into existing logic devices. Primarily the success of Spintronics will depend upon the choice of magnetic materials utilized and their compatibility with metals and semiconductors presently in use. It is further believed that the choice of ferromagnetic materials for Spintronics applications will also boast the feature of high spin polarization such that the relative ratio of spin up to spin down electrons in the ferromagnetic material will be maximized. These ferromagnetic materials will come in a variety of forms being either a ferromagnetic metal [10], a ferromagnetic semiconductor [11] or even a ferromagnetic oxide [12]. Regardless of the type of material and specifics of the device, the degree of spin polarization ultimately determines the usefulness of the spin dependent transport.

Spin polarization is a basic intrinsic quantity that defines the ratio of the number of spin up to spin down electrons at the Fermi energy in a ferromagnetic material. The spin polarization follows the simple relation:

$$P = \frac{n_{\uparrow} - n_{\downarrow}}{n_{\uparrow} + n_{\downarrow}} \tag{1}$$

where $n_{\uparrow}$ is the number of spin up electrons and $n_{\downarrow}$ is the number of spin down electrons. It is clear from equation 1 that P can be zero when the number of $n_{\uparrow}$ and $n_{\downarrow}$ are equivalent as in the case of non-magnetic materials. For ferromagnetic materials P takes on a non-zero value. For example, recent experimental measurements have placed the achievable limit on P in transition metal (Fe, Co and Ni) ferromagnets and their alloys below 0.5 [13]. However for spin transport applications, even higher values of spin polarization is desired and other types of ferromagnets need to be considered as alternative materials to the transition metal ferromagnets. Notice a special case results when the spin polarization takes on a value of 1. In this instance at the Fermi energy one of the spin bands has a finite number of electrons while the other spin channel is empty thus creating a gap between the valence and conduction bands. A magnetic material in this case is labelled as "half-metallic" since one spin band exhibits metallic behavior while the other spin band acts as a semiconductor but most importantly both bands exist in a single material. Simply put, half-metallic ferromagnets are materials that possess conduction electrons being 100% spin polarized at the Fermi energy.

There are many ferromagnetic materials that are predicted to be half-metallic. Some common examples are CrO_2 [14], $FeCoS_2$ [15], Fe_3O_4 [16] as well as some Heusler alloys. Of these materials, the oxides require considerable effort to fabricate in thin film form. Furthermore CrO_2 has a Curie temperature around room temperature, making it difficult to realize the use of these materials in a thermally stable device application. On the other hand, Fe_3O_4 has a relatively high Curie temperature but suffers from multiple crystalline phases for the same Fe-O stoichiometry. In general half metallic oxides can suffer from oxygen non-stoichiometry effects and in this case the multiple phases detract the benefits of the high spin polarization phase. For these reasons we argue for the development and understanding of certain Heusler alloys as high spin polarized thin film candidates.

Heusler alloys are ferromagnetic ternary alloys that contain manganese [17]. They have the following chemical formula: A_xMnB where A is a transition metal and B either a transition metal or a semiconductor. The value of x can be 1 or 2 designating the half or full Heusler alloys respectively. Ferromagnetism in Heusler alloys is critically dependent upon both the magnetic and chemical ordering of the Mn atom. Furthermore many of the Heusler alloys posses very high Curie temperatures (>700°C) along with large magnetic moments (>3.5 μ_B/formula unit) [18] that removes any thermal instability limitations. While there are a variety of Heusler alloys, only a select number have been predicted to be half metallic from electronic band structure calculations. The half-Heusler alloys NiMnSb and PtMnSb have been the most widely studied both theoretically [19] and experimentally [20]. Other full-Heusler alloy structures, for example Co_2MnGe and Co_2MnSi, have also been predicted theoretically to posses conduction electrons to be 100% spin polarized [21].

Electronic structure calculations play an important role in predicting and understanding highly spin polarized materials [22]. These calculations are most intriguing but the results are limited to the ideal magnetic ground state for T $= 0K$. Thermal excitations are expected to naturally mix the majority and minority spins leading to a deviation from the half metallic state at some finite temperature [23]. The simplest comparison of the theoretical findings with experiment relies on magnetometry measurements that suggest a particular half-metallic material should possess an integer moment at low temperature [24]. Even the most careful magnetic moment measurements can only give an accuracy within 5 to 10%. Other important electronic structure results suggest a small amount of substitutional disorder can lead to a significant reduction in the spin polarization [25]. However characterization of numerous Heusler thin film samples using x-ray and electron diffraction suggests high quality films. Lastly transport measurements that indicate a linear temperature dependence in the sample's resistivity have also been suggestive of half-metallic behavior based upon a one magnon scattering process [26]. While theoretical calculations advocate high spin polarization in a variety of materials and characterization indicates high quality samples, a true determination of the spin polarization is needed and will ultimately influence the extent of use in an actual spin transport based device.

There are numerous ways to determine the spin polarization of a ferromagnetic material. Throughout the years techniques such as photoemission [27], which probes surface states, and tunneling spectroscopy [28], which samples the Zeeman splitting, have been used. Each technique requires careful preparation of the sample surface or sophisticated fabrication into a trilayer device. Recently a much simpler technique, Point-Contact Andreev Reflection (PCAR) has been developed [29] in which a measurement of the conductance between a superconductor in point contact with a ferromagnet can determine the spin polarization. By properly modeling the change in conductance below the superconducting gap [30], an accurate measurement of the spin polarization can be made at low temperature. Note, one must differentiate between the results of the fore mentioned techniques since each measures slightly different things. First photoemission measures the actual density of states of the material that can be directly compared with the calculated band structure while the tunneling and PCAR techniques measure the amount of spin polarization in the spin current. The two measurements can be easily related by the appropriate Fermi velocity dependence for ballistic or diffuse transport. In any case, these techniques do not involve an actual device and may need to be made at low temperature due to the use of a conventional superconductor.

Finally we mention that high spin polarization measurements using photoemission on the manganate $LaCaMnO_3$ [12] and low temperature measurements of CrO_2 using PCAR [31] have been reported. However, the interpretation and accuracy of the photoemission data remains unclear due to the techniques low energy resolution near the Fermi energy. Furthermore the PCAR measurements can only be made at low temperature and an accurate value of spin polarization at room temperature remains elusive. Therefore to remove

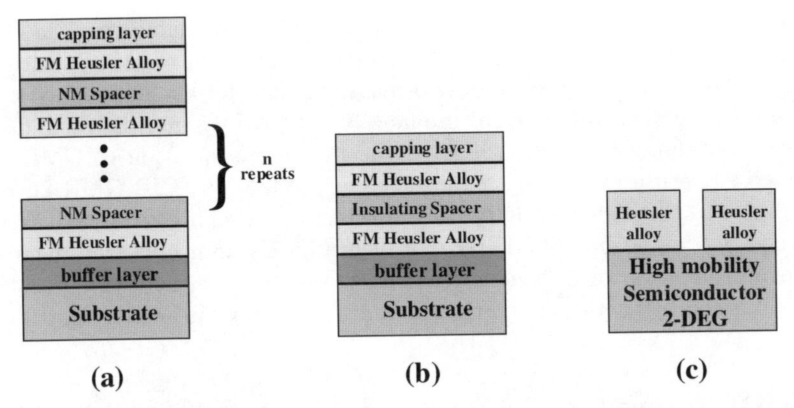

Fig. 1. Schematic representation of the 3 types of spin transport systems utilizing a ferromagnetic Heusler alloy layer: (**a**) giant magnetoresistance, (**b**) tunneling magnetoresistance and (**c**) spin injection

all doubt of the existence of highly spin polarized ferromagnets will ultimately involve either a Current Perpendicular to the Plane (CPP) GMR measurement in a multilayer film or tunnel junction device containing half metallic ferromagnetic layers and showing a large magnetoresistance effect [32]. A spin injection device (both inject and detect) [33] having half metallic ferromagnetic pads in contact with a high mobility semiconductor or 2DEG material with a 100% injection efficiency will also satisfy this criterion. A schematic diagram of these suggested experiments and devices using a Heusler alloy ferromagnet is shown in Fig. 1. These are generic Spintronic-based devices that would also work with regular ferromagnetic materials however their performance would not be optimized. While the choice of materials used in these experiments remains open for debate, to date, no clear demonstration of room temperature high spin polarization behavior has been reported. These results are unfortunate but indicate the rather complicated nature of incorporating half metallic ferromagnetic materials into new devices.

In this chapter we concentrate on the magnetotransport properties of Co_2MnGe thin films. We have selected this Heusler alloy for its predicted half metallic properties and ease of thin film growth by various deposition methods. We further select this full Heusler alloy compared to the half Heusler alloys due to a lower probability of anti-site disorder. We describe both experimental and theoretical results for Co_2MnGe in a variety of thin film layered structures. In Sect. 2 we first discuss the characterization and epitaxial growth of single crystal Co_2MnGe thin films on GaAs (001) substrates using Molecular Beam Epitaxy (MBE). Such growth can be utilized in spin injection devices where clean and abrupt interfaces are needed. Next in Sect. 3 we focus on the characterization and growth of Co_2MnGe prepared by magnetron sputtering. We compare the properties of the polycrystalline sputtered films with those of the single crystal MBE grown films. Giant magnetoresistance is then presented in Sect. 4 in the context of a "band engineering" methodology that requires the matching of the energy bands of Co_2MnGe with the energy bands of an appropriate non-magnetic Heusler-like spacer layer. We show room temperature current in plane (CIP) GMR in multilayers and spin valves containing Co_2MnGe. This is the first demonstration of GMR in a layered structure containing a Heusler alloy. While our CIP GMR effect is small due to the high resistivity materials chosen, it serves as a model system for design, development and study of future highly spin polarized magnetotransport. Lastly we utilize the Hall effect to characterize and understand the poor conductivity apparently inherent to the Co_2MnGe Heusler alloy.

2 MBE Growth of Co_2MnGe on GaAs (001)

2.1 Molecular Beam Epitaxy and Reflection High Energy Electron Diffraction

Molecular Beam Epitaxy (MBE) is a sophisticated form of vacuum evaporation in which atomic or molecular beams are directed at a carefully prepared single crystal substrate in a very clean ultra high vacuum (UHV) system. The resulting epitaxially grown single crystal film mimics the crystal structure of the underlying substrate, being free of grain boundaries with a minimized number of defects. Because of these unique and nearly perfect film qualities, MBE has grown into prominence as an important research tool having clean and controlled deposition conditions, which result in a variety of metal and semiconductor systems possessing enhanced electrical and magnetic properties.

MBE has made its greatest contribution in the semiconductor field. The classic example system is a multilayer structure of GaAs and AlGaAs in which a layer of GaAs is sandwiched between thicker layers of AlGaAs. If the thickness of the GaAs is less than an electron's mean free path, it then behaves as a two-dimensional quantum well containing a free electron gas. This quantum well structure can show extraordinarily high electron mobilities and unusual effects such as the Quantum Hall Effect [34] that was first discovered in ultraclean Si-MOSFET structures. Other MBE achievements also include the development of the semiconductor laser and its incorporation into many consumer products.

However in the last 15 years MBE has become an essential tool for metal film deposition as well. The development of high temperature sources for transition metal evaporation and the important finding of the epitaxial relationship between Fe and GaAs were paramount. For the first time, a single crystal magnetic thin film could be incorporated onto a semiconducting substrate. This breakthrough further lead to the discovery of interlayer coupling [35] and subsequently the giant magnetoresistance (GMR) effect [7], which were both initially measured in single crystal Fe/Cr superlattices grown using MBE on a

GaAs substrate. These two discoveries were not only important fundamental physics contributions but they were also responsible for revolutionizing magnetic recording by prompting the development of the spin valve reader [36]. With all the accolades for MBE, magnetron sputtering still remains the preferred thin film deposition technique. Sputtering has found a common place in both research and production facilities while MBE has often found refuge in only universities and basic research laboratories. The machine expense and slow deposition rates have stalled its acceptance as a production tool. However, today MBE production machines do exist, allowing for a simple transfer of thin film technology from the research level to products and furthermore these machines have similar throughput as compared to production sputtering systems. What makes MBE most attractive though, is that it offers many advantages over magnetron sputtering deposition. First of all, MBE growth takes place in an ultrahigh vacuum environment without the need for sputter gas, which may become incorporated into the film during growth. MBE also offers precise layer thickness control where deposition rates can vary over a broad range from angstroms per minute to microns per hour. To achieve a comparable deposition range with sputtering would require low power deposition, which generally results in unstable plasma. In a MBE system the atoms are two orders of magnitude less energetic as compared to sputtered atoms. The lower energy atoms reduce the number of film defects and coupled with the low deposition rate allows for sharp and abruptly clean interfaces. Lastly MBE can grow single crystal material of near perfect quality. For such films, both the magnetic and transport properties can be directly compared with theoretical predictions without the difficulty in modelling effects such as random crystal orientations, grain boundaries and vacancy defects.

MBE also offers opportunities for *in situ* monitoring of film growth using a standard technique of Reflection High Energy Electron Diffraction (RHEED). A schematic diagram of a simple RHEED system is shown in Fig. 2. The RHEED setup consists of a high voltage electron gun which aims a beam of electrons, ranging from 10 to 30 keV of energy, that impinge at a grazing incidence upon the sample surface. The high energy electrons interact with the top most atoms and are diffracted toward a phosphorous-based fluorescent screen producing a pattern that is a reciprocal space representation of the sample surface. Because of the low incident angle (less than 3°) the electrons only interact with the first few monolayers of the sample. Note RHEED is typically found in MBE type deposition systems due to the low vacuum pressure during film growth. Other deposition techniques such as sputtering involve higher vacuum pressures during film growth and will cause scattering of the electron beam before it impinges upon the film surface. And finally, the power of RHEED not only extends to monitoring the arrangement of atoms during deposition but it also provides an indication of the substrate starting surface quality before film growth as well.

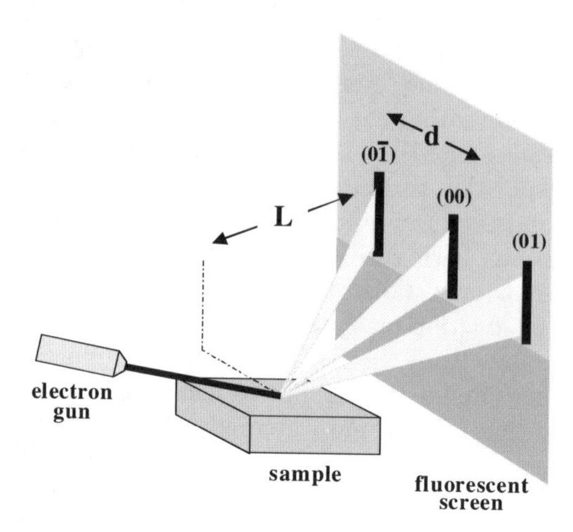

Fig. 2. A schematic view of Reflection High Energy Electron Diffraction (RHEED) setup used to characterize the top most surface atoms in a thin film

The generated RHEED pattern provides a wealth of information regarding sample crystalline quality. First of all, a pattern consisting of spots indicate a single crystal surface while a circular pattern indicates a polycrystalline surface. Furthermore amorphous surfaces are identified by a diffuse RHEED pattern. Flat surfaces are indicated by streaks and the appearance of Kikuchi lines suggests high quality crystalline material. Other attributes such as surface reconstruction and island growth also have distinct features that can be identified in a RHEED pattern. The authors suggest for further reading an excellent text by Prutton [37]. Lastly, analysis of a RHEED pattern can provide a direct measurement of the in-plane lattice parameter by determining the distance between diffraction spots (for example d is defined as the distance between the specular spot (00) and the $(0\bar{1})$ diffraction spot as shown in Fig. 2) and the distance between the sample and the fluorescent screen (L) using the following relation:

$$a_0 = \frac{(h^2 + k^2)^{1/2} \lambda L}{d} \tag{2}$$

where λ is the wavelength of the electrons and h and k are the Miller indices for cubic symmetry. The accuracy in the determination of the lattice parameter a_0 depends upon how precisely the distance between spots can be measured.

2.2 Epitaxial Registry between Co_2MnGe and GaAs (001)

GaAs is a widely used direct band gap semiconductor having a zinc blende or rock salt (NaCl) crystal structure and a lattice parameter of 5.654 Å. For this material, the zinc blende unit cell is made up of 2 interpenetrating face centered cubic (fcc) sublattices of Ga and As respectively. GaAs can be selectively doped with a variety of elements, varying the carrier concentration over a large range of densities reaching well beyond 10^{19} atoms/cm^3, to provide insulating to semi-insulating properties being of either p or n type conduction. Today's high speed logic and dynamic random access memory (DRAM) are mostly GaAs based. Furthermore GaAs is reasonably inexpensive to manufacture in large substrate form. The three principle crystallographic surfaces (001), (011) and (111) can easily be prepared for epitaxial growth using simple chemical etching and/or thermal means. Thus GaAs has many versatile uses in semiconductor devices, but it serves as a platform for thin film growth as well.

Magnetic film growth on GaAs has a long and distinguished history. The first single crystal ferromagnetic thin film grown on a GaAs (001) surface was body centered cubic (bcc) Fe [38]. Fe has a lattice parameter of 2.866 Å and the unit cell has almost a 2 to 1 correspondence to GaAs such that 2 Fe cell edges fit in registration to 1 cell edge of GaAs with a lattice mismatch of only 1.4%. As a general rule of thumb, epitaxial growth has a high occurrence between two different materials when the lattice mismatch is below a few percent. Using RHEED, the epitaxial growth of single crystal Fe on GaAs has been confirmed. Numerous experiments to study the effects of growth conditions, such as crystallographic orientation, substrate temperature and growth rate, on the magnetic properties of Fe have been reported throughout the literature. The first few monolayers of Fe grown on GaAs have been shown to be non-magnetic [39]. Other studies suggest the chemical bonding and atomic arrangement between Fe and zinc blende structures, such as GaAs, play an important role in the origin of the unique uniaxial anisotropy [40]. Understanding the chemical nature of the Fe/GaAs interface is also crucial to the success of high spin injection efficiency needed for Spintronic based devices [9]. Thus Fe growth on GaAs has become a model system of study of magnetic thin film growth on a semiconductor substrate.

The epitaxial growth of single crystal Fe on GaAs has also prompted investigation of other transition metal ferromagnets and their possible epitaxial growth on semiconducting substrates as well. One such example is the single crystal growth of bcc Co on GaAs. In bulk, the low temperature equilibrium structure of Co is hexagonal close packed (hcp) while the high temperature phase is fcc. However a non-equilibrium bcc phase can be stabilized by epitaxial growth on GaAs (110) and (001) surfaces [41,42]. The bcc Co phase is similar to the bcc Fe phase with a slightly smaller lattice parameter of 2.825 Å and also grows in similar registry to the GaAs surface. It has been argued that the presence of an As rich region at the substrate surface stabilizes the bcc Co phase [43]. A cross sectional transmission electron microscopy (TEM) study has shown that the metastable bcc Co layer can grow to about 150 Å before undergoing a transition to an hcp phase. Above this critical thickness, both bcc and hcp Co phases coexist [44].

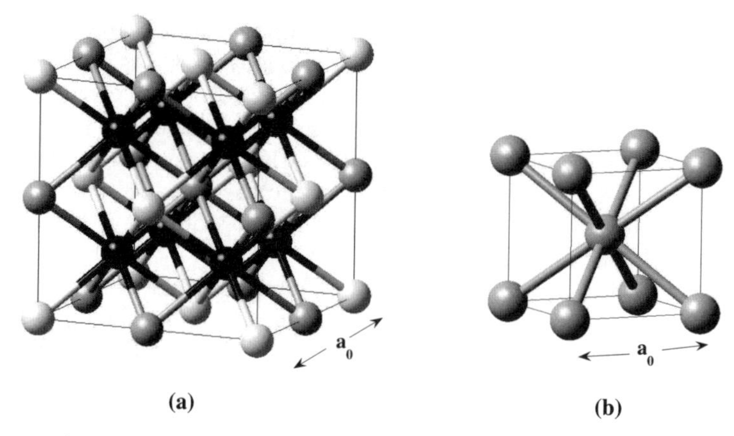

Fig. 3. A comparison of the (**a**) Heusler alloy crystal structure $L2_1$ made up of four interpenetrating fcc sublattices and (**b**) the bcc Fe or bcc Co unit cell. Note the lattice parameter for Co_2MnGe is 5.743 Å and almost twice that of Fe

The bcc Co phase is important because it serves as a guidline for thin film growth of the Co_2MnGe Heusler alloy on GaAs. Co_2MnGe is a full Heusler alloy having a $L2_1$ crystal structure and a lattice parameter of 5.743 Å. This Heusler alloy unit cell is shown in Fig. 3. The $L2_1$ structure is classified as a fcc-like crystal made up of four interpenetrating fcc sublattices. In contrast, the half-Heusler unit cell has a $C1_b$ structure. In the half-Heusler alloy, the probability of anti-site disorder is higher compared to the full-Heusler alloy due to empty sites in the complicated unit cell. Therefore film growth of the full Heusler alloy may be easier and of higher quality due to a smaller percentage of anti-site disorder which can reduce the value of the material's spin polarization.

The epitaxial growth of Co_2MnGe on GaAs (001) proceeds naturally due to a similar growth scheme of both bcc Fe and bcc Co [45]. The lattice mismatch between Co_2MnGe and GaAs is only 1.6%. In this case the edges of the unit cells of the Heusler alloy and the GaAs directly match in an almost one to one correspondence. In Fig. 4 the exact arrangement of the individual Heusler atoms with respect to the GaAs (001) surface is shown where chemical bonds exist between the Mn and Ge atoms with the As atoms.

2.3 Epitaxial Growth of Single Crystal Co_2MnGe and GaAs (001)

The Heusler alloy Co_2MnGe was grown on GaAs (001) substrates using a commercial MBE system with a base pressure of 9×10^{-10} Torr using three separately calibrated Knudsen cell sources. The flux from each source was carefully monitored using a quadrupole mass analyzer and the overall growth rate was approximately 2Å/min. The films were grown on chemically prepared (001) GaAs substrates using a sulphuric acid etch [41]. The substrate

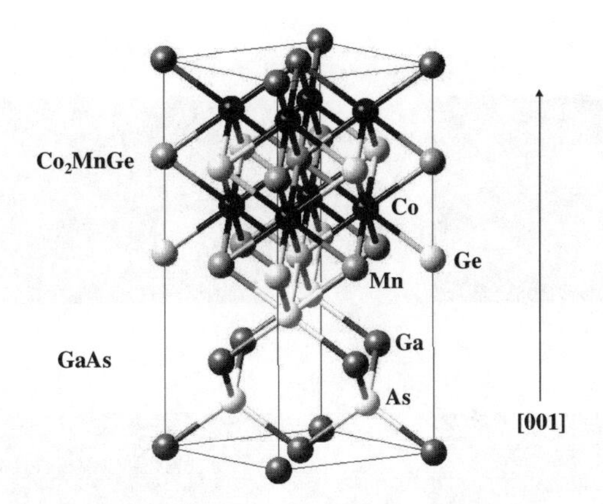

Fig. 4. View along the [001] direction of the Co_2MnGe Heusler alloy crystal and the epitaxial relationship with respect to the GaAs (001) substrate surface

was first annealed to 625°C for 15 min. to allow for surface reconstruction. Afterwards the temperature was lowered to 175°C for growth. The film quality was characterized *in situ* using RHEED and Auger spectroscopy. After film growth, the stoichiometry was also confirmed by *ex situ* x-ray fluorescence measurements.

In Fig. 5 the RHEED patterns obtained at various times in the deposition of a Heusler alloy film are shown. The patterns are presented in two columns denoting the $\langle 110 \rangle$ and the $\langle 100 \rangle$ directions with respect to the GaAs substrate respectively. Figure 5a shows the obtained patterns for the (4×6) reconstructed GaAs (001) surface before film growth. We also note along the $[1\bar{1}0]$ direction, the reconstruction shows six lines as opposed to only four lines along the [110]. Immediately upon deposition of the Heusler alloy film, the faint reconstruction lines in the substrate RHEED pattern disappear and bright spots begin to emerge in registry with the former GaAs streaks. At about 6 Å thickness (roughly 1 unit cell) the underlying substrate RHEED pattern has completely disappeared and the new RHEED pattern is solely from the Heusler film. As an example, in Fig. 5(b), the RHEED pattern obtained at 25 Å thick is shown. For the $\langle 110 \rangle$ direction, along with the three bright spots are faint (2×2) reconstruction lines. Streaking has also begun to be observed indicating a rather flat surface. By carefully measuring the distance between the brighter streaks and applying equation 2, a lattice parameter of 5.77 Å has been calculated which is close to the bulk value.

As mentioned previously, a metastable bcc Co phase has been shown to epitaxially grow on a GaAs (001) surface [41]. For comparison the RHEED patterns obtained for a 25 Å Co_2MnGe film and a 25 Å Co film along both the $\langle 110 \rangle$ and $\langle 100 \rangle$ directions are shown in Fig. 6. The 25 Å layer thickness is chosen to be below the critical bcc Co thickness such that no hcp diffraction spots are visible. The RHEED patterns of the two materials look similar in

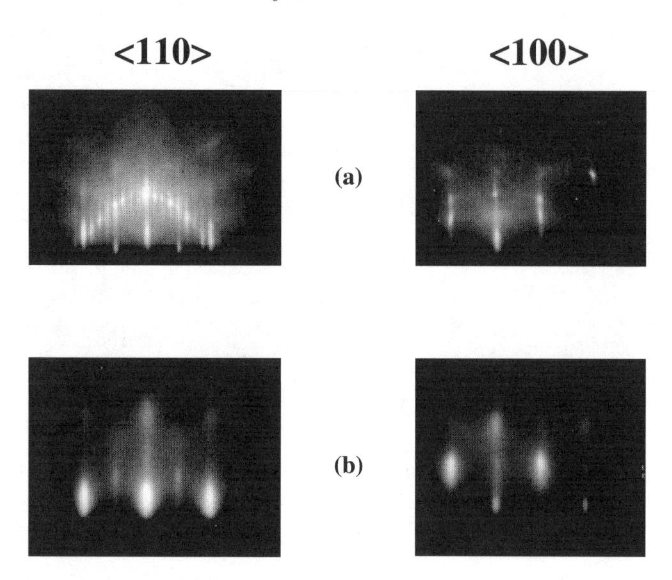

Fig. 5. RHEED patterns obtained during various stages of growth of a Co_2MnGe Heusler alloy film. Note that the two directions, the ⟨110⟩ and the ⟨100⟩ are with respect to the GaAs substrate. Shown are the (**a**) reconstructed (4 × 6) GaAs substrate starting surface and (**b**) a 25 Å thick film RHEED patterns

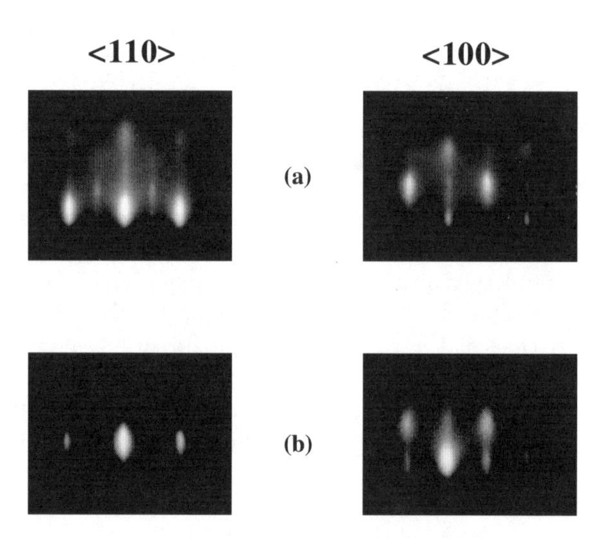

Fig. 6. A comparison of RHEED patterns of (**a**) 25 Å thick $L2_1$ Co_2MnGe Heusler alloy and (**b**) 25 Å thick bcc Co on GaAs (001). Note that the two directions, the ⟨110⟩ and the ⟨100⟩ are with respect to the GaAs substrate

terms of distance between diffraction spots except for the additional (2×2) reconstruction lines found only in the Co_2MnGe film shown in Fig. 6a. The pure Co film may also have a rougher surface due to the lack of streaking RHEED lines. However these RHEED patterns clearly demonstrate an almost identical growth nature of both the Co_2MnGe Heusler alloy and bcc Co on GaAs (001). It is important to remember that the metastable bcc Co phase has a critical thickness of 150 Å such that Co resorts to a hcp structure [44] while in contrast the $L2_1$ Co_2MnGe can be grown to very large film thickness [45].

Finally Co_2MnGe can also be grown on other zinc blende surfaces [46]. For example ZnSe is a well known zinc blende semiconductor that epitaxially grows on GaAs where the lattice mismatch is less than 0.4%. This material can be grown quite thick and can provide an extremely smooth starting surface with a small amount of roughness typically less than 3 Å. ZnSe is also highly insulating which is desirably for use as a buffer layer in a CIP GMR device because of a low amount of current shunting. An Fe seed layer only 3 Å thick provides the proper template to ensure well-ordered and epitaxial growth of the Co_2MnGe Heusler alloy on the ZnSe surface [46].

2.4 Magnetic Properties of Single Crystal Co_2MnGe on GaAs

In Fig. 7 the magnetization curves for a 250 Å thick film measured at 298K is shown. The magnetic field is applied in the plane of the film along the principal crystallographic axes. The three hysteresis curves have been plotted over ± 100 Oe range for detail clarity. The $\langle 100 \rangle$ direction is rather hard as seen from the rounded hysteresis curve shown in Fig. 7a. A magnetic field slightly greater than 500 Oe is required to saturate the magnetization when the field is applied along this direction.

Interestingly, the two $\langle 110 \rangle$ directions, while both relatively easy, are inequivalent. In Fig. 7b, the easier of the two axes, labeled [110] gives the expected square hysteresis loop. When the magnetic field is applied perpendicular to the [110] axis, along the $[1\bar{1}0]$, the resultant hysteresis curve is no longer square but has additional features as shown in Fig. 7c. At large fields, the magnetization M lies parallel to H. As H is reduced below 20 Oe, the favored direction of M lies closer to the easier $\langle 110 \rangle$ axis so that near zero fields, M lies almost perpendicular to the applied field direction H. Upon field reversal, M returns to the field direction in a similar manner together with hysteresis. Note that a similar hysteresis loop shape to the one shown in Fig. 7c has also been observed for both thin films of Fe on GaAs [38] and Co on GaAs [41]. This clearly indicates that the resultant hysteresis loop arises from epitaxial growth on zinc-blende substrates and *not* from antiferromagnetically coupled magnetic sublattices in the Heusler alloy unit cell. This multistep magnetization switching has been extensively studied and is due to combined cubic magnetocrystalline and uniaxial anisotropies [47].

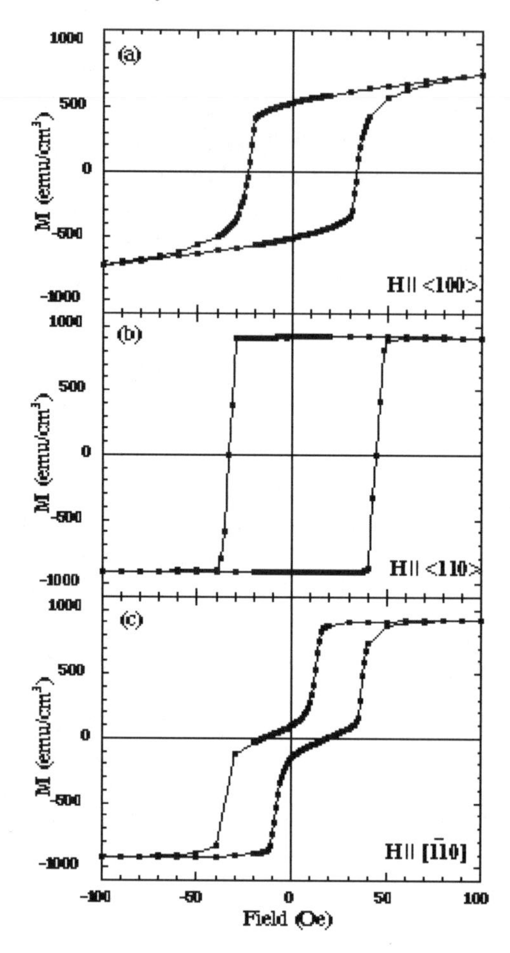

Fig. 7. Hysteresis curves measured at room temperature for a 25 nm thick Co_2MnGe film with the magnetic field applied in the plane of the film along the **(a)** $\langle 100 \rangle$ **(b)** $\langle 110 \rangle$ and **(c)** $\langle 1\bar{1}0 \rangle$ directions

The Heusler alloy film was also examined by in-plane angle-dependent ferromagnetic resonance (FMR) measurements at 35 GHz. The angular dependence of the resonance field is shown in Fig. 8 where ϕ represents the angle between the applied magnetic field and the hard axes. In this configuration ϕ is equal to 0 along the [100] direction and the plotted resonance fields repeat outside the range $\pm 90^o$. The fourfold anisotropy is clearly seen from this figure where the resonance fields complete two periods within 180°. The easy $\langle 110 \rangle$ directions are observed at $\phi = \pm 45^o$ where the resonance fields are at their minimum values. The difference in the magnitude of these two resonance fields, shows the inequivalence in the $\langle 110 \rangle$ directions that was

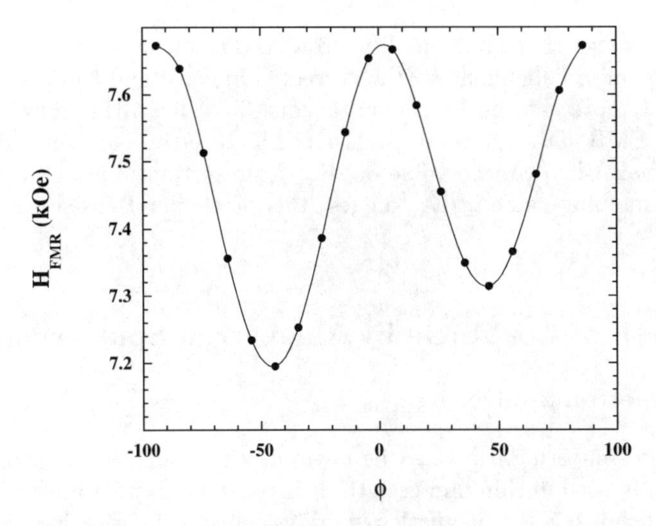

Fig. 8. Angular dependence of the FMR resonance field vs. ϕ where ϕ is the angle between the applied magnetic field and the hard axis direction [100]. The solid line is a fit using the parameters of Table 1

Table 1. Magnetization and FMR parameters for a 250 Å-thick film of Co_2MnGe on GaAs (001)

$4\pi M = 12.5 \pm 0.3$kG
$K_1/M = -0.133 \pm 0.01$ kOe
$K_u/M = 0.035 \pm 0.01$ kOe
$4\pi M + 2K_\perp/M = 10.1 \pm 1.0$ kOe
$g = 2.2 \pm 0.1$

observed in the hysteresis curves shown in Figs. 7b and 7c and indicates the existence of an in-plane uniaxial anisotropy along a $\langle 110 \rangle$ axis.

The FMR parameters determined for the 250 Å thick Co_2MnGe film from the data in Fig. 8 are given in Table 1. These parameters were determined from the free energy density given by

$$E = K_1(\alpha_1^2\alpha_2^2 + \alpha_2^2\alpha_3^2 + \alpha_3^2\alpha_1^2) + K_u \cos^2(\phi - \pi/4) + K_\perp \cos^2\theta - H \cdot M + 2\pi M^2 \cos^2\theta \tag{3}$$

where K_1 is the usual cubic anisotropy, α_i are the direction cosines for cubic axes, K_u is the thickness dependent (surface) uniaxial in-plane anisotropy that distinguishes the two $\langle 110 \rangle$ directions in the (001) surface, and $K_\perp$ is an anisotropy perpendicular to the film. The polar angles are θ and ϕ with the z axis along the surface normal. The K_u term is characteristic of metallic magnetic films grown on a (001) zinc blende substrate [40]. The 35 GHz FMR linewidth was measured to be only about 110 Oe for H along the easy directions suggesting good quality growth. The deduced value of $4\pi M + 2K_\perp/M$ is quite dependent on the correct value of g and those values given in Table 1 represent the best simultaneous fit of 9.5 GHz and the 35 GHz extremal FMR data. Note the relatively large error bars for the $4\pi M + 2K_\perp/M$ make the deduced value of $2K_\perp/M$ uncertain to a factor of 2. Even so, the remaining value of $K_\perp$ suggest this particular Heusler alloy is quite sensitive to strain.

3 Growth of Co_2MnGe by Magnetron Sputtering

3.1 Magnetron Sputtering

Magnetron sputtering is a versatile physical vapor deposition technique that is commonly used in thin film growth. It is considered the primary deposition method of choice in the magnetic recording industry. Fabrication of recording heads and media rely heavily on sputtered films. Sputtering provides the ability to grow films with a high growth rate, typically a few Å/s, having good thickness control, uniformity and reproducibility. Unlike MBE, sputtering is less sophisticated, requiring only a simple vacuum chamber and appropriate target material cathodes. The sputter deposition process occurs when ionized gas particles impact on a target surface and transfer energy to the target lattice. This results in some of the lattice atoms acquiring a sufficient amount of energy to be ejected or sputtered from the target surface. These atoms are collected on a prepared substrate and films of varying thickness can be easily obtained. In addition, the inclusion of permanent magnets in the cathode housing improves the sputtering yield by increasing the ionization efficiency and confining the plasma close to the target surface to allow for sputtering to occur at low gas pressure. The amount of material which is sputtered from the target cathode onto the substrate depends primarily on the sputter gas pressure, target voltage and current, the geometry of the deposition, the mean free path of the gas in the vacuum chamber and the sticking coefficients between the target atoms and the substrate [48].

The versatility of sputtering lies in the fact that almost any material can be sputtered as long as it is in solid form and does not degrade the vacuum system due to a high vapor pressure. More importantly, both metallic and insulating materials can be sputtered. A DC voltage source is typically used for metallic targets while a rf power source is used for insulating targets to circumvent the large voltage drop within the sputter target. Sputter gas incorporation in films can be minimized using a low pressure growth mode reducing defects to improve film qualities such as resistance and strain effects. These are clearly obvious differences between sputter deposition and MBE deposition. Nonetheless high quality sputtered thin films can be grown with comparable electrical and magnetic properties to films grown by MBE deposition.

3.2 Co_2MnGe Prepared by Magnetron Sputtering

There are two basic approaches to fabricating thin films of Co_2MnGe using magnetron sputtering. The first approach involves a single sputter target with the appropriate volume concentrations of Co, Mn and Ge respectively. However using this method, one may encounter difficulty in obtaining films with the correct stoichiometry due to different sputter yields of the atomic constituents. A better approach is to start with individual sputter targets of Co, Mn and Ge such that the growth rates for all three materials can be individually adjusted by varying the target power and thus films with precise stoichiometry can be obtained. The work presented here utilizes this methodology to control film composition to match the Co_2MnGe stoichiometry alloy without any additional atoms to promote other impurity phases.

To achieve such $L2_1$ crystal structure growth, the substrate temperature during deposition becomes an important parameter of control. The substrate temperature increases the atom's mobility thereby promoting crystalline order and/or film texture which is necessary for growth of the Heusler alloys. While the MBE films utilized the epitaxial growth onto a lattice matched GaAs surface, optimal growth conditions required a substrate temperature of 175°C during deposition. For the deposition of the sputtered films, amorphous starting substrates can be used. It is expected that the substrate temperature during growth is much higher compared to the MBE grown films. In this case epitaxy helps reduce the amount of thermal energy added to the system to obtain highly ordered Heusler alloys.

All the Heusler alloy films described hereafter were fabricated using the following deposition conditions. The Heusler alloys were grown on thermally oxidized Si (001) substrates in a commercial sputter system with a base pressure of 4×10^{-10} Torr. Films were fabricated using magnetron sputtering from elemental targets with a 2mTorr Ar sputter gas pressure and a substrate temperature of 400°C. The power to each sputter cathode was adjusted to a predetermined value to ensure the proper stoichiometry of the various Heusler alloy films. The 400°C substrate temperature was optimized and necessary to ensure an ordered alloy crystal structure. Heusler alloy films grown at room temperature using the same deposition conditions were amorphous.

The samples were characterized using x-ray fluorescence (EDAX) and x-ray diffraction to check film stoichiometry and crystal structure. In Fig. 9 an asymmetric grazing incidence x-ray diffraction scan [49] is shown for a 250 Å thick Co_2MnGe film. A polycrystalline diffraction pattern is observed for this Heusler alloy film where the (220) peak as expected has the highest intensity. All other peaks have been identified as part of the $L2_1$ crystal structure and no impurity phases are present. As a comparison, the MBE grown Heusler film would have only a single (200) peak around $2\theta = 64.9°$ due to the epitaxial relationship with the underlying GaAs substrate. Thus the Co_2MnGe Heusler alloy can be fabricated as a single crystal and polycrystalline film by both MBE deposition and magnetron sputtering deposition respectively.

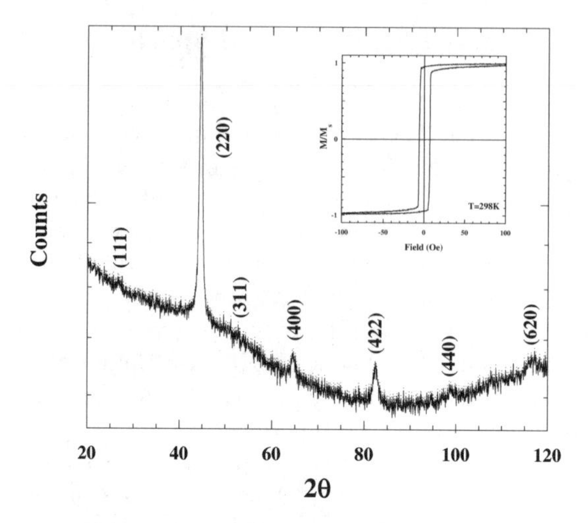

Fig. 9. Asymmetric grazing incidence x-ray diffraction scan of a 250 Å thick polycrystalline Co_2MnGe film on SiO_2/Si substrate. Insert: The corresponding room temperature hysteresis loop

Lastly a comparison between the magnetic and transport properties of MBE and sputtered deposited films can be made. The insert in Fig. 9 shows a room temperature hysteresis loop for the polycrystalline film. This magnetization curve looks similar to the curve shown in Fig. 7b. for the MBE grown film. The unique four fold symmetry present in the single crystal film (Fig. 7) is obviously absent from the sputtered film. This illustrates the effect of epitaxial growth on a zinc blende surface as compared to polycrystalline growth onto an amorphous surface and how the magnetic properties of the same material can be modified by an interfacial anisotropy.

In Fig. 10 the resistivity for both the single crystal and polycrystalline Co_2MnGe films are shown as a function of temperature. The absolute resistivity values have been calculated from the sheet resistance using a standard 4-point probe Van der Pauw technique, which minimizes the effects of the sample geometry [50], and the measured film thickness from x-ray reflectivity. At room temperature the resistivity of both films match at a relatively high value of $115\mu\Omega$-cm. As a comparison, transition metal ferromagnets such as Fe and Co have a room temperature resistivity around $10\mu\Omega$-cm. However, a higher resistivity value is expected for a ternary alloy containing a component which is a semiconductor in pure form.

The resistivity for a metallic thin film will decrease linearly as the temperature is reduced. At low temperatures the effect of phonon scattering is suppressed and scattering of electrons by impurities and defects becomes dominant. In both films the resistivity flattens out below 50K and the extrapolated 0K values differ by $6\mu\Omega$-cm where the sputtered film has the higher residual resistivity. In the polycrystalline films defects arise primarily

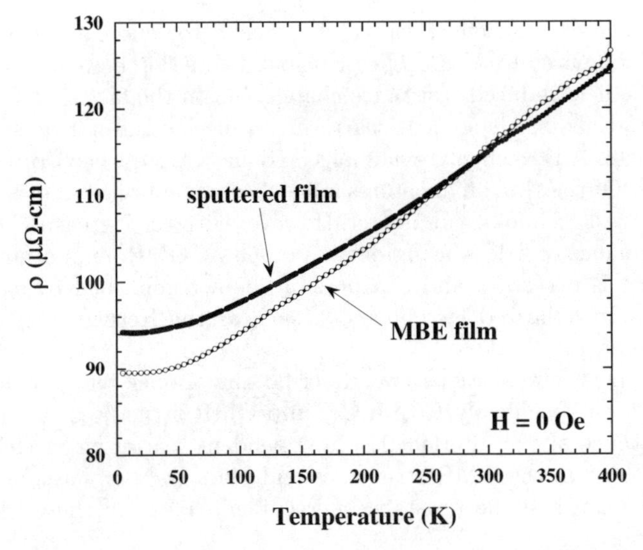

Fig. 10. Resistivity vs. temperature for both the single crystal MBE grown and the polycrystalline sputtered Co_2MnGe films. Both films were 250 Å thick

from grain boundaries and Ar gas impurities incorporated during the sputter process as compared to single crystal films. The difference in slope of the linear part of the resistivity for both films must be addressed. The slope in the resistivity as a function of temperature is an intrinsic property of the material. The additional defects provided by sputtering may add to the slope difference. Other effects such as film stress and off stoichiometry can also contribute to these differences.

4 GMR in Co_2MnGe Layered Structures

4.1 Introduction to Magnetoresistance

Magnetoresistance is a well-known phenomenon observed as a variation in the resistance upon the application of a magnetic field. In particular, non-magnetic transition metals exhibit an ordinary magnetoresistance (OMR) due to asymmetric energy bands or non-spherical Fermi surfaces [51]. Moreover, ferromagnetic materials can also exhibit an anisotropic magnetoresistance (AMR) effect that is related to the angle between the magnetization and current directions. Still further, materials can be engineered to exhibit additional magnetoresistive effects such as the giant magnetoresistance (GMR), first discovered in MBE grown thin films of alternating layers of Fe and Cr grown on GaAs [7]. GMR was shown to be the result of the change in the relative alignment of the magnetizations in adjacent ferromagnetic layers in a multilayer stack. The effect was coined "giant" to denote a larger resistance change as compared to AMR. The principal distinction however, is not in the value of the field induced resistance change, but in the fact that GMR stems from interfacial spin dependent scattering, a mechanism not present in bulk ferromagnets. An excellent review of GMR has already been provided elsewhere [52]. Of the three magnetoresistive effects mentioned above, GMR has had the largest technological impact. However with the increasing demand on the performance of field sensing devices based on GMR, enhancement of the GMR effect is necessary and can possibly be accomplished by substitution of highly spin polarized ferromagnets, such as the Heusler alloys, into the multilayer stacks.

In the next few sections, we describe the proper way to incorporate Heusler alloys, specifically Co_2MnGe, into GMR structures. We emphasize the importance of the interfacial spin dependent scattering mechanism and band matching in the GMR structures and place less emphasis on the term "giant" in regard to the resistance percentage change in these layered systems.

4.2 GMR and the Importance of Band Matching

GMR has been observed in numerous multilayer systems containing transition metals [53]. To date, there has been no conclusive demonstration of room temperature GMR in layered structures containing Heusler alloys [20,54,55]. It has been suggested that interfacial alloying, off stoichiometry [55] and atomic disorder leading to a lower spin polarization [25], are the reasons for poor GMR in NiMnSb/Cu/NiMnSb structures. However, PCAR measurements have indicated the spin polarization of NiMnSb is around 60% [29] which is higher than Fe or Co. This value of polarization is significantly smaller than one may anticipate from theoretical band structure calculations but still larger than spin polarization measured for 3d ferromagnets. Other Heusler alloy containing multilayer systems have also shown no GMR effect [56]. In their work, high quality multilayers of Co_2MnX where X = Si, Ge or Sn intercalated with either Au or V have been fabricated without any substantial magneto-transport. While it is reasonable to consider that the Heusler alloys may somehow intrinsically prevent GMR, careful examination of the published results commonly show the use of transition metals (Au, Cu, Mo and V) as spacer layer materials. In this section, we show the importance of the choice of the spacer layer to achieve spin dependent transport across the interface and thus the inappropriate selection of the non-magnetic spacer layer is the fundamental limitation.

A GMR multilayer consists of two parts; a ferromagnetic (FM) layer that supplies spin polarized conduction electrons and a non-magnetic (NM) spacer layer that acts to mediate exchange and transport between the FM layers. According to the Mott theory, the conductance of any FM material is given by the conductance for spin up ($\rho^{\uparrow}$) and spin down ($\rho_{\downarrow}$) channels [57]. The difference between $\rho^{\uparrow}$ and $\rho_{\downarrow}$ also known as conduction electron spin asymmetry, reflects the spin dependent scattering characteristic of homogenous magnetic materials. In the case of artificially structured GMR multilayers, the properties of the non-magnetic spacer become equally important. The spacer layer plays a critical role as a spin discriminator by creating strongly spin dependent scattering potentials at the FM/NM interfaces. This is accomplished by the NM spacer selectively matching its electronic energy bands with one of the spin-split bands in the ferromagnetic layer [58]. If other scattering contributions such as dislocations and grain boundaries effects are relatively small, band matching becomes the decisive factor for magneto-transport and controls the GMR characteristics. Unfortunately, while most of the focus centers on incorporating half-metallic FM materials into GMR structures, the choice of the spacer layer material is often overlooked.

To illustrate the importance of band matching, consider the following simplified trilayer structure of Co/Cu/Co. This GMR trilayer is made up of two FM layers of fcc Co sandwiching a NM spacer layer of fcc Cu. In zero magnetic field and appropriately chosen layer thickness, anti-parallel alignment of the two Co layer magnetizations results from strong exchange coupling between the FM layers and a high resistance state is measured. By applying a magnetic field the magnetizations of both Co layers can be aligned parallel to each other and a low resistance is measured. GMR is observed in this trilayer system because the electronic structure of Cu has a similar electronic character to the spin up bands of Co. A comparison of the electronic structure of both materials is shown in Fig. 11 where the energy bands along the high symmetry lines in the Brillouin zone are shown. The Fermi energy is defined at 0 eV. From Fig. 11 around the Fermi energy, it is clear that the energy bands of Cu and the spins up energy bands of Co have a similar slope and energy for one spin channel. These band matching conditions must be met, or electrons will be scattering at the interface without the desirable spin asymmetry increase as shown by Tsymbal [59] and references therein.

The forementioned GMR effect in Co/Cu multilayers is unique magneto-transport for spin up conduction electrons due to the choice of the Cu spacer layer. Conversely Co/Re multilayers also exhibit GMR [60], however in this case, the Re band structure matches the spin down bands of Co. From this example, it is clear by choosing an appropriate spacer one can select the spin states (either up or down) to be transmitted across the FM/NM interface. This reasoning also works for other GMR multilayer systems such as Fe/Cr [7] with spin down transport and Fe/Pd [61] with spin up transport. As for both Fe and Co, spin up transport requires a NM spacer possesing a higher number of valence d electrons (ex. Pd and Cu) and for spin down transport requires a NM spacer possesing a lower number of d valence electrons (ex. Cr and Re). This simple rule for spacer layer material selection can be applied to all transition metal ferromagnets and is graphically summarized in Fig. 12. As

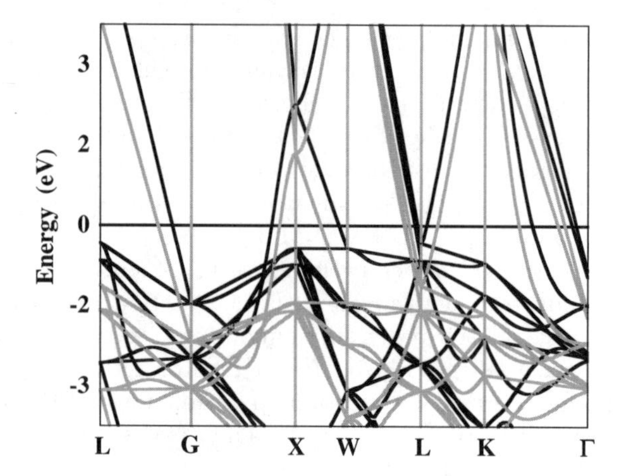

Fig. 11. The energy bands for fcc Co spin up (*black*) and the energy bands for fcc Cu (*grey*) in the first Brillouin zone are shown. The Fermi energy is at 0 eV. Good band matching (similar energy and similar slope) is evident around the Fermi energy

a first best guess, the spacer layer that falls two columns to the right of the ferromagnet in the periodic table promotes majority spin transport and the spacer layer that fall two columns to the left of the ferromagnet promotes minority spin transport [62].

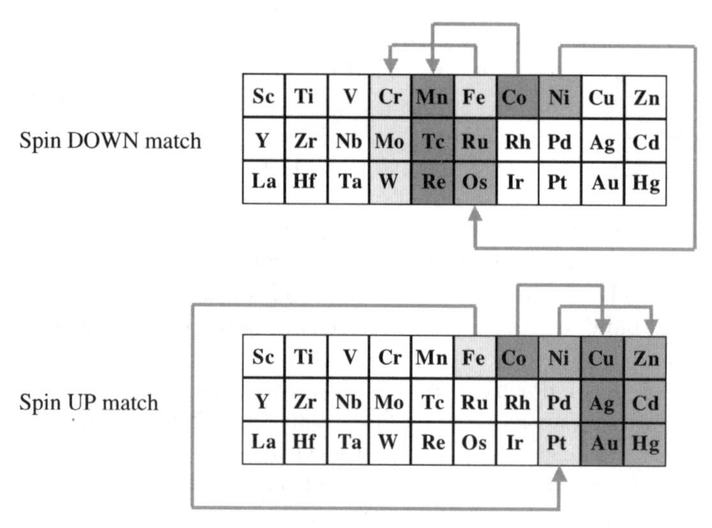

Fig. 12. A graphical representation of the transition metal spacer layer selection to promote spin up and spin down transport in Fe, Co and Ni

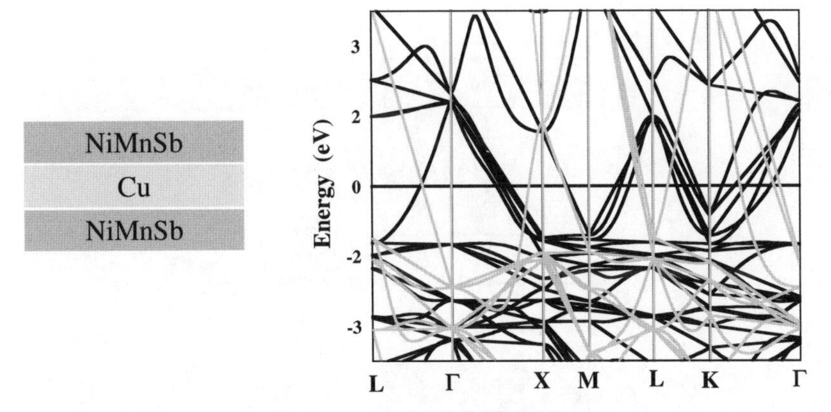

Fig. 13. The energy bands for the spin up NiMnSb half-Heusler alloy (*black*) and the energy bands of Cu (*grey*) are shown. The Fermi energy is at 0 eV. Band mismatch is clearly evident around the Fermi energy

Lastly we turn our attention to the failure to observe of GMR in systems containing Heusler alloys. From the previous discussions the selection of the NM spacer layer must have similar energy bands compared to one of the spin split FM layer energy bands. It is also well-known that the Heusler alloys have complicated band structures compared to the transition metals [19,63,64] and that the probability of band matching between these materials is unlikely. As an example we illustrate the large band mismatch between the half-Heusler alloy NiMnSb and fcc Cu shown in Fig. 13 where the electronic structure for the spin up bands of NiMnSb and Cu are plotted. We have chosen NiMnSb and Cu in this example because of several publications exploring the combination of these materials as a viable GMR stack where no GMR has been convincingly demonstrated [20, 55, 65]. Around the Fermi energy complete band mismatch is evident where the bands of the two materials of the same energy intersect with dissimilar slopes. Most importantly, this mismatch prohibits preferable transmission of the majority spin electrons across the FM/NM interface. The dissimilar energy bands also suggest predominantly different carrier types for the two materials as well. Since Cu is a well known n-type conductor [66], NiMnSb is therefore p-type as follows from the band structure plot in Fig. 13. Recent Hall effect measurements of the carrier type have confirmed p-type conduction in NiMnSb [67] and the results are in agreement with the calculated band structure.

To properly match a spacer layer for the Heusler alloys requires a material that not only has a similar electronic structure but also a similar crystal structure, lattice parameter, resistivity and Hall mobility. The spacer material must also be a ternary alloy containing a transition metal component and a semiconductor component. The transition metal atoms will be chosen to promote either spin up or spin down transport respectively. For completeness,

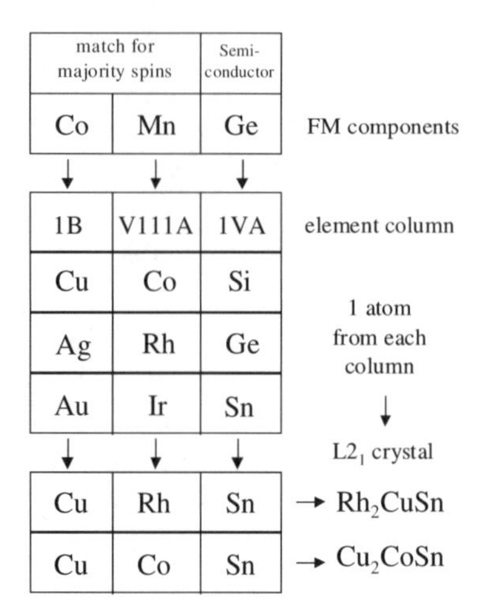

Fig. 14. Choice of appropriate spacer layer selection for the spin up channel of the Co_2MnGe Heusler alloy. The two resultant $L2_1$ spacer layers are Rh_2CuSn and Cu_2CoSn

the semiconductor atom will be chosen from the same periodic table column as the semiconductor atom in the Heusler alloy. The addition of the semiconductor atom in the spacer material alloy is most important since it drastically modifies the electronic band structure of the transition metal components [19].

4.3 "Band Engineering" of an All-Heusler GMR Stack

Using the rules outlined in the previous section we suggest an appropriate spacer layer for the Co_2MnGe Heusler alloy. In Fig. 14, a block diagram explaining the selection procedure is graphically shown. Electronic band structure calculations have identified the minority spin bands possessing a gap in electron states at the Fermi level in Co_2MnGe [21]. For magneto-transport to occur, the NM spacer layer *must* match the majority spin bands in this Heusler alloy. Therefore we "band engineer" a combination of atoms for the spacer layer material to match the majority spin bands for Co and Mn as well as a group IV semiconductor to match Ge. For a Co spin up match, Cu, Ag or Au satisfy our criterion. Similarly Co, Rh and Ir match the spin up bands for Mn. We argue Mn inclusion is vital since these atoms carry a substantial magnetic moment in all Heusler alloys [68]. Lastly Si, Ge or Sn are group IV semiconductors needed to complete the spacer alloy structure. Our appropriate non-magnetic spacer layer will be a ternary alloy made up of one atom from each of the 3 columns in Fig. 14, have a $L2_1$ crystal structure and similar lattice spacing compared to Co_2MnGe. There are numerous alloys that fit these criteria and we suggest Rh_2CuSn and Cu_2CoSn as viable candidates. Recall the strict definition of a Heusler alloy is a ferromagnetic ternary alloy that contains Mn. We loosely deviate from this definition and call both Rh_2CuSn and Cu_2CoSn, Heusler alloys as well.

We propose that the multilayer structure of (Co_2MnGe/Rh_2CuSn) will satisfy the selective band matching and correspondingly the spin-dependent interfacial scattering requirement. Note this type of GMR multilayer is completely different from previously studied transition metal systems that contain Heusler alloys [20, 54, 56]. Obviously, our selection process of the NM spacer is simplistic and without rigorous confirmation. We compare the calculated band structures from first principles for Co_2MnGe and Rh_2CuSn in a similar way as the NiMnSb/Cu case shown in Fig. 13. In Fig. 15(**a**) a schematic representation of our suggested all-Heusler alloy (Co_2MnGe/Rh_2CuSn) multilayer is shown. Correspondingly in Fig. 15(b) the calculated electronic band structure for the majority spin up channel of Co_2MnGe and the electronic band structure of Rh_2CuSn are shown with the Fermi levels at 0 eV. Our band structure calculations were done using a full-potential linear muffin-tin orbital methodology that allows an accurate description of the electronic band-structure within the local spin density approximation [69]. Band matching is clearly evident as electron states at the Fermi energy have similar energy and slopes.

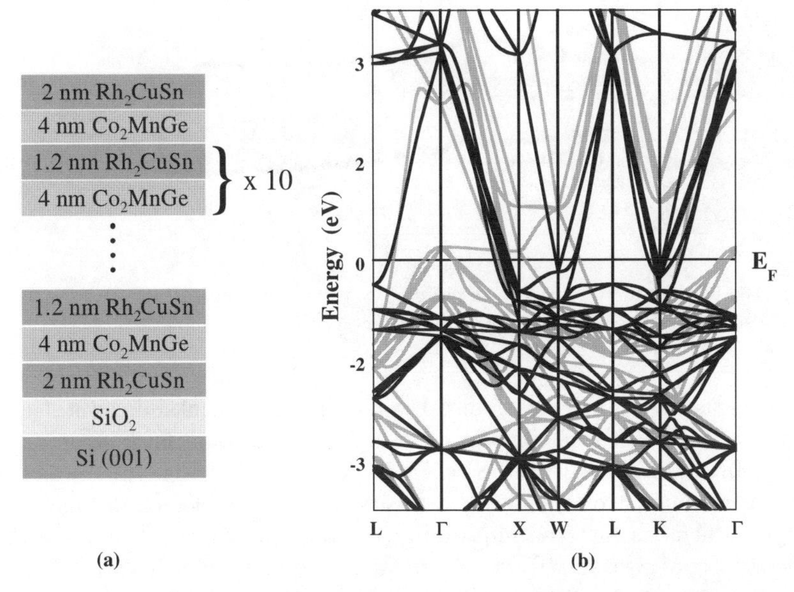

Fig. 15. (**a**) Multilayer deposition design for all-Heusler GMR structure $(Co_2MnGe/Rh_2CuSn)_{10}$ and (**b**)a comparison of the band structure for spin up Co_2MnGe (black) and Rh_2CuSn (grey). The Fermi energy is at 0 eV

4.4 GMR in (Co_2MnGe/Rh_2CuSn) Multilayers

Our electronic band structure calculations have identified two non-magnetic full Heusler alloys, Rh_2CuSn and Cu_2CoSn which have been reported in bulk form only, as potential spacer layer candidates that provide good band matching to Co_2MnGe for spin polarized transport. The bulk lattice parameters for Rh_2CuSn and Cu_2CoSn are 6.146 Å and 5.982 Å respectively. These Heusler alloys mismatch Co_2MnGe by less than 10%. We have fabricated these potential spacer layer candidate materials into thin film form using the same sputter deposition conditions and characterized them using the same techniques described in Sect. 3.2. In Fig. 16 the asymmetric grazing incidence x-ray diffraction scans for the 3 Heusler alloys is shown. All the x-ray peaks have been identified to belong to the $L2_1$ crystal phase for each Heusler alloy.

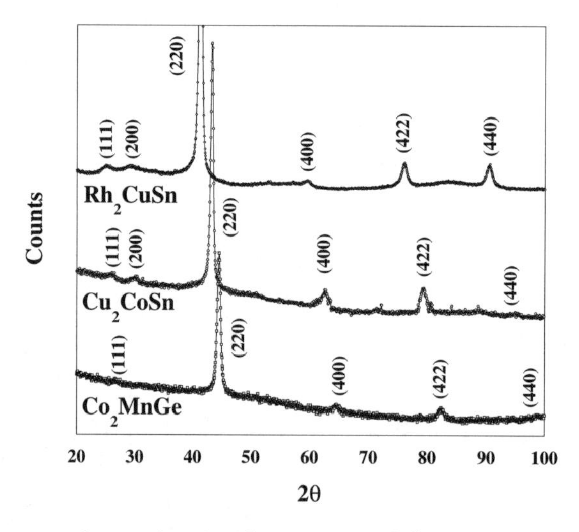

Fig. 16. Asymmetric grazing incidence x-ray diffraction scan for Co_2MnGe, Cu_2CoSn and Rh_2CuSn. The curves have been offset for clarity

To measure a GMR effect in a layered structure containing the Heusler alloy Co_2MnGe as the ferromagnetic component, the multilayer structure (40 Å Co_2MnGe/12 Å $Rh_2CuSn)_{10}$ schematically shown in Fig. 15(a) was fabricated using magnetron sputtering deposition. We have selected this multilayer design to demonstrate the importance of matching the electronic structure between Co_2MnGe and Rh_2CuSn in a high quality multilayer stack. The thickness of each layer was chosen for the best possible film growth quality and not for GMR optimization.

For magnetoresistance measurements we have employed a CIP geometry as opposed to a CPP geometry to remove all post processing constraints. Unfortunately, the CIP geometry limits the observation of a large GMR effect due to the incorporation of highly spin polarized FM materials as suggested from a simple two-current model. However the fabrication of CPP stacks requires extensive effort and making such devices without any substantiated proof of proper selection of materials is unwise. Therefore our main focus is to demonstrate a room temperature CIP GMR effect in new and untested materials first, before building CPP devices.

The multilayer sample quality was characterized using both low angle reflectivity and asymmetric glancing incidence diffraction and the results are shown in Fig. 17. In low angle reflectivity, the length scales studied are greater than the lattice spacing of the constituent layers, such that x-ray scattering arises from chemical modulations in the layered structure. The low angle scan in Fig. 17(a) exhibits a large number of peaks with decreasing intensity as the detection angle, 2θ, increases. The spacing between each peak is related to the overall film thickness. Also present are peaks with an additional increase in intensity around $2\theta = 2.0$, 3.9 and 5.8 degrees. These peaks arise from the bi-layer thickness modulation, designated as Λ, in the layered structure.

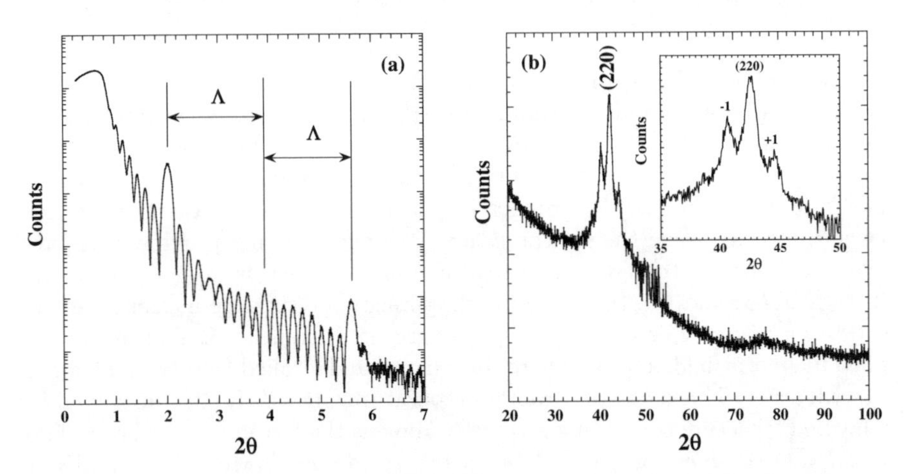

Fig. 17. (**a**)Low angle x-ray reflectivity and (**b**) asymmetric glancing incidence scan for (40 Å Co_2MnGe/ 12Å $Rh_2CuSn)_{10}$. The insert in (**b**) shows the high angle superlattice peaks

The asymmetric glancing incidence scan for the all-Heusler alloy multilayer is shown in Fig. 17(b). The x-ray scan shows a major peak which corresponds to the (220) crystallographic planes. Notice superlattice peaks from the multilayer structure are also observed around the (220) peak. A more detailed view is provided in the insert of Fig. 17(b). The appearance of superlattice peaks indicates a high quality multilayer sample with abrupt interfaces.

The room temperature magneto-transport properties in the all-Heusler alloy multilayer were measured using a Vibrating Sample Magnetometer (VSM)

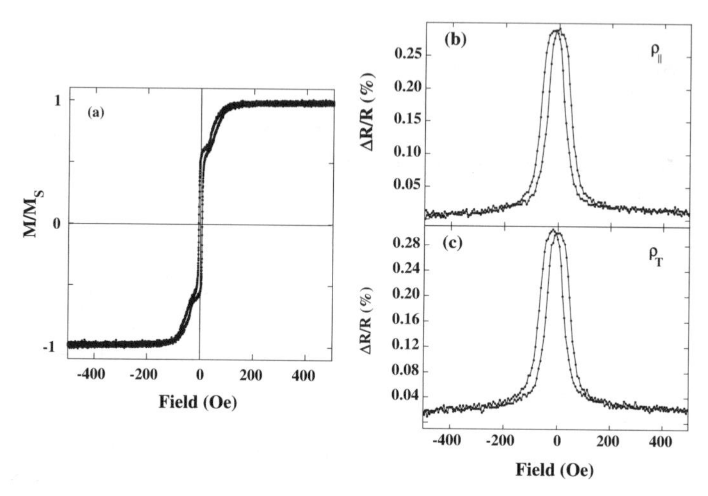

Fig. 18. Room temperature (**a**) magnetization, (**b**) $\rho_{||}$ and (**c**) ρ_T versus field for the (40 Å Co_2MnGe/ 12 Å Rh_2CuSn/$)_{10}$ multilayer

and a 4-point probe in line resistivity setup. The magnetic field is applied in the sample plane and the results are shown in Fig. 18. A weak interlayer exchange coupling between Co_2MnGe layers is observed in the hysteresis loop data shown in Fig. 18(a). By contrast, transition metal multilayers such as Fe/Cr and Co/Cu exhibit stronger interlayer exchange for a similar spacer layer thickness. Nevertheless the weak exchange coupling provides antiparallel alignment of the adjacent Co_2MnGe layers which is needed for GMR. In Figs. 18(b) and (c), the corresponding magnetoresistance measurement is shown for two sample geometries, $\rho_{||}$, where the current is parallel to the applied magnetic field, and ρ_T where the current and applied field lie orthogonal to each other. This is the first room temperature CIP GMR effect observed in a layered structure containing a Heusler alloy as the ferromagnetic layer. The GMR value is small, only 0.26% but it is certainly not AMR. Unlike AMR, a multilayer film that exhibits GMR must satisfy the condition that $\rho_{||}$ and ρ_T be equivalent. The GMR curves shown in Fig. 18(b) and (c) clearly indicate this condition is satisfied.

We have also fabricated spin valve structures containing Co_2MnGe and Rh_2CuSn. The spin valve allows for abrupt switching of one FM layer magnetization by pinning one of the FM layers with an antiferromagnet [36]. The spin valve design facilitates controlled antiparallel alignment of the FM layers in small fields. A schematic diagram of our spin valve device is shown in Fig. 19(a) where FeMn is the antiferromagnet to pin the top Co_2MnGe layer. In Fig. 19(b) and (c) we plot the mangetization and GMR effect for the $\rho_{||}$ versus field. In our system, $\rho_{||}$ is defined along the same direction as the unidirectional (UD) anisotropy axis arising from the exchange bias effect between FeMn and Co_2MnGe layers. In the $\rho_{||}$ geometry, 180° alignment of

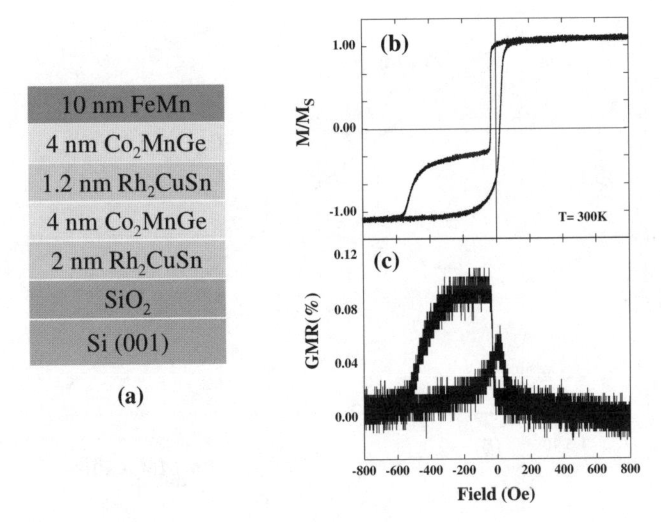

Fig. 19. (**a**) Schematic diagram of an all-Heusler spin valve structure containing Co_2MnGe and Rh_2CuSn layers. The corresponding room temperature magnetoresistance:(**b**) $\rho_{\|}$ and (**c**) ρ_T versus field

the magnetization of the FM layers has been achieved and a GMR effect of 0.10% is measured. While the GMR is smaller than in the multilayer, the resistance curve still shows a negative GMR effect and abrupt switching of the magnetization states.

It should be mentioned that there are reports on the observation of CIP GMR in NiMnSb/Mo/NiMnSb spin valve structures made at low temperature [54]. These results are quite surprising since Mo is an ordinary transition metal having a small probability of properly matching the energy bands in the NiMnSb Heusler alloy. However, upon close examination of the published data one notices that the magnetoresistance curves shows no abrupt resistance change, as shown in Fig. 19(b) and (c), that is a signature condition for a spin valve. Furthermore the GMR response suspiciously resembles GMR found in granular systems [70, 71].

Finally we address the amplitude of the GMR effect observed in our multilayers and spin valves. It has been shown that CIP GMR in transition metal multilayer systems can be as large as 150% at room temperature [72] while the CIP GMR for the all-Heusler alloy multilayer is less than 0.5%. In a CIP GMR geometry, the relevant length scale is the mean free path (λ_{eff}). If the mean free path of the NM spacer material is small, the corresponding GMR effect will also be small. The Heusler alloys are materials with much larger resistivities as compared to the individual transition metal atoms and therefore have for a similar electron density and mass, small mean free paths.

To prove our point, we have used the Hall effect to determine the mobility and carrier concentration in our proposed Heusler alloys. In Fig. 20

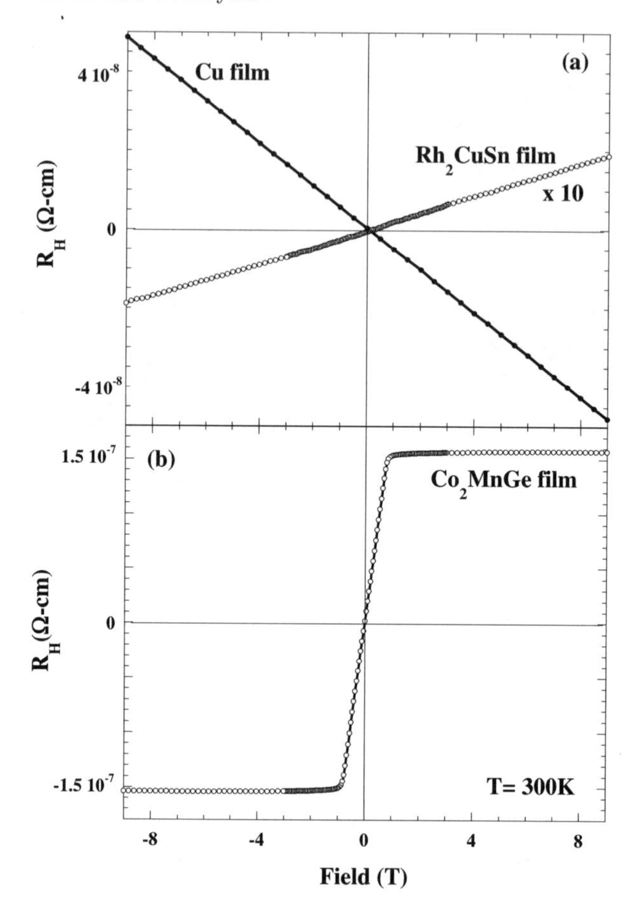

Fig. 20. Room temperature Hall Effect for (**a**) Cu and Rh_2CuSn and (**b**) Co_2MnGe films. The data for the Rh_2CuSn has been multiplied by a factor of 10 for clarity

we plot the room temperature Hall resistance vs. magnetic field for Cu and Rh_2CuSn (Fig. 20a) and Co_2MnGe (Fig. 20b). In NM materials the Hall resistance varies linearly with field. One immediately notices the difference in the resistivity slope between Cu and Rh_2CuSn which indicates predominant n-type conduction in Cu and predominant p-type conduction in Rh_2CuSn. An extraordinary Hall component is observed for the FM Co_2MnGe where as expected, the Hall resistivity curve resembles a hard axis hysteresis loop as shown in Fig. 20(b). At fields above the $4\pi M$ value of Co_2MnGe, where the extraordinary Hall effect is negligible, a positive slope is observed similar to the Rh_2CuSn. Our results show that both Heusler alloys are p-type conductors similar to previous Hall results on the NiMnSb Heusler alloy [67].

Combining our Hall coefficient results with accurate determination of the film resistivities using the Van de Pauw technique, the mean free path can be

Table 2. Room temperature Hall Coefficients for Cu and the Heusler alloys

	fcc Cu	Rh_2CuSn	Co_2MnGe	Cu_2CoSn
a_0 (Å)	3.61	6.146	5.743	5.982
R_0 (10^{-5}cm^3/C)	−5.36	1.99	0.599	2.65
R_E (10^{-5}cm^3/C)	–	–	222	–
ρ_{xx} ($\mu\Omega$-cm)	2.43	115.5	113.4	91.6
μ(cm^2/Vs)	−22.1	0.15	0.06	0.29
n (10^{22}/cm^3)	11.6	31.3	16.6	23.6
λ_{eff}(Å)	2200	2.41	3.36	14.9

determined using a simple Drude model approximation. We use the following relation of $\lambda_{eff} = \hbar(3\pi n)^{1/3}/(ne^2\rho_{xx})$ where n is the carrier concentration determined from the Hall data and ρ_{xx} is the absolute resistivity. Our results for the Heusler alloys and for Cu are tabulated in Table 2. We have determined quite small values of λ_{eff} for the Heusler alloys, most specifically the Rh_2CuSn NM spacer layer has a mean free path almost 1000 times smaller than ordinary Cu. Thus the reason for a small CIP GMR effect in the (Co_2MnGe/Rh_2CuSn) layered structures may be due to an intrinsically small mean free path in the Rh_2CuSn.

5 Conclusions

In this chapter we have investigated the growth and magneto-transport properties of the Co_2MnGe Heusler alloy. We have discussed various fabrication methods for thin film deposition of Co_2MnGe as well as the thin film deposition of Rh_2CuSn and Cu_2CoSn. We have stressed the importance of band matching of the electronic structure of both the FM and NM materials in a GMR stack, a feature that has thus far been absent in previous GMR studies of Heusler alloys. Using a "band engineering" methodology we have demonstrated room temperature CIP GMR in the (Co_2MnGe/Rh_2CuSn) multilayer stack. Our CIP GMR results are small due to the high resistivities and small mean free path inherent to these Heusler alloys. Our results create a starting point for utilizing Heusler alloys in GMR applications where ultimately the largest effect will be realized in a CPP measuring geometry.

Acknowledgements

The authors are indebted to J. J. Krebs, G.A Prinz, K. B. Hathaway, R.J. M. van de Veerdonk, T.J Klemmer, J. Rosenberg and P. Ohodnicki for their valuable technical contributions.

References

1. R. Wood: IEEE Trans. Magn. **36**, 36 (2000)
2. J. Daughton: Thin Solid Films **216**, 162 (1992)
3. A. Ney, C. Pampuch, R. Koch, and K.H. Ploog: Nature **425**, 485 (2003)
4. R.M. Bozorth. In: *Ferromagnetic Materials* (IEEE Press, New York 1993)
5. O.A. Ivanov, L.V. Solina, V.A. Demshima, and L.M. Maget: Phys. Met. Metallogr. **35**, 81 (1973)
6. R.M. Bozorth: Rev. Mod. Phys. **25**, 42 (1953)
7. M.N. Baibich, J.M. Broto, A. Fert, F. Nguyen Van Dau, F. Petroff, P. Etienne, G. Creuzet, A. Friedrich and J. Chazelas: Phys. Rev. Lett. **61**, 2472 (1988)
8. J.S. Moodera, L.R. Kiner, T.M. Wong, and R. Meservey: Phys. Rev. Lett **74**, 3273 (1995)
9. S.A. Wolf, D.D. Awschalom, R.A. Buhrman, J.M. Daughton, S. von Molnár, M.L. Roukes, A.Y. Chtchelkanova, and D.M. Treger: Science **294**, 1488 (2001)
10. G.A. Prinz: Science **282**, 1660 (1998)
11. H. Ohno: Science **281**, 951 (1998)
12. J.-H. Park, E. Vescovo, H.-J. Kim, C. Kwon, R. Ramesh, and T. Venkatesan: Nature **392**, 794 (1998)
13. D.J. Monsma and S.S.P. Parkin: Appl. Phys. Lett. **77**, 720 (2000)
14. K. Schwarz: J. Phys. F: Met. Phys. **16**, L211 (1986)
15. I.I. Mazin: Appl. Phys. Lett. **77**, 3000 (2000)
16. Y.S. Dedkov, U. Rudiger, and G. Guntherodt: Phys. Rev. B**65**, 64417 (2002)
17. F. Heusler: Verh. Dtsch. Phys. Ges. **5**, 219 (1903)
18. P.J. Webster: J. Phys. Chem. Solids **32**, 1221 (1971)
19. R.A. de Groot. F.M. Mueller, P.G. van Engen, and K.H.J. Buschow: Phys. Rev. Lett. **50**, 2024 (1983)
20. J.A. Caballero, Y.D. Park, J.R. Childress, J. Bass, W.-C. Chiang, A.C. Reilly, W.P. Pratt Jr., and F. Petroff: J. Vac. Sci. Technol. A**16**, 1801 (1998)
21. S. Fujii, S. Sugimura, S. Ishida, and S. Asano: J. Phys. Condens. Matter **43**, 8583 (1990)
22. W.E. Pickett and D.J. Singh: Phys. Rev. B**53**, 1146 (1996)
23. R. Skomski and P.A. Dowben: Euro Phys. Lett. **58**, 544 (2002)
24. W.E. Pickett: Phys. Rev. Lett. **77**, 3185 (1996)
25. D. Orgassa, H. Fujiwara, T.C. Schulthess, and W.H. Butler: Phys. Rev. B**60**, 13237 (1999)
26. J.S. Moodera and D.M. Mootoo: J. Appl. Phys. **76**, 6101 (1994)
27. See, for example: *Polarized Electrons in Surface Physics*, ed by R. Feder (World Scientific, Singapore 1985)
28. P.M. Tedrow and R. Meservey: Phys. Rep. **238**, 173 (1994)
29. R.J. Soulen Jr., J.M. Byers, M.S. Osofsky, B. Nadgorny, T. Ambrose, S.F. Cheng, P.R. Broussard, C.T. Tanaka, J. Nowak, J.S. Moodera, A. Barry, and J.M.D. Coey: Science **282**, 85 (1998)
30. G.J. Strijkers, Y. Ji, F.Y. Yang, C.L. Chien, and J. M Byers: Phys. Rev. B**63**, 104510 (2001)
31. Y. Ji, G.J. Strijkers, F.Y. Yang, C.L. Chien, J.M. Byers, A. Anguelouch, G. Xiao, and A. Gupta: Phys. Rev. Lett. **86**, 5585 (2001)
32. A.M. Bratkovsky: Phys. Rev. B**56**, 2344 (1997)
33. S. Datta and B. Das: Appl. Phys. Lett. **56**, 665 (1990)

34. K. von Klitzing, G. Dorda, and M. Pepper: Phys. Rev. Lett. **45**, 494 (1980)
35. P. Grunberg, R. Schreiber, Y. Pang, M.B. Brodsky, and H. Sowers: Phys. Rev. Lett. **57**, 2442 (1986)
36. B. Dieny, V.S. Speriosu, S.S.P. Parkin, B.A. Gurney, D.R. Wilhoit, and D. Mauri: Phys. Rev. B**43**, 1297 (1991)
37. M. Prutton. In: *Introduction to Surface Physics* (Oxford University Press 1994)
38. J.R. Waltrop and R.W. Grant: Appl. Phys. Lett. **34**, 630 (1979)
39. T.J. Mc Guire, J.J. Krebs, and G.A. Prinz: J. Appl. Phys. **55**, 2505 (1984)
40. J.J. Krebs, B.T. Jonker, and G.A. Prinz: J. Appl. Phys. **61**, 2596 (1987)
41. G.A. Prinz: Phys. Rev. Lett. **54**, 1051 (1985)
42. Y.U. Izerda, W.T. Elam, B.T. Jonker, and G.A. Prinz: Phys. Rev. Lett. **62**, 2480 (1989)
43. A.Y. Liu and D.J. Singh: Phys. Rev. B**47**, 8515 (1993)
44. M.A. Mangan, G. Spanos, T. Ambrose, and G.A. Prinz: Appl. Phys. Lett. **75**, 346 (1999)
45. T. Ambrose, J.J. Krebs, and G.A. Prinz: Appl. Phys. Lett. **76** 3280 (2000)
46. T. Ambrose, J.J. Krebs, and G.A. Prinz: J. Appl. Phys. **89**, 7522 (2001)
47. F.Y. Yang, C. H. Shang, C. L. Chien, T. Ambrose, J.J. Krebs, G.A. Prinz, V.I. Nikitenko, V.S. Gornakov, A.J. Sharpiro, and R.D. Shull: Phys. Rev B**65** 174410 (2002)
48. J. Vossen and W. Kern. In: *Thin Film Processes* (Academic Press, New York 1978)
49. M.F. Toney and S. Brennan: J. Appl. Phys. **65**, 4763 (1989)
50. L.J. van der Pauw: Philips Res. Repts. **13**, 1 (1958)
51. N.W. Ashcroft and N.D. Mermin. In: *Solid State Physics,* (Saunders Company 1976)
52. See, for example: *Magnetic Multilayers and Giant Magnetoresistance*, ed by U. Hartmann (Springer Verlag 1999)
53. S.S.P. Parkin: Phys. Rev. Lett. **67**, 3598 (1991)
54. C. Hordequin, J.P. Nozières, and J. Pierre: J. Magn. Magn. Mater. **183**, 225 (1998)
55. P.R. Johnson, M.C. Kautzky, F.B. Mancoff, R. Kondo, B.M. Clemens, and R.L. White: IEEE Trans. Mag. **32**, 4615 (1996)
56. K. Westerholt, U. Geirsbach, and A. Bergmann: J. Magn. Magn. Mater. **253**, 239 (2003)
57. N.F. Mott. In: *Metal Insulator Transitions*, (Taylor and Francis, London 1997)
58. S. Zhang, P.M. Levy, and A. Fert: Phys. Rev. B**45**, 8689 (1992)
59. E.Y. Tsymbal and D.G. Pettifor. In: *Solid State Physics*, Vol. 56, ed by H. Ehrenreich and F. Sapen (Academic Press 2001)
60. T. Charlton and D. Lederman: Phys. Rev. B**63**, 94404 (2001)
61. J.R. Childress, O. Durand, A. Schuhl, and J.M. George: Appl. Phys. Lett. **63**, 1996 (1993)
62. K.B. Hathaway. In: *Ultrathin Magnetic Structures II*, ed by B. Heinrich and J.A.C. Bland (Springer-Verlag, Berlin 1994)
63. I. Galanakis, S. Ostanin, M. Alouani, H. Dreyssé, and J.M. Wills: Phys. Rev B**61**, 4093 (2000)
64. S. Picozzi, A. Continenza, and A.J. Freeman: Phys. Rev. B**66**, 94421 (2002)
65. J. Caballero, A.C. Reilly, Y. Hao, J. Bass, W.P. Pratt, F. Petroff, and J.R. Childress: J. Magn. Magn. Mater. **199**, 55 (1999)

66. W.W. Schulz, P.B. Allen, and N. Trivedi: Phys. Rev. B**45**, 10886 (1992)
67. M.C. Kautzky, F.B. Mancoff, J.F. Bobo, P.R. Johnson, R.L. White, and B.M. Clemens: J. Appl. Phys. **81**, 4026 (1997)
68. H.H. Potter: Proc Phys. Soc. **41**, 135 (1929)
69. M. Methfessel: Phys. Rev. B**38**, 1537 (1988); M. Methfessel, C.O. Rodriguez, and O.K. Andersen: Phys. Rev. B**40**, 2009 (1989); A.T. Paxton, M. Methfessel, and H. Polatoglou: Phys. Rev. B**41**, 8127 (1990)
70. A.E. Berkowitz, J.R. Mitchell, M.J. Carey, A.P. Young, S. Zhang, F.E. Spada, F.T. Parker, A. Hutten, and G. Thomas: Phys. Rev. Lett. **68**, 3745 (1992)
71. J.Q. Xiao, J.S. Jiang, and C.L. Chien: Phys. Rev. Lett. **68**, 3749 (1992)
72. E.E. Fullerton, M.J. Conover, J.E. Mattson, C.H. Sowers, and S.D. Bader: Appl. Phys. Lett. **63**, 1699 (1993)

Surface Segregation and Compositional Instability at the Surface of Half-Metal Ferromagnets and Related Compounds

Hae-Kyung Jeong[1], Anthony Caruso[1], and Camelia N. Borca[2]

[1] Department of Physics, University of Nebraska-Lincoln, USA
hjeong@unlserve.unl.edu and acaruso1@bigred.unl.edu
[2] Institute of Applied Optics, Swiss Federal Institute of Technology, Switzerland
cborca@jilau1.colorado.edu

Abstract. Interface engineering in order to exploit the possibilities of the interface electronic structure may be a route to a good spin injector. Nonetheless, for many potential half metallic systems, the stoichiometric surface is generally not in thermodynamic equilibrium with the bulk and the surface Debye temperature is often quite low. Several such classes of materials are currently under investigation as potential high spin polarization materials. These include the half metallic systems such as the semi-Heusler alloys, the manganese perovskites and the "simpler" oxides like chromium dioxide and magnetite. But all these materials suffer from a tendency of the surface to adopt a different composition than the bulk. Surface segregation and surface composition for three classes of potential half metallic systems: NiMnSb(100), $La_{0.65}(M{=}Ca,Pb,Sr)_{0.35}MnO_3(100)$ and CrO_2 will be explored. The surface compositional stability and different surface terminations will be compared with the results from recent electronic structure calculations.

1 Introduction

The phenomenon of surface segregation, which induces surface compositional instability in alloys, is the preferential enrichment of one component of a multi-component system at a boundary or interface. Half-metallic ferromagnets are defined by an electronic structure, which should exhibit conduction by charge carriers of one spin direction exclusively. Consequently, the spin polarization of the conduction electrons should theoretically be 100%. In reality this complete spin polarization is not observed, often due to the presence of surface segregation and compositional instabilities in the surface region of half-metals, as outlined in the present review.

Surface segregation is a strong indication that the surface enthalpy differs significantly from the bulk, in the context of standard statistical models. The difference in total free energy of the surface with respect to the bulk is a consequence of the surface truncation to vacuum and the resultant breaking of symmetry. Besides surface segregation, the free energy difference drives other phenomena in order to minimize the total free energy [1], such as surface relaxations and surface reconstructions. All these phenomena related to changes in crystalline order in the surface result in the creation of different electronic and magnetic properties of the surface with respect to the bulk. The present review focuses on the interplay between surface composition and surface electronic structure of several half-metallic materials and related compounds, which can explain the reduced experimental values of the spin-polarization found in the literature at the surface or interface of these materials.

It is well known that surface segregation plays a crucial role in affecting the half-metallic character. The half-metallic character in which the density of states of one spin channel is zero (insulating), but finite in the other spin channel (metallic) at the Fermi energy, is often considered as very important for many spintronic applications and is strongly influenced by composition [2–7] and surface termination [8–11]. Evidence that surface composition affects spin polarization of half-metallic systems abounds [2, 3, 5, 12–15].

One of the best examples of reduced spin-polarization due to surface compositional instabilities is presented in Fig. 1. The three panels show the spin-integrated as well as the spin-resolved unoccupied (above the Fermi level)

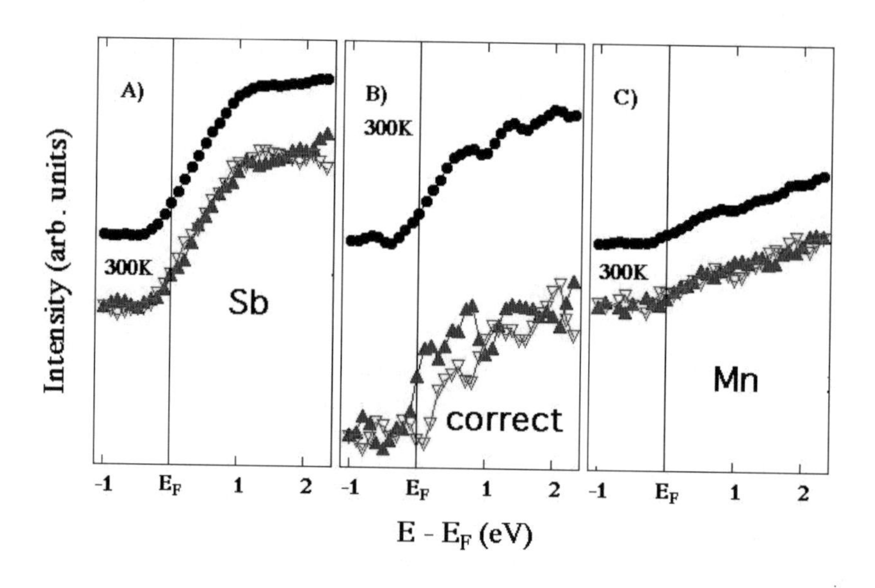

Fig. 1. Spin-polarized unoccupied electronic bands from spin polarized inverse photoemission spectroscopy at normal incidence for NiMnSb (100) surface. The spin-integrated inverse photoemission is shown for comparison (•). Three cases are shown: (**a**) Sb-rich NiMnSb surface, (**b**) stoichiometric (clean) NiMnSb surface and (**c**) Mn-rich NiMnSb surface. Up triangles (▲) indicate spin majority, while down triangles (▽) indicate spin minority. Adapted from [2]. Note that spin-resolved and spin-integrated curves are offset for clarity

bands of NiMnSb half-metal alloy for three different surfaces, all recorded at room temperature. The technique used to measure the unoccupied electronic bands is spin polarized inverse photoemission spectroscopy, which is sensitive to the first 2–3 atomic layers from the surface towards the bulk. The difference between the spin-up and spin-down bands gives the spin-polarization or the asymmetry value near the Fermi level. Figure 1a shows that for a Sb-rich NiMnSb surface, the spin-asymmetry is almost zero, as is the case in Fig. 1c where spin-polarization is rapidly reduced to zero due to Mn segregation to the surface region during sample preparation [2]. The spin-polarization at the Fermi level is as high as 100% above the background level in case of a clean and stoichiometric NiMnSb surface, as shown in Fig. 1b. As discussed in [2], the stoichiometric surface measured in Fig. 1b is very fragile with a short "life-time" of the order of 4–5 hours, after which the surface becomes Mn-rich with lower spin-asymmetry. Similar work, done by Bona, showed 50% spin polarization from in situ cleaved NiMnSb using spin polarized photoemission spectroscopy and it was suggested that this discrepancy of a much higher value was a consequence of Sb surface segregation [3]. This is consistent to 30% at 0.25 K and 18% at 10 K of spin polarization of MnSb, respectively [16]. It is therefore clear that the different composition of surface from bulk reduces the spin polarization of the supposed half-metal.

Even the most useful tools for the investigation of the spin polarization have only rarely found values of polarization close to 100% as in the case for $La_{0.7}Sr_{0.3}MnO_3$ by spin polarized photoemission spectroscopy at 42 K [17] and spin polarized inverse photoemission spectroscopy at 100 K [18]. However, Soulen et al. measured 78% of the spin polarization of $La_{0.7}Sr_{0.3}MnO_3$ by Andreev scattering at 1.6 K [19], and the result corresponds to the spin polarization obtained by TMR junctions at around 5 K [20–22]. Even less polarizations, 25% at a few K [23], 50% at 70 K and 10% at 300 K [24], were obtained by the TMR measurements [24]. For NiMnSb, several measurements were performed to detect the spin polarization. Soulen et al. measured 58% of spin polarization using the Andreev-reflection method at 1.6 K and compared to values between 42% and 46.5% for Fe, Ni and Co [19]. Bona [3] and Zhu [25] obtained 50% and 40% of the spin polarization at 300 K by spin polarized photoemission spectroscopy. Higher spin polarization for NiMnSb at 300 K was observed using spin polarized inverse photoemission, but still less than 100% [26]. TMR measurements give 28% [27], 20% [28] and 10% [29, 30] for the spin polarization of NiMnSb respectively. Much higher spin polarizations were obtained for CrO_2 by Andreev scattering, namely $98.4 \pm 4\%$ [31], 97% [32], $96.4 \pm 4\%$ [33], $90 \pm 3.6\%$ [19] and $81 \pm 3\%$ [34] at 1.6 K. Spin- polarized photoemission gives higher spin polarization, 95%, at even 300 K for CrO_2 [25, 35], but $20 \pm 10\%$ at 300 K by scanning tunneling spectroscopy [36]. Two magnetoresistive measurements also have found this contradiction between $85 \pm 3\%$ at 4.2 K [37] and 10% at 77 K [38] for CrO_2. Spin polarized photoemission spectroscopy experiments also showed

−80% [39] and −40% [40] for Fe_3O_4, and 15% [41] for $La_{0.7}Pb_{0.3}MnO_3$, while 80% [5] for $La_{0.7}Pb_{0.3}MnO_3$ by spin polarized inverse photoemission.

Knowing how the surface composition differs from the bulk composition of half metal alloys is essential in order to understand the presence of surface/interface states as well as to engineer the necessary properties, such as high values of spin-asymmetry, for building reliable spin-valves, spin-tunnel junctions and other spin dependent devices and structures.

2 Characterization of Surface Composition

The question "what is a surface?" can be answered in many ways depending on the significance of the respective property to be studied. In a strictly physical sense, it is the outermost, top monatomic layer of a solid. Since it is now well known that equilibrium segregation is confined to one – or in some cases – to a few atomic layers, its study is consistent with the general goal of surface analysis, for which the basic requirements can be stated as follows: (a) Information depth of the order of one or a few atomic layers and, (b) Detection of all elements, quantitatively and with high sensitivity for any matrix material.

At present, these requirements are met to a large extent by analysis techniques based on electron and ion spectroscopies [42, 43]. This brief consideration is confined to the most popular techniques for which commercial equipment is available. These are Auger Electron Spectroscopy (AES) [44], Angle-Resolved X-ray Photoelectron Spectroscopy (ARXPS or ESCA = Electron Spectroscopy for Chemical Analysis) [44], Ion Scattering Spectroscopy (ISS) [45] and Secondary Ion Mass Spectrometry (SIMS) [46]. The principal instrumentation of any of these techniques consists of a vacuum chamber, which contains the radiation source for excitation [electrons (AES), X-rays (ARXPS), ions (ISS, SIMS)], the sample and the analyzer/detector system [energy (AES, XPS, ISS) or mass (SIMS) analyzer]. The electron spectroscopies need an auxiliary ion gun for sample cleaning and sputter depth profiling. In the present work we focus on ARXPS to characterize the surface composition of half-metals and related compounds.

In ARXPS, surface composition can be determined with great accuracy since the effective probing depth becomes shorter as the emission angle is increased with respect to the surface normal. The experimental core level intensities for any two components from a multi-component alloy are acquired at several emission angles, usually from 0° (normal emission) to 60° (off-normal emission). Then, a linear background contribution is systematically subtracted from each raw spectrum. The peak intensities are further normalized by the corresponding differential cross-section for emission and by the analyzer transmission function [1, 12–15]. The experimentally normalized intensity ratio for any two elements A and B is thus given by:

$$R^{\exp}(\theta) = \frac{I(A)/\sigma_A\, T(E_A)}{I(B)/\sigma_B\, T(E_B)} \tag{1}$$

where $I(A)$ and $I(B)$ are the measured core level intensities for elements A and B, σ_A and σ_B are the cross sections, and $T(E_A)$ and $T(E_B)$ are the transmission functions of the analyzer for elements A and B as a function of the corresponding photoelectron kinetic energies E_A and E_B and are based on the measured transmission functions for each analyzer. For example, the transmission function of PHI 10–360 Precision Energy Analyzer is $T(E_A) = \sqrt{E_A}$ [47]. For the Gamma Scienta SES-100 analyzer with 0.8 mm and 1.6 mm slits [48], the corresponding transmission functions are $T_{0.8mm}(E_A) = 0.01833 + 1.0023(\frac{E_A}{E_{pass}})^{-0.439}$ or $T_{1.6mm}(E_A) = -0.02805 + 2.3467(\frac{E_A}{E_{pass}})^{-0.352}$.

Over the past decades, the phenomenon of surface segregation attracted quite a lot of theoretical interest. Up to now, various theories have been developed in order to account for the enrichment at a binary alloy surface. First-principles approach [49], embedded-atom method (EAM) model [50], Finnis-Sinclair (FS) potential [51], and Bozzolo-Ferrante-Smith (BFS) method [52] have been used to simulate the surface segregation for fcc type random and ordered alloys, while for fcc and hcp elements the modified analytic embedded-atom method (MAEAM) many-body potential is used to theoretically study the surface segregation phenomena of binary alloys [53]. Unfortunately, to our knowledge, there are no detailed theoretical studies of the surface segregation of multi-component half-metallic alloys. We use here a simplified model to fit the experimental intensity ratios obtained from ARXPS.

The comparison between theory and experiment is accomplished through considerations of the theoretical normalized intensity ratio of element A and B as given by:

$$R^{theor}(\theta) = \frac{\sum_{j=0}^{\infty} f_j(A) \exp\left[\frac{-jd}{\lambda_A^j} \cos(\theta)\right]}{\sum_{j=0}^{\infty} f_j(B) \exp\left[\frac{-jd}{\lambda_B^j} \cos(\theta)\right]} \tag{2}$$

where λ_A^j and λ_B^j are the inelastic mean free paths of the core electrons generated from elements A and B respectively and passing through the material contained in layer j. The inelastic mean free paths can be adopted from previously published methodologies [54] contained within the NIST Electron Inelastic Mean Free Path program [55]. The atomic fraction of element A (chosen as the element which segregates to the surface) in the jth layer below the surface is given by:

$$f_j(A) = b + \delta \exp(-jd/G) \tag{3}$$

where b is the bulk fraction of element A, δ and G are fitting parameters representing the extent of the segregation and the segregation depth respectively, and d is the distance between atomic layers. These two quantities are

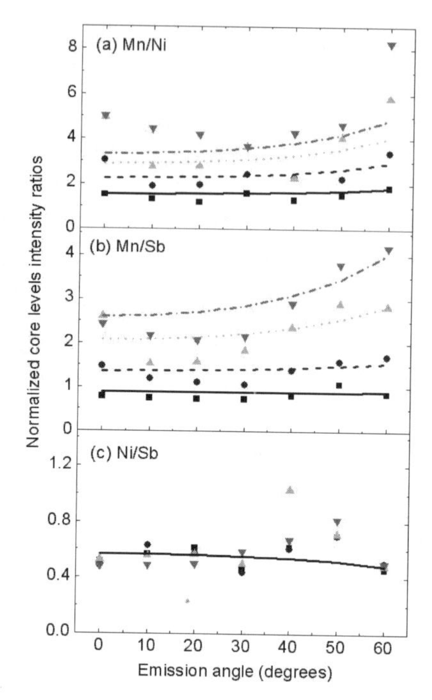

Fig. 2. The normalized x-ray photoemission intensity ratios of the Ni ($2p_{3/2}$), Mn ($2p_{3/2}$), and Sb ($3d_{5/2}$) of NiMnSb as a function of the emission angles. The intensities are normalized using their differential cross section for emission and the analyzer transmission function according to equation 1. Data denoted by the symbol (■) are representative of the stoichiometric surface (*fitted by a solid line*) following removal of the Sb capping layer, while (•) indicates data obtained after 10 min, (▲) 30 min., and (▼) 60 min of sputtering and annealing, respectively. The data (■•▲▼) are fitted by lines using a segregation model in which Mn atoms migrate to both vacancy sites as well as replace Sb and Ni atoms at their lattice sites in the surface region except for the first monolayer which is MnSb. Adapted from [12]

also the fitting parameters when comparing the model with experimental values. From the profile form $f_j(A)$ one can calculate the apparent surface concentration (or relative intensity) of element A for a particular core level.

Figure 2 shows the experimental normalized intensity ratio and a theoretical model fit of the NiMnSb surface with different annealing and sputtering times. The experimental core level intensity corresponding to each element is normalized using its differential cross section for emission and the analyzer transmission function according to (1). As the annealing time increases, the intensity ratios of Mn/Ni and Mn/Sb also increase, which indicates that Mn segregates to the surface. The best result of the fitting was from Mn segregation to the surface and to the vacancy sites within the Ni layer, and the best fitting parameters are $G = 8.55$ Å with $\delta = 13$, 22 and 27% for 10, 30 and

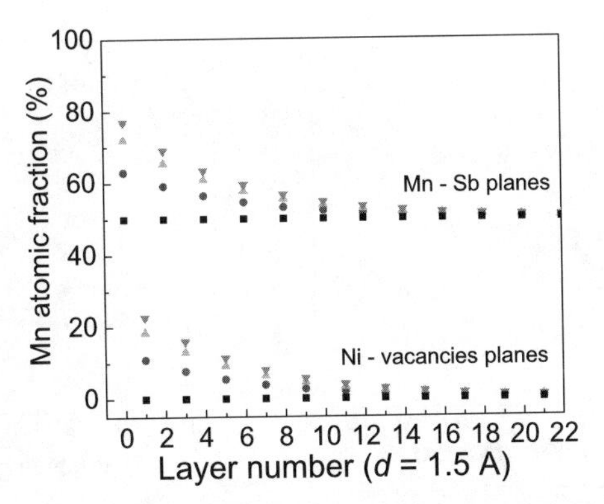

Fig. 3. The concentration of Mn in the near surface region, obtained from fits to the angle resolved x-ray photoemission data are shown. The Mn concentration profile symbols (■•▲▼) are fits (*by dashed lines*) to the data shown by similar symbols in Fig. 2. The ■ indicates an as-deposited surface, • is after a 10 min., ▲ is after 30 min., and ▼ is after 60 min. of sputtering and annealing. Adapted from [12]

60 min of sputtering and annealing. From the comparison of the normalized core level intensities ratios, both experimentally and theoretically, Mn segregation was indicated with an MnSb terminal layer [2], and different sample preparation times were shown to affect the segregation depth δ [12]. Using (3) with the best fitting parameters, the concentration of Mn in the near surface region was mapped out in Fig. 3 as a function of the layer thickness. Mn segregation in the near surface region was shown as the dominant component in the NiMnSb alloy.

Surface composition of $La_{0.65}Pb_{0.35}MnO_3$ was also thoroughly investigated [15]. Depending on the annealing treatments, the thin film surface exhibited wildly different behavior and composition. Figures 4(a) and (b) show core level intensity ratios of Pb $(4f)$/La $(3d)$ and Mn $(2p)$/[La $(3d)$+Pb$(4f)$]. The solid lines represent the optimum fits, which assumed a model like the schematic pictures shown as inset to Fig. 4. With a gentle annealing procedure (up to 250°C), the surface is dominated by appreciable Pb segregation (Fig. 4c), whereas a heavily annealed surface (up to 520°C) undergoes an irreversible restructuring into a Ruddlesden-Popper phase $(La_{1-x}Pb_x)_2MnO_4$.

For $La_{1-x}Ca_xMnO_3$ ($x = 0.1$ and 0.35), the terminal layer is predominately Mn-O for $x = 0.35$, while for $x = 0.1$, the majority of the surface is La/Ca-O terminated according to analysis of surface composition using the core level intensity ratios [13]. Figure 5(a) shows normalized core level intensity ratios (•) of Ca$(2p)$ and La$(3d)$ together with a fitting that assumes Ca surface segregation (∘) for $x = 0.35$ and 0.1, respectively. For $La_{0.65}Ca_{0.35}MnO_3$, the fitting parameters are $\delta = 0.24$ and $G = 0.9$ layers, while for $La_{0.9}Ca_{0.1}MnO_3$, $\delta = 0.82$ and $G = 0.4$ layers. From those parameters the Ca atomic fraction as a function of depth was mapped out as

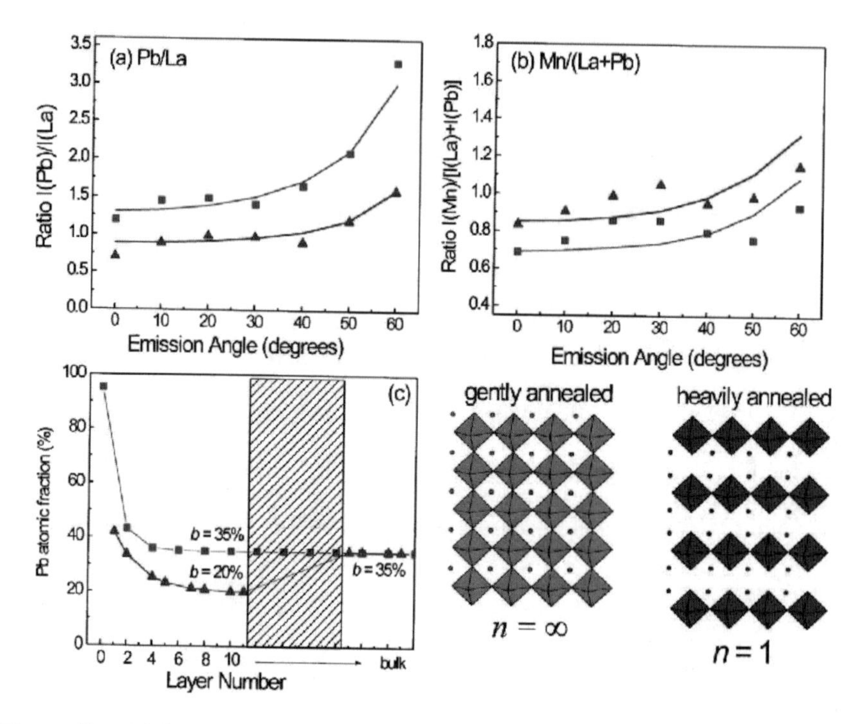

Fig. 4. Panel (**a**) shows the normalized level core intensity ratios of Pb($4f$) peaks to that of La($3d$) peaks and the best fits (–) for the gently annealed surface (■) and for the heavily annealed surface (▲), as a function of photoelectron emission angle (θ). Part (**b**) shows the corresponding experimental and calculated relative core level intensity ratios of Mn/(La+Pb) for gently (■) and heavily (▲) annealed surfaces. Panel (**c**) displays the resulting theoretical Pb atomic fractions as a function of layer number-the shaded area shows the transition region between the surface layers with doping level $b = 20\%$ and the bulk layers with $b = 35\%$. The inset shows schematically the layers stacking sequence used to obtain the compositional fits in panels (**a**) and (**b**) for the gently (*left*) and heavily (*right*) annealed surfaces. Adapted from [15]

illustrated in Fig. 5(b). Both surfaces exhibit a large enhancement in the Ca concentration compared to the bulk.

For $La_{0.65}Sr_{0.35}MnO_3$, the surface chemical composition was also investigated by angle resolved x-ray photoelectron spectroscopy [14]. In Fig. 6, the experimental core level intensity ratios of Sr/La, Mn/Sr and Mn/La are presented together with two model fits assuming a $La_{0.65}Sr_{0.35}MnO_3$ (blue) stacking or $(La_{0.65}Sr_{0.35})_2MnO_4$ (red) one. From these fits we conclude that Sr atomic fraction shows an appreciable Sr surface segregation causing a major restructuring of the surface region characterized by the formation of a Ruddlesden-Popper phase $(La,Sr)_{n+1}Mn_nO3_{n+1}$ with $n = 1$.

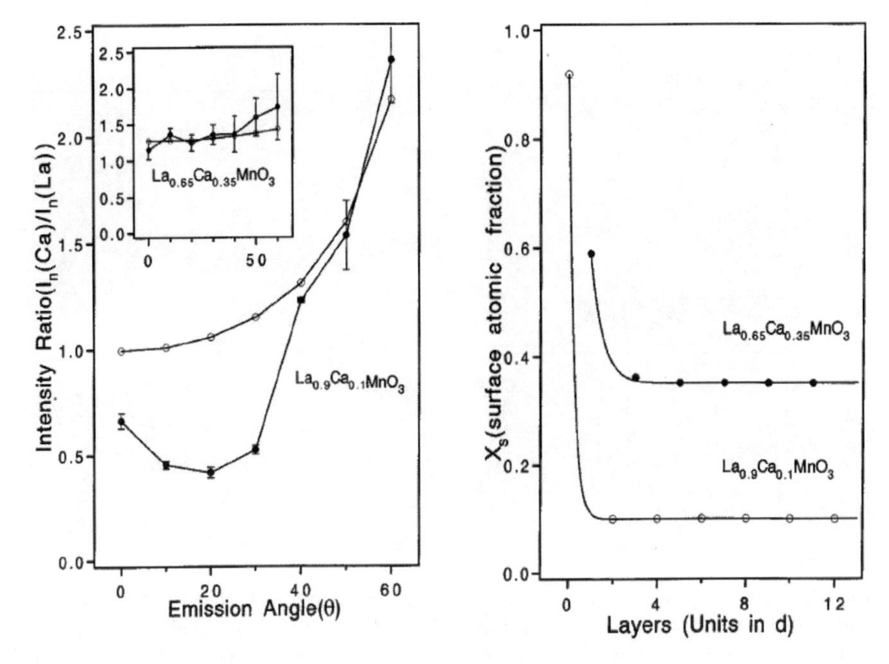

Fig. 5. The experimental normalized core level intensity ratios (•) of Ca(2*p*) and La(3*d*) as a function of the emission angle and with solid line a model fit (○) assuming Ca segregation to the surface (**a**). The resulting theoretical Ca atomic fractions as a function of layer number are presented for $x = 0.1$ and 0.35 (**b**). Adapted from [13]

Recently it has been also found that the surface composition of CrO_2 is not CrO_2 [56–58]. The same methods as described above show that a surface layer of insulating Cr_2O_3 is covering the metallic CrO_2 (Fig. 7). The segregation and restructuring in the surface region, therefore, is common in the half metals and will have a crucial impact on the electronic and magnetic properties at the created surfaces/interfaces. Such interfaces will degrade the performance of potential spintronic devices.

3 Free Energy in Segregation

Due to the truncation of the bulk, surface has a larger free energy than that of the bulk before the segregation process. In order to minimize the free energy difference, surface segregation, surface reconstruction or/and surface relaxation is likely to occur. The presence of surface segregation is an indicator of a difference in free energy (chemical potential) between the surface and the bulk [1]. The substantial surface segregation implies a different surface enthalpy than the bulk. This energy difference has been calculated from the

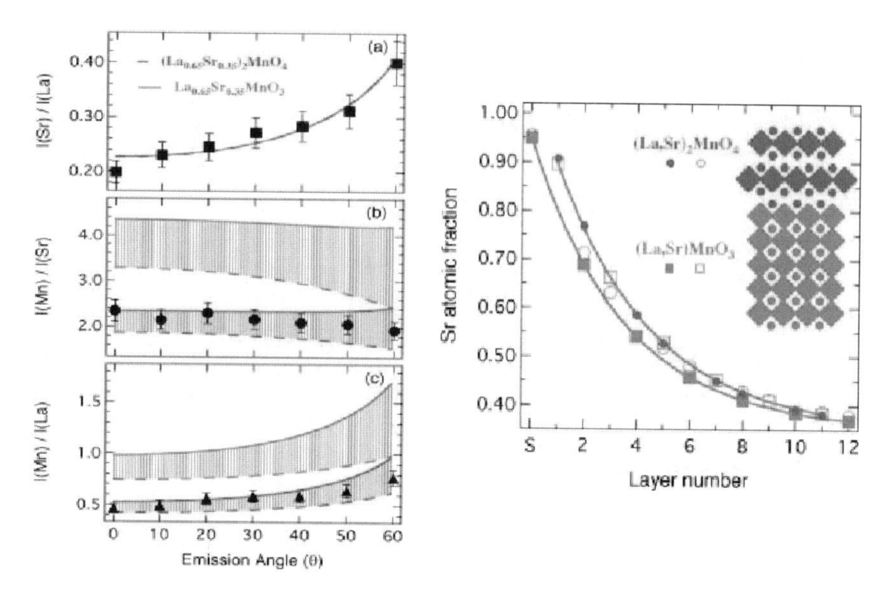

Fig. 6. Panel (**a**) shows the experimental core level intensity ratios (■) of Sr($3d$) and La($3d$) together with the optimum fits for the possible stacking sequences of $La_{0.65}Sr_{0.35}MnO_3$ structure (*blue dashed line*) and that of $(La_{0.65}Sr_{0.35})_2MnO_4$ structure (*red dashed line*). Panel (**b**) and (**c**) represent the normalized core level intensity ratios of Mn/Sr (•) and Mn/La (▲) with the corresponding calculated ratios of Mn/Sr and Mn/La for $La_{0.65}Sr_{0.35}MnO_3$ (blue hatched region) and $(La_{0.65}Sr_{0.35})_2MnO_4$ (red hatched region), respectively. The corresponding Sr atomic fraction is also shown as a function of layer for $La_{0.65}Sr_{0.35}MnO_3$ with La/SrO(MnO_2) termination in blue solid (*open*) squares and $(La_{0.65}Sr_{0.35})_2MnO_4$ with La/SrO (MnO_2) termination in red open (*solid*) circles. Adapted from [14]

segregation profiles using the approximation:

$$f_j(A)/f_j(B) = [f_b(A)/f_b(B)] \exp(-\Delta H/kBT) \tag{4}$$

where $f_b(A)$ and $f_b(B)$ are the bulk layer concentrations of constituents A and B, and T is the annealing temperature [59]. Figure 8 shows the resulting enthalpy difference between the surface and the bulk based on the most extensive Mn segregated surface of NiMnSb as well as for the Pb segregated surface of $La_{0.65}Sr_{0.35}MnO_3$. Since the enthalpy is smaller at the surface region compared to the bulk after surface segregation, the free energy may be minimized as a result of the Mn or Pb surface segregation. Also, different surfaces show different values of free energy in the surface and selvedge regions, depending on the amount of segregation.

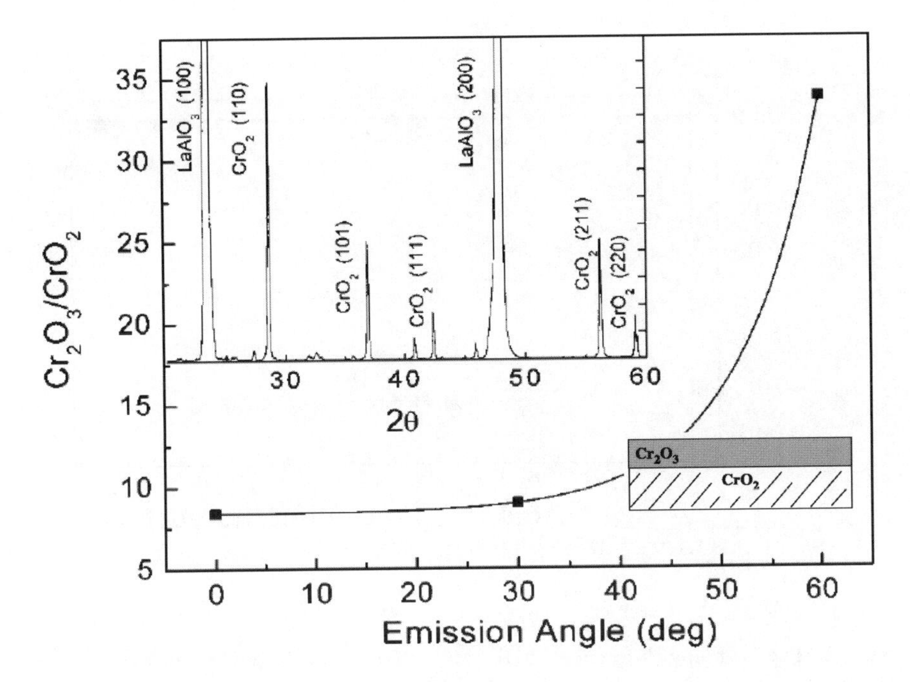

Fig. 7. The room temperature core level intensity ratio of Cr_2O_3/CrO_2 vs. different emission angles. The x-ray diffraction data and the schematic model of the sample are shown in the inset. Adapted from [56–58]

4 Magnetic Ordering in Segregation

All half-metals inherently possess magnetically ordered components. Most half-metals show stoichiometric composition in the bulk, but almost always a non-stoichiometric surface or interface when exposed to vacuum or joined with another material. The root of this non-stoichiometry is not entirely clear using classical arguments for the mechanisms, which cause or allow for a reduction in the total free energy of the system. The standard arguments for segregation or migration have been typically limited to the atomic size induced stress, lattice mismatch, solubility considerations and free enthalpies. Only in few instances has the argument of magnetism or moment ordering as a driving force in lowering the total free energy been considered [60].

During a study in the early 1980's on binary alloys, Moran-Lopez [60] found that they could predict between two metals, which would act as the solvent and which as a solute. Of the many binary systems which they considered, it was only those containing transition metal elements, which were known to order magnetically (ferromagnetically, anti-ferromagnetically, etc.) that did not fit their predicted model and empirical data.

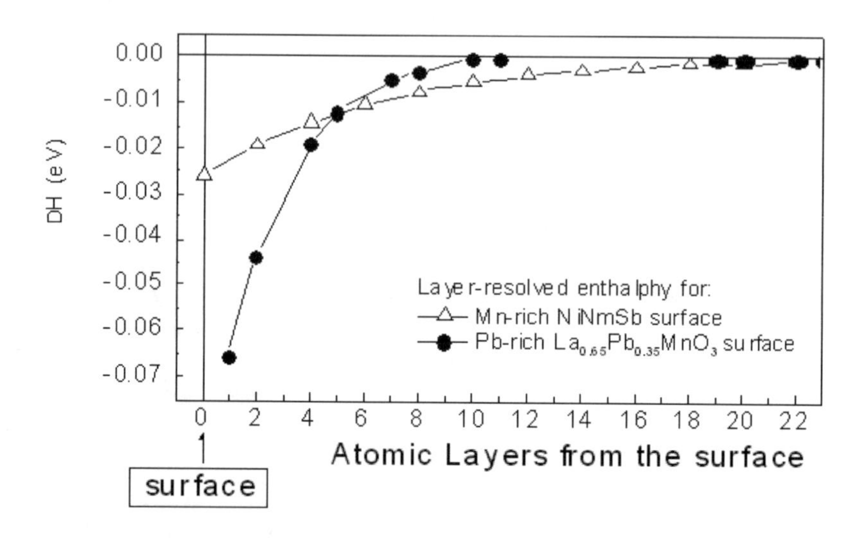

Fig. 8. The enthalpy difference (ΔH) between the surface and the bulk based on the most extensive Mn segregated surface of NiMnSb

In 2001 it was found [61] that for two very similar ternary semi-Heusler alloys, NiMnSb and TiCoSb, the resultant segregation could not be predicted from classical mechanism arguments. In the case of NiMnSb, it was found that mostly Mn enriched the surface layers whereas for TiCoSb, Co and Sb dominated the surface and sub-surface regions, respectively.

Figure 9 shows the normalized intensity ratios of the Ti ($2p_{3/2}$), Co ($2p_{3/2}$), and Sb ($3d_{5/2}$) core levels of TiCoSb. Several models have been explored in order to fit the experimental results, including the stoichiometric model fit (shown with black line). One model that agrees with the data obtained following very extensive annealing of TiCoSb, is based on the conjecture that a surfactant layer of $CoSb_3$ is a probable surface termination with Sb dominating the sub-surface layers. Due to extensive Sb sub-surface enrichment, a quantitative characterization of the segregation could not be obtained.

The most obvious origin for segregation is that size-induced stress distorts the lattice, causing strain, which is then alleviated by sending the misfit solute atoms to the surface [1]. This origin for segregation favors Sb segregation in the case of NiMnSb and TiCoSb (or any similar Heusler alloy, including NiMnSb) as Sb $\approx$ 1.45 Å possesses the largest atomic radius with respect to the other constituents Ni $\approx$ 1.35Å, Mn $\approx$ 1.40Å, Ti $\approx$ 1.40Å and Co $\approx$ 1.35Å, as taken from Slater [62]. The thermodynamic parameters, such as those that

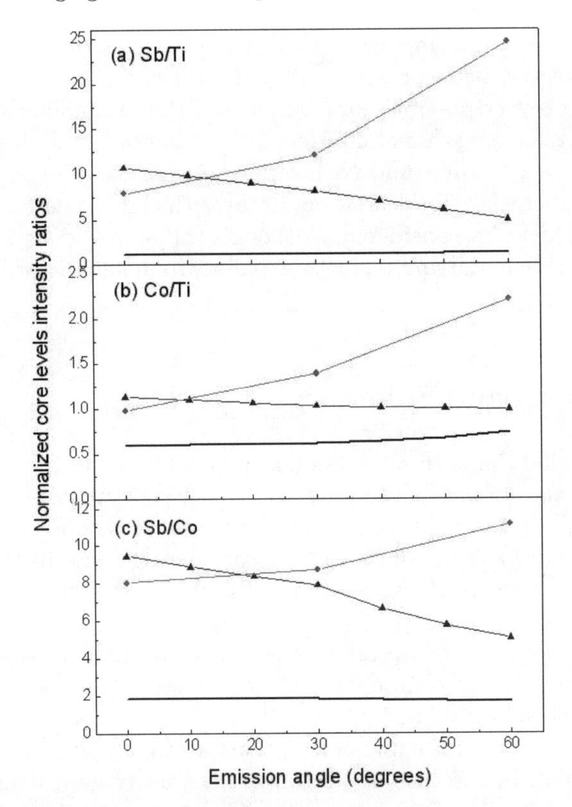

Fig. 9. The normalized intensity ratios of the Ti ($2p_{3/2}$), Co ($2p_{3/2}$), and Sb ($2d_{5/2}$) core levels of TiCoSb normalized as in Fig. 2. Panels (**a**), (**b**), and (**c**) denote a similar surfactant situation (as segregated Mn in NiMnSb) where in this case, Sb dominates the surface by replacing Ti and Co, and filling the vacant sites, with the first monolayer consisting of $CoSb_x$. The symbol (blue ▲) indicates data obtained from the surface using a procedure that should have removed the Sb capping layer and (•) is from a surface after extensive annealing. The solid black line represents the stoichiometric model fit. Adapted from [61]

lead to predictions of segregation based on the type of melting curve for each alloy [63], could favor segregation of Sb that includes a surfactant, such as the proposed $CoSb_3$. However, if one tries to consider the Gibbs free energy alone, where the system wants to go toward its state of greatest disorder, we find that no simple numbers or argument exists. For the NiMnSb, neither the thermodynamic or size arguments explain why Mn segregation appears to be wholly appropriate. We therefore considered magnetic ordering.

In all cases of transition metals, one must go back to basic ideas to identify what magnetic ordering would occur and justify why such ordering is energetically favorable. For Cr and Mn alone in bulk or thin layers, their spin coupling leads to an anti-ferromagnetic alignment; although, depending on crystal orientation, uncompensated spins can exist at the surface or buried interface. However, if Cr or Mn is alloyed with another element, ferromagnetic ordering is very possible, for example in $BiMnO_3$ and CrO_2. For Fe, Co, and Ni in bulk form, each metal orders ferromagnetically with Fe having the greatest long-range order, and Ni having the weakest. The consideration of the Curie temperature T_C decreasing as layer thickness decreases (finite size scaling) must also be considered. Although the above ideas are simplistic, they help provide a starting point to explain why magnetic ordering may be energetically favorable.

5 Discussions and Summary

The half metallic character at a boundary or interface is affected by several factors. First and foremost, the presence of surface segregation decreases the high spin-polarization value, as we discussed above (see Fig. 1). If the surface of such half-metallic alloys is different than the one in the bulk, then the electronic structure changes and resultantly affects the spin polarization value.

A second factor is the atomic disorder [64–70], namely the exchange of two components or two components at empty sites. For example, Co atoms present in anti-sites in Co_2MnGe results in defect-induced states near the Fermi level from first-principle calculations so that half-metallic character is destroyed [64]. In the case of NiMnSb, the spin-resolved density of states were calculated by first-principles with different types of atomic disorders, namely Mn-Ni disorder, Ni-Mn at empty sites and Mn-Sb at empty sites, respectively [65, 66]. The minority-spin band gap and the spin polarization at the Fermi energy is significantly reduced with increasing atomic disorders. The total magnetic moment of NiMnSb is reduced from 3.94 μ_B to 3.15 μ_B with increasing disorder. The reduction of the moment is caused by the disorder induced minority-spin states, which also cause the reduction of the spin polarization for disorder levels above 1% [66, 67]. Other half metals such as $Co_2Cr_{1-x}Fe_xAl$ [68] and Co_2MnSi [69, 70] showed also considerable reduction of the spin polarization due to Co-Cr and Co-Mn types of disorder, respectively.

Furthermore, the value of the effective Debye temperature, which is indicative of the dynamic motion of the lattice normal to the surface, can play an important role. Most half metals have low Debye temperatures. The effective Debye temperatures of $La_{0.65}Pb_{0.35}MnO_3$ are 71–77 K in surface region and 310–352 K in bulk region [15]. The big difference of the Debye temperature between surface and bulk is due to the presence of Pb surface segregation. For NiMnSb, the effective surface Debye temperature using low energy electron diffraction, inverse photoemission, and core-level photoemission is 145 ± 13 K, which is much smaller than the bulk Debye temperature of 312 ± 5 K obtained from wave vector dependent inelastic neutron scattering [71]. Half metals with low Debye temperature are not good enough to use in spintronic devices because of the high diffusivity of their components and the surface segregation. Since the vibrational structure at both interfaces and the bulk can have a profound effect on the polarization [72] and polarization injection [73], low Debye temperatures, particularly at interfaces, are generally undesirable for spin filters and spin injection.

Another factor that influences half metallic behavior is the temperature. Spin fluctuation, magnon-phonon coupling and population of spin minority states near the Fermi energy increase as the temperature increases, which results in lower spin polarization values [78,79]. Spin disorder (spin fluctuation) at finite temperatures can destroy the full spin polarization [72,73,80,81]. Transverse and longitudinal optical modes couple to spin-wave modes and reduce the net magnetization [80]. Generally the magnons and the transverse optical modes couple in near the Brillouin zone edge [72,73,81].

A final point is introduced by the presence of interface and/or surface states in a minority-semiconducting gap, which reduces the value of the minority gap [8–11,82–86]. Galanakis considered surface states in ab initio calculations based on the full-potential version of the screened KKR Green function method with different surface terminations [8,84]. For NiMnSb, CoMnSb and PtMnSb, MnSb terminated surface has surface states due to Mn d and Sb p orbitals within the minority gap so that the spin polarization is reduced to 38%, 46% and 46%, respectively. Ni, Co and Pt terminated surfaces, however, have different spin polarizations, which are 0, -22 and $+22$% respectively. For Co_2MnGe, Co_2MnSi and Co_2CrAl, Co and MnGe terminated surfaces introduce different surface states, which destroy the minority gap differently. $La_{0.6}Sr_{0.4}O$ or MnO_2 terminated $La_{0.6}Sr_{0.4}MnO_3$ (LSMO) thin film shows that the surface peak of Sr($3d$) is shifted differently [9,10]. A possible recrystallization of Sr at the surface may seriously affect the electronic and magnetic properties of these perovskite oxides. The core levels and electronic band structure exhibit strong dependence on surface morphology of $La_{1-x}Sr_xMnO_3$ thin films [11].

Bulk properties of half metals are very different from surface and interface properties, which are of very crucial importance in spintronic applications. Surface states also become interface states but their coupling to the bulk of the overlayer is minimal and predominantly confined to the boundary of Brillouin zone [82,83,87–89]. Therefore, controlled surface and interface engineering as well as fundamental investigation of surface and interface states with discovery of new materials will be required for the use of half metals as a spin injector and spin source for the future.

Acknowledgements

It is a pleasure to thank Professor Peter Dowben in cooperation with whom a large part of the work discussed here was done. Also his comments and advice concerning the present review were of much help. The authors are grateful to Takashi Komesu, Delia Ristoiu and Jean-Pierre Nozieres with whom part of the discussed work was carried out. The support of the NSF "QSPINS" MRSEC (DMR 0213808) and the Office of Naval Research are gratefully acknowledged.

References

1. Surface Segregation Phenomena, ed by P.A. Dowben and A. Miller, (CRC Press, Boston 1990) p. 145
2. D. Ristoiu, J.P. Nozières, C.N. Borca, T. Komesu, H.-K. Jeong, and P.A. Dowben: Europhys. Lett. **49**, 624 (2000)
3. G.L. Bona, F. Meier, and M. Taborelli: Sol. St. Commun. **56**, 391 (1985)
4. W. Zhu, B. Sinkovic, E. Vescovo, C. Tanaka, and J.S. Moodera: Phys. Rev. B **64**, *R*060403 (2001)
5. C.N. Borca, B. Xu, T. Komesu, H.-K. Jeong, M.T. Liu, S.-H. Liou, S. Stadler, Y. Idzerda, and P.A. Dowben: Europhys. Lett. **56**, 722 (2001)
6. L. Ritchie, G. Xiao, Y. Ji, T.Y. Chen, C.L. Chien, M. Chang, C. Chen, Z. Liu, G.Wu, and X.X. Zhang: Phys. Rev. B **68**, 104430 (2003)
7. A. Plecenik, K. Frohlich, J.P. Espinos, J.P. Holgado, A. Halabica, M. Pripko, and A. Gilabert: Appl. Phys. Lett. **81**, 859 (2002)
8. I. Galanakis: J. Phys.: Condens. Matter **14**, 6329 (2002)
9. H. Kumigashira, K. Horiba, H. Ohguchi, M. Oshima, N. Nakagawa, M. Lippmaa, K. Ono, M. Kawasaki, and H. Koinuma: J. Mag. Mag. Mat. **272–276**, 1120 (2004)
10. H. Kumigashira, K. Horiba, H. Ohguchi, K. Ono, M. Oshima, N. Makagawa, M. Lippmaa, M. Kawasaki, and H. Koinuma: Appl. Phys. Lett. **82**, 3430 (2003)
11. M. Oshima, D. Kobayashi, K. Horiba, H. Ohguchi, H. Kumigashira, K. Ono, N. Nakagawa, M. Lippmaa, M. Kawasaki, and H. Koinuma: J. Elec. Spec. Relat. Phen. **137–140**, 145 (2004)
12. D. Ristoiu, J.P. Nozières, C.N. Borca, B. Borca, and P.A. Dowben: Appl. Phys. Lett. **76**, 2349 (2000)
13. J. Choi, J. Zhang, S.-H. Liou, P.A. Dowben, and E.W. Plummer: Phys. Rev. B **59**, 13453 (1999)
14. H. Dulli, P.A. Dowben, S.-H. Liou, and E.W. Plummer: Phys. Rev. B **62**, R14629 (2000)
15. C.N. Borca, Bo Xu, T. Komesu, H.-K. Jeong, M.T. Liu, S.-H. Liou, and P.A. Dowben: Surf. Sci. Lett. **512**, L346 (2002)
16. A.F. Panchula, C. Kaiser, A. Dellock, and S.S.P. Parkin: Appl. Phys. Lett. **83**, 1812 (2003)
17. J.-H. Park, E. Vescovo, H.-J. Kim, C. Kwon, R. Ramesh, and T. Venkatesan: Nature **392**, 794 (1998)

18. R. Bertacco, M. Protalupi, M. Marcon, L. Duo, F. Ciccacci, M. Bowen, J.P. Contour, and A. Barthelemy: J. Magn. Magn. Mater. **242**, 710 (2002)
19. R.J. Soulen Jr., J.M. Byers, M.S. Osofsky, B. Nadgorny, T. Ambrose, S.F. Cheng, P.R. Broussard, C.T. Tanaka, J. Nowak, J.S. Moodera, A. Barry, and J.M.D. Coey: Science **282**, 85 (1998)
20. J.M. De Teresa, A. Barthelemy, A. Fert, J.P. Contour, R. Lyonnet, F. Montaigne, P. Seneor, and A. Vaures: Phys. Rev. Lett. **82**, 4288 (1999)
21. J.Z. Sun, L. Krusin-Elbaum, P.R. Duncombe, A. Gupta, and R.B. Laibowitz: Appl. Phys. Lett. **70**, 1769 (1997)
22. Yu Lu, X.W. Li, G.Q. Gong, and G. Xiao: Phys. Rev. B **54**, R8357 (1996)
23. H.Y. Hwang, W.-W. Cheong, N.P. Ong, and B. Batlogg: Phys. Rev. Lett. **77**, 2041 (1996)
24. J.M. De Teresa, A. Barthelemy, A. Fert, J.P. Contour, F. Montaigne, and P. Seneor: Science **286**, 507 (1999)
25. K.P. Kamper, W. Schmitt, and G. Guntherodt: Phys. Rev. Lett. **59**, 2788 (1987)
26. C.N. Borca, T. Komesu, H.-K. Jeong, P.A. Dowben, D. Ristoiu, Ch. Hordequin, J.P. Nozieres, J. Pierre, S. Stadler, and Y.U. Idzerda: Phys. Rev. B **64**, 052409 (2001)
27. C.T. Tanaka, J. Nowak, and J.S. Moodera: J. Appl. Phys. **86**, 6239 (1999)
28. J.A. Caballero, A.C. Reilly, Y. Hao, J. Bass, W.P. Pratt, F. Petroff, and J.R. Childress: J. Magn. Magn. Mat. **198-199**, 55 (1999)
29. K.E.H.M. Hanssen, P.E. Mijnarends, L.P.L.M. Rabou, and K.H.J. Buschow: Phys. Rev. B **42**, 1533 (1990)
30. C.T. Tanaka, J. Nowak, and J.S. Moodera: J. Appl. Phys. **81**, 5515 (1997)
31. A. Anguelouch, A. Gupta, G. Xiao, G.X. Miao, D.W. Abraham, S. Ingvarsson, Y. Ji, and C. L. Chien: J. Appl. Phys. **91**, 7140 (2002)
32. J.S. Parker, S.M. Watts, P.G. Ivanov, and P. Xiong: Phys. Rev. Lett. **88**, 196601 (2002)
33. Y. Ji, G.J. Strijkers, F.Y. Yang, C.L. Chien, J.M. Byers, A. Anguelouch, G. Xiao, and A. Gupta: Phys. Rev. Lett. **86**, 5585 (2001)
34. W.J. DeSisto, P.R. Broussard, T.F. Ambrose, B.E. Nadgorny, and M.S. Osofsky: Appl. Phys. Lett. **76**, 3789 (2000)
35. Yu.S. Dedkov, M. Fonine, C. Konig, U. Rudiger, and G. Guntherodt: Appl. Phys. Lett. **80**, 4181(2002)
36. R. Wiesendanger, H.-J. Guntherodt, G. Guntherodt, R.J. Gambino, and R. Ruf: Phys. Rev. Lett. **65**, 247 (1990)
37. J.J. Versluijs, M.A. Bari, and J.M.D. Coey: Phys. Rev. Lett. **87**, 6601 (2001)
38. A. Barry, J.M.D. Coey, and M. Viret: J. Phys.: Condens. Matter **12**, L173 (2001)
39. Yu.S. Dedkov, U. Rudiger, and G. Guntherodt: Phys. Rev. B **65**, 064417 (2002)
40. S.A. Morton, G.D. Waddill, S. Kim, I.K. Schuller, S.A. Chambers, and J.G. Tobin: Surf. Science **513**, L451 (2002)
41. S.F. Alvarado, W. Eib, P. Munz, H.C. Siegmann, M. Campagna, and J.P. Remeika: Phys. Rev. B **13**, 4918 (1976)
42. H.W. Werner and R.P.H Garten: Rep. Prog. Phys. **47**, 221 (1984)
43. S. Hofmann: Surf. Interface Anal. **9**, 3 (1986)
44. Practical Surface Analysis by XPS and AES, ed by D. Briggs and M.P. Seah, (Wiley, Chichester 1983)

45. W.L. Baun: Surf. Interface Anal. **3**, 243 (1981)
46. Secondary Ion Mass Spectrometry, ed by S.A. Benninghoven, F.G. Rudenauer, and H.W. Werner, (Wiley, New York 1987)
47. M.P. Seah, M.E. Jones, and M.T. Anthony: Surface and Interface Analysis **6**, 242 (1984); M.P. Seah: Surface and Interface Analysis **20**, 243 (1993)
48. H.-K. Jeong, T. Komesu, C.-S. Yang, P.A. Dowben, B.D. Schultz, and C.J. Palmstrom: Mater. Lett. **58**, 2993 (2004)
49. R. Monnier: Phil. Mag. B **75**, 67 (1997)
50. S.M. Foiles: Phys. Rev. B **32**, 7685 (1985)
51. M.W. Finnis and J.E. Sinclair: Phil. Mag. A **50**, 45 (1984)
52. G. Bozzolo, J. Ferrante, R.D. Noebe, B. Good, F.S. Honecy, and P. Abel: Comput. Mater. Sci. **15**, 169 (1999)
53. H. Deng, W. Hu, X. Shu, and B. Zhang: Appl. Surf. Sci. **221**, 408 (2004)
54. S. Tanuma, S. Ichimura, and K. Yoshihara: Appl. Surf. Sci. **101**, 47 (1996)
55. NIST Electron Inelastic Mean Free Path program ver. 1.1 (NST std. ref. Database 71), http://www.nist.gov.srd
56. R. Cheng, B. Xu, C.N. Borca, A. Sokolov, C.-S. Yang, L. Yuan, S.-H. Liou, B. Doudin, and P.A. Dowben: Appl. Phys. Lett. **79** 3122 (2001)
57. N. Pilet, C. Borca, A. Sokolov, E. Ovtchenkov, Bo Xu, and B. Doudin: Mater. Lett. **58**, 2016 (2004)
58. R. Cheng, T. Komesu, H.-K. Jeong, L. Yuan, S.-H. Liou, B. Doudin, and P.A. Dowben: Phys. Lett. A **302**, 211 (2002)
59. S.Y. Liu and H.H. Kung: Surf. Sci. **110**, 504 (1981)
60. J.L. Moran-Lopez: Surf. Sci. **188**, L742 (1987)
61. A.N. Caruso, C.N. Borca, D. Ristoiu, J.P. Nozieres and P.A. Dowben: Surf. Sci. **525**, L109 (2003)
62. J.C. Slater: J. Chem. Phys. **39**, 3199 (1964)
63. J.J. Burton and E.S. Machlin: Phys. Rev. Lett. **37**, 22 (1976)
64. S. Picozzi, A. Continenza, and A.J. Freeman: Phys. Rev. B **69**, 094423 (2004)
65. D. Orgassa, H. Fujiwara, T.C. Schulthess, and W.H. Butler: Phys. Rev. B **60**, 13237 (1999)
66. R.B. Helmholdt, R.A. de Groot, F.M. Müller, P.G. van Engen, and D.H.J. Buschow: J. Magn. Magn. Mater. **43**, 249 (1984)
67. D. Orgassa, H. Fujuwara, T.C. Schulthess, and W. H. Butler: J. Appl. Phys. **87**, 5870 (2000)
68. Y. Miura, K. Nagao, and M. Shirai: Phys. Rev. B **69**, 144113 (2004)
69. M.P. Raphael, B. Ravel, Q. Huang, M.A. Willard, S.F. Cheng, B.N. Das, R.M. Stroud, K.M. Bussmann, J.H. Claassen, and V.G. Harris: Phys. Rev. B **66**, 104429 (2002)
70. L.J. Singh, Z.H. Barber, Y. Miyoshi, Y. Bugoslavsky, W.R. Branford, and L.F. Cohen: Appl. Phys. Lett. **84**, 2367 (2004)
71. C.N. Borca, T. Komesu, H.-K. Jeong, and P.A. Dowben: Appl. Phys. Lett. **77**, 88 (2000)
72. R. Skomski and P.A. Dowben: Europhys. Lett. **58**, 644 (2002)
73. H. Itoh, T. Oshsawa, and J. Inoue: Phys. Rev. Lett. **84** 2501 (2000)
74. T. Komesu, H.-K. Jeong, J. Choi, C.N. Borca, P.A. Dowben, B.D. Schultz, and C.J. Palmstrom: Phys. Rev. B **67**, 035104 (2003)
75. H.-K. Jeong, T. Komesu, P.A. Dowben, B.D. Schultz, and C.J. Palmstrom: Phys. Lett. A. **302**, 217 (2002)

76. K. Pohl. J.H. Cho, K. Terakura, M. Scheffler, and E.W. Plummer: Phys. Rev. Lett. **80**, 2853 (1998)
77. C.F. Walters, K.F. McCarthy, E.A. Soares, and M.A. van Hove: Surf. Sci. **464**, L732 (2000)
78. P.A. Dowben and R. Skomski: J. Appl. Phys. **95**, 7453 (2004)
79. P.A. Dowben and S.J. Jenkins: review article "The Limits to Spin-Polarization in Finite-Temperature Half-Metallic Ferromagnets" unpublished
80. P.A. Dowben and R. Skomski: J. Appl. Phys. **93**, 7948 (2003)
81. A.H. MacDonald, T. Jungwirth, and M. Kasner: Phys. Rev. Lett. **81**, 705 (1998)
82. S.J. Jenkins and D.A. King: Surf. Sci. **494**, L793 (2001)
83. S. Picozzi, A. Continenza, and A.J. Freeman: J. Appl. Phys. **94**, 4723 (2003)
84. I. Galanakis, P.H. Dederichs, and N. Papanikolaou: Phys. Rev. B **66**, 174429 (2002)
85. G. Lee, I.G. Kim, J.I. Lee, and Y.-R. Jang: Phys. Stat. Sol (b) **241**, 1435 (2004)
86. W.R. Branford, S.K. Clowes, Y.V. Bugoslavsky, S. Gardelis, J. Androulakis, J. Giapintzakis, C.E.A. Grigorescu, S.A. Manea, R.S. Freitas, S.B. Roy, and L.F. Cohen: Phys. Rev. B **69**, R201305 (2004)
87. S.J. Jenkins and D.A. King: Surf. Sci. **501**, L185 (2002)
88. G.A. Wijs and R.A. de Groot: Phys. Rev. B **64**, R020402 (2001)
89. S. Picozzi, A. Continenza, and A.J. Freeman: J. Phys. Chem. Solids **64**, 1697 (2003)

Heusler Alloyed Electrodes Integrated in Magnetic Tunnel-Junctions

Andreas Hütten, Sven Kämmerer, Jan Schmalhorst, and Günter Reiss

Department of Physics, University of Bielefeld, P.O. Box 100131, D-33501 Bielefeld, Germany
huetten@physik.uni-bielefeld.de

Abstract. As a consequence of the growing theoretically predictions of 100% spin polarized half- and full-Heusler compounds over the past 6 years, Heusler alloys are among the most promising materials class for future magnetoelectronic and spintronic applications. We have integrated Co_2MnSi as a representative of the full-Heusler compound family as one magnetic electrode into technological relevant magnetic tunnel junctions. The resulting tunnel magnetoresistance at 20 K was determined to be 95% corresponding to a Co_2MnSi spin polarization of 66% in combination with an AlO_x barrier thickness of 1.8 nm. For magnetic tunnel junctions prepared with an initially larger Al layer prior to oxidation the tunnel magnetoresistance at 20 K increases to about 108% associated with a Co_2MnSi spin polarization of 72% clearly proving that Co_2MnSi is already superior to 3*d*-based magnetic elements or their alloys. The corresponding room temperature values of the tunnel magnetoresistance are 33% and 41%, respectively. Structural and magnetic properties of the Co_2MnSi / AlO_x – barrier interface have been studied with X-ray diffraction, electron and X-ray absorption spectroscopy and X-ray magnetic circular dichroism and it is shown that the ferromagnetic order of Mn and Co spins at this interface is only induced in optimally annealed Co_2MnSi layer. The underlying atomic ordering mechanism responsible for achieving about its theoretical magnetic moment could be assigned to the elimination of Co-Si antisite defects whereas the reduction of Co-Mn antisite defects results in large tunnel magnetoresistance. The presence of a step like tunnel barrier which is already created during plasma oxidation while preparing the AlO_x tunnel barrier has been identified as the current limitation to achieve larger tunnel magnetoresistance and hence larger spin polarization and is a direct consequence of the oxygen affinity of the Co_2MnSi - Heusler elements Mn and Si.

1 Introduction

Spin electronic devices such as read heads, magnetic field sensors or memory cells in magnetic random access memories (MRAMs) are based on two exciting physical effects – the Giant Magnetoresistance (GMR) and the Tunnel Magnetoresistance (TMR) – which both have triggered enormous research activities on magnetic thin films over the past two decades. The discoveries of antiferromagnetic coupling in Fe/Cr multilayers [1] and the spin dependent scattering on magnetic interfaces within Co/Cu multilayers [2] led to the manifestation of the GMR-effect. Although the first measured TMR was already reported in 1975 [3] it took another 20 years to realize reliably large TMR at room temperature [4, 5]. The effective amplitude of both types of magnetoresistance in simple sandwiches consisting of two ferromagnetic metals which are separated either by thin nonmagnetic but well conducting layers in case of GMR or by a thin insulating layers serving as the tunnel barrier in case of TMR can identically be expressed as:

$$\frac{\Delta R}{R}|_{GMR\ or\ TMR} = \frac{R_A - R_P}{R_P} \tag{1}$$

where R_A and R_P represent the resistance of the two ferromagnetic layers when their magnetizations are aligned antiparallel or parallel to each other. Furthermore, Julliere's model [3] for the TMR of such a sandwich structure which is also called magnetic tunnel junction (MTJ) predicts that the resulting TMR-effect amplitude is linked to the effective spin polarization P_i of each ferromagnetic layer as is formulated by:

$$TMR = \frac{\Delta R}{R} = \frac{R_A - R_P}{R_P} = \frac{2P_{layer\ 1}P_{layer\ 2}}{1 - P_{layer\ 1}P_{layer\ 2}} \tag{2}$$

Consequently, huge TMR-effect amplitude could be achieved as is shown in Fig. 1 employing high spin polarized ferromagnetic electrodes on both sides of the MTJ.

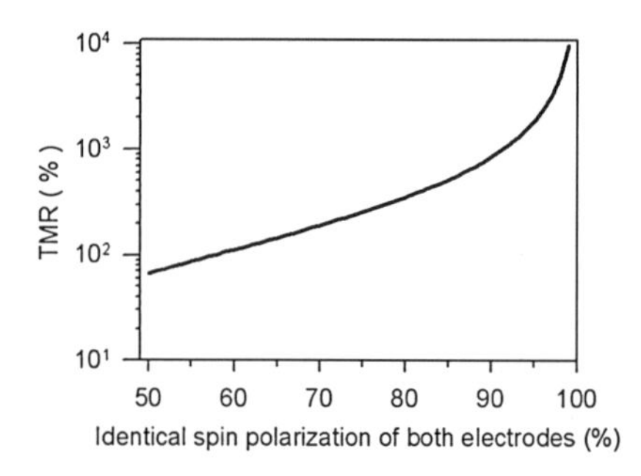

Fig. 1. Estimated TMR-effect amplitude using Julliere's model of a magnetic tunnel junction consisting of two identical ferromagnetic electrodes separated by an insulating tunnel barrier as a function of their spin polarization

Hence, one of the crucial issues today in the development of spin electronic devices is the search for new materials that exhibit large carrier spin polarizations [6]. Potential candidates include half-metallic ferromagnets oxides [7] or ferromagnetic half-metals such as Heusler alloys [8]. Characteristic for these magnetic materials named "half-metallic" [9] is their unusual band structure with only one spin band being metallic. Electrons of the opposite spin have a gap in their density of states (DOS) at the Fermi level (E_F) and hence are

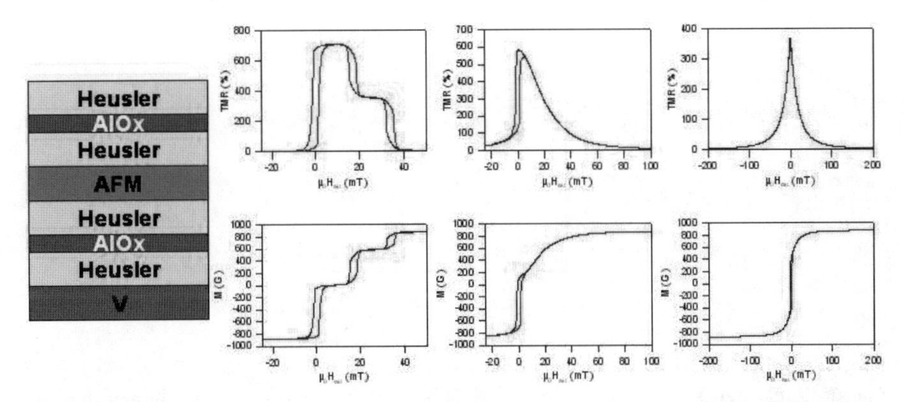

Fig. 2. Numerically calculated TMR characteristics and corresponding magnetization curves of two combined MTJ – Heusler-1 (10 nm) / AlO_x (1.8 nm) / Heusler-2 (10 nm) and Heusler-3 (20 nm) / AlO_x (1.8 nm) / Heusler-4 (20 nm) – made of generic Heusler alloyed layers assuming an identical spin polarization of 80%. Second and third Heusler layer are coupled to a thin adjacent antiferromagnetic $Mn_{83}Ir_{17}$ (10 nm) layer material employing the phenomenon of exchange biasing [17,18]. (**a**) aligning Heusler-2 and Heusler-3 via the antiferromagnetic layer parallel to the reference line results in a TMR- sensor characteristic with three applicable field ranges of high sensitivity at about zero field, at about 20 mT and at nearly about 40 mT. (**b**) aligning Heusler-2 and Heusler-3 via the antiferromagnetic layer at 45° of the reference line results in a TMR- sensor characteristic with one sharp applicable field range at about zero field and a linear field range up to about 40 mT. (**c**) aligning Heusler-2 and Heusler-3 via the antiferromagnetic layer at perpendicular to the reference line results in a hysteresis-free TMR- sensor characteristic. Changing the individual Heusler layer thickness and/or modifying there magnetic anisotropy by incorporating soft or hard magnetic materials layers leaves much room for tailoring the TMR characteristics

insulating. Consequently with only one spin band present at E_F half-metallic ferromagnets are 100% spin-polarized at the Fermi level and allow the transport of only one spin carrier across an interface or tunnel barrier into an adjacent material. Therefore, the spectacular theoretical prediction of 100% spin polarization in an entire class of materials, the half-Heusler XYZ [9–12] as well as full-Heusler X_2YZ [13–16] alloys is currently the driving force for evaluating the potential of MTJs with at least one magnetic electrode made of a Heusler alloy. The technological consequence of huge TMR-effect amplitudes will enable to realize TMR-sensors with incredible sensitivities in different field ranges as is exemplary summarized in Fig. 2. Besides detecting the amplitude of external magnetic fields a combination of the these TMR-sensors will provide the detection of the direction of the applied external field. Furthermore, the large TMR-effect amplitude is directly related to large working distances of the sensors in operation and will open up new fields

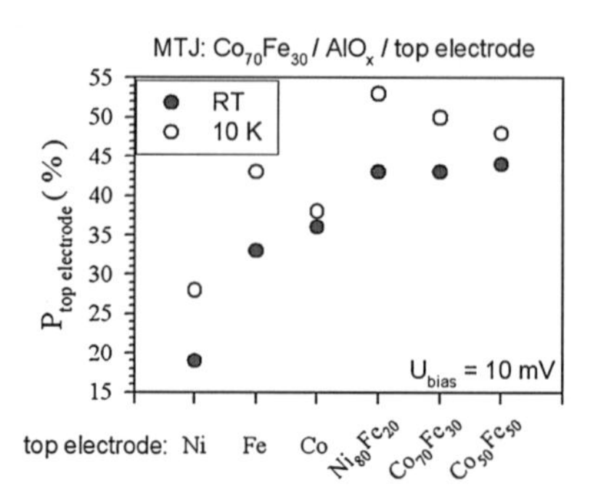

Fig. 3. Spin polarization data at room temperature and at 10 K for a variety of magnetic elements and their alloys [21]. The spin polarization was determined at 10 mV bias voltage employing magnetic tunnel junctions with the relevant TMR – building block being $Mn_{83}Ir_{17}$ (15 nm) / $Co_{70}Fe_{30}$ (3 nm) / AlO_x (1.8 nm) / top electrode (4 nm)

of applications one of which being the automotive developments concerning steering by wire and monitoring the engine performance.

However, the earlier experimental efforts to realize MTJs did not show any evidence for a true enhancement of the TMR-effect amplitude when using Heusler alloys as magnetic electrodes. The spin polarization of the half-Heusler compound NiMnSb which has been integrated in a MTJ [19] was measured to be 25% at 4.2 K and is related to a TMR-effect amplitude of 19.5%. The corresponding TMR value at room temperature (RT) was found to be 9% only. Using a full-Heusler alloy of type $Co_2Cr_{0.6}Fe_{0.4}Al$ as one of the magnetic electrodes in a MTJ slightly larger TMR values – 16% at RT and 26% at 4.2K – could be realized [20].

For comparison, a variety of effective spin polarizations of different ferromagnetic transition elements and their alloys which have been determined at RT and 10 K [21] are compiled in Fig. 3. At RT a maximum spin polarization of 44% was found for $Co_{50}Fe_{50}$, whereas that of $Ni_{80}Fe_{20}$ reached 53% at 10 K.

Nevertheless, the list of promising half- and full-Heusler alloys is long [12, 16]. One interesting candidate is Co_2MnSi which is also characterized by a 100% spin polarization as was predicted [13] from band structure calculations. In addition, Co_2MnSi possesses with $T_C = 985$ K [22] the highest Curie temperature among all full Heusler alloys identifying it to be an excellent candidate for technological applications. This is because it was found that the temperature dependence of the spin polarization scales with the corresponding magnetic moment of the material [23] and is described by the Bloch $T^{3/2}$ dependence [24]. Consequently, materials with large Curie temperatures should have a high remnant spin polarization at RT.

Recently, the spin polarization of Co_2MnSi was determined to be 54% at 4.2 K using point contact Andreev reflection spectroscopy [25] which is fairly comparable with the spin polarization of the 3*d*-based magnetic elements or their alloys given in Fig. 3. By that time the potential of Co_2MnSi integrated as one ferromagnetic electrode in technological relevant magnetic tunnel junctions had yet to be proven. To cope with this challenging task magnetic tunnel junctions had been fabricated [26] consisting of one $Co_{70}Fe_{30}$ and one Co_2MnSi electrode separated by a very thin insulating AlO_x barrier so as to determine the spin polarization of Co_2MnSi and the resulting TMR-effect amplitude. Currently, our maximum TMR-effect amplitude achieved in these MTJs with a AlO_x barrier thickness of 1.8 nm is 94.6% at 20 K applying a bias voltage of 1 mV [27]. This corresponds to a 65.5% spin polarization of Co_2MnSi which clearly exceeds that of the 3*d*-based magnetic elements or their alloys but is also well below the predicted 100%.

From a fundamental physics point of view the magnetoresistance of these MTJs depends not only on the electronic states of Co_2MnSi but also on the AlO_x tunnel barrier itself. Hence, the resulting measured spin polarization of the Co_2MnSi may not be taken as a purely intrinsic materials parameter. Nevertheless, using a MTJ so as to quantify the spin polarization of Co_2MnSi has certain advantages. Firstly, the tunnel junction area preparation is precisely controllable and so is the magnetic orientation of the two magnetic electrodes in contrast to point contacts. Secondly, cooling the MTJs allows to determine the temperature dependence of TMR-effect amplitude of Co_2MnSi which cannot be done by employing superconducting tunnel junctions or point contacts. Thirdly, the experimentally determined TMR-effect amplitude is based on all electrons which are directly involved in the tunneling process probing the relevant electronic states of Co_2MnSi without any energy resolution limitations as being present in photoemission experiments. And lastly, the spin polarization measured via a MTJ is the realistic limit important for spin electronics applications and proves the current experimental know-how to realize the Co_2MnSi / AlO_x combination.

Although our TMR results clearly prove that Co_2MnSi integrated as one of the magnetic electrodes is already superior to 3*d*-based magnetic elements or their alloys, the key issue in fabricating MTJ with integrated Heusler alloyed electrodes is the control of the Heusler alloy / tunnel barrier interface quality which will be discussed in the following sections.

2 Preparation of MTJs with one Integrated Co_2MnSi – Electrode

As was recently shown [28] textured Co_2MnSi thin films can successfully be grown onto V seed layers at RT but the resulting saturation magnetization of Co_2MnSi remains very low indicating that this Heusler phase is only weakly ordered. Post annealing of up to 600°C of these films initiates atomic ordering which leads to a saturation magnetization close to the value of 5.1 μ_B per formal unit for bulk Co_2MnSi [22]. Consequently, for synthesizing MTJs

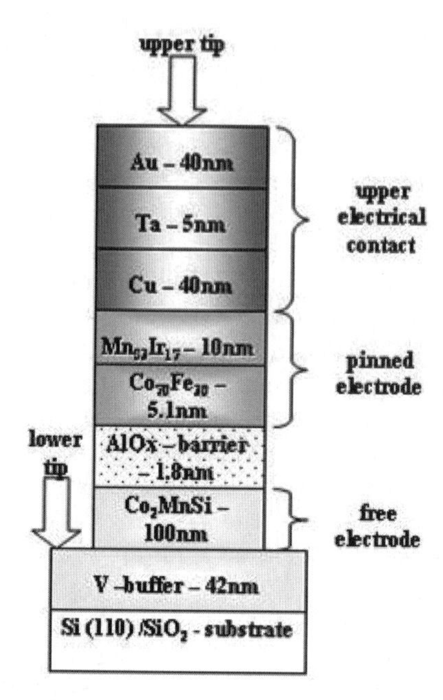

Fig. 4. Sketch of the MTJ layer stack with a Co_2MnSi layer as the lower magnetic free electrode

containing a Co_2MnSi layer as one magnetic electrode one would deduce the following strategy: firstly, to prepare a magnetically optimized Co_2MnSi layer onto a V-buffer layer, then successively continuing to build the AlO_x barrier, then the upper ferromagnetic $Co_{70}Fe_{30}$ electrode which is exchanged biased to an antiferromagnetic $Mn_{83}Ir_{17}$ layer and finally the upper current leads to complete the total layer stack. However, this strategy is still limited by the layer deposition technique in use. Employing our magnetron sputtering tools limited to a base pressure of 1×10^{-7} mbar requires to adjust our preparation strategy so as to avoid the presence of impurities at the Co_2MnSi/AlO_x – as well as at the $AlO_x/Co_{70}Fe_{30}$ – interfaces. Without braking the vacuum eight preparation steps lead to realize the MTJ layer stack sketched in Fig. 4:

1. Successively DC-magnetron sputtering of the V-buffer layer, the Co_2MnSi-Heusler layer and a 1.4 nm thick Al layer.
2. Oxidizing the Al layer in a pure oxygen plasma for 150 s. This oxidation procedure is identical to that used for the preparation of AlO_x-barriers in standard MTJs [29] with conventional 3*d* ferromagnets as ferromagnetic electrodes.
3. Annealing the present layer stack for 40 min at 500°C to induce texture and atomic ordering into the Co_2MnSi-layer and to homogenize simultaneously the AlO_x-barrier.

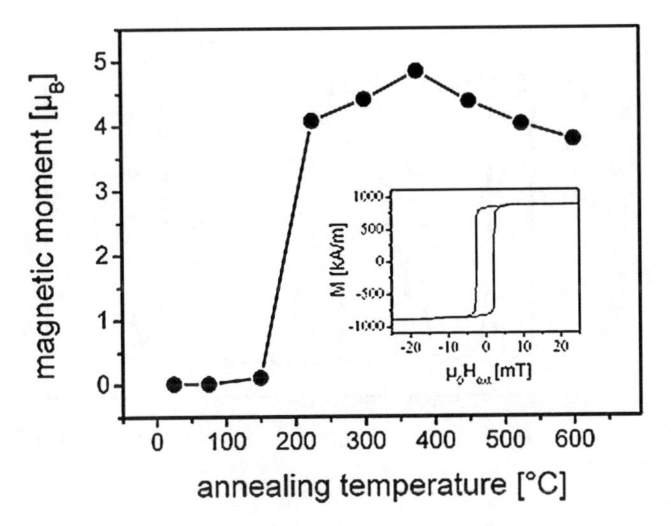

Fig. 5. Magnetic moment of 100 nm thick Co_2MnSi layers deposited at RT onto a 42 nm V-buffer layer and subsequently annealed at different temperatures for 60 min

4. Oxidizing the AlO_x-barrier layer a second time in a pure oxygen plasma for 50 s so as to clean its surface.
5. DC-magnetron sputtering of the upper magnetic $Co_{70}Fe_{30}$ layer and subsequent RF-magnetron sputtering of the $Mn_{83}Ir_{17}$ antiferromagnet.
6. DC-magnetron sputtering of the upper current Cu/Ta/Au multilayered lead.
7. Annealing the total layer stack for 1 h at 275°C in an external magnetic field of 100 mT so as to establish the exchange bias between the antiferromagnetic $Mn_{83}Ir_{17}$ and the upper ferromagnetic $Co_{70}Fe_{30}$ electrode and finally,
8. Patterning the whole layer stack by optical lithography and ion beam etching to quadratic MTJs with 100 μm – 300 μm length.

That the annealing procedure of step (3) is already sufficient to develop an ordered Co_2MnSi-layer was confirmed by the evolution of the magnetic moment vs. annealing temperature of single Co_2MnSi-layers deposited onto V-buffer shown in Fig. 5. Annealing for 60 min above 225°C already results in a magnetic moment per formal unit which is larger than 4 μ_B and a maximum of 4.7 μ_B was measured at 375°C. The hysteresis loop of a Co_2MnSi-layer annealed at 300°C, which was measured at RT and shown as inset, is almost rectangular and reveals a saturation magnetization of 881 kA/m. The corresponding X-ray diffraction pattern is given in Fig. 6 clearly showing the (110)-texture of the Co_2MnSi-layer developed during annealing. Hence it can be concluded that the combination of the V-buffer together with the RT magnetron sputtering of the Co_2MnSi-layer followed by subsequent

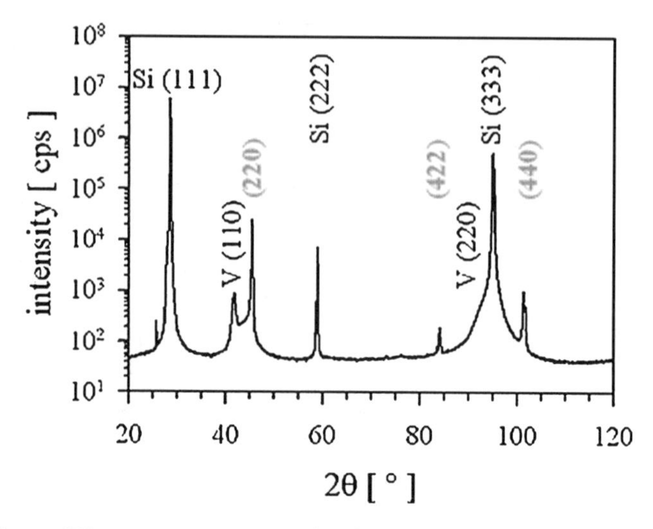

Fig. 6. X-ray diffraction pattern of a (110) textured 100 nm Co_2MnSi layer deposited at RT onto 42 nm V-buffer layer and subsequently annealed at 300°C for 60 min. The (hkl)-peaks belonging to Co_2MnSi are indicated in gray

annealing provides a proper method so as to obtain magnetically well defined Co_2MnSi-layers within the MTJ layer stack.

3 The Resulting TMR-Effect Amplitude and Spin Polarization of MTJs with Integrated Ferromagnetic Co_2MnSi

The resulting TMR-effect amplitude of such a completely processed SiO_2-substrate// V (42 nm) / Co_2MnSi (100 nm) / AlO_x (1.8 nm) / $Co_{70}Fe_{30}$ (5.1 nm) / $Mn_{83}Ir_{17}$ (10 nm) / Cu (40 nm) / Ta (5 nm) / Au (40 nm) MTJ layer stack is shown in Fig. 7 as a function of temperature. As can be seen, the TMR-effect amplitude clearly exceeds that of the two reference systems with identically made tunnel barriers below 100 K. Applying Julliere's model [3] the spin polarization of the Co_2MnSi layer can be deduced at 10 K. This analysis is made possible since the oxidation procedure of the AlO_x-barrier is the same for all three systems shown. Using the spin polarization for $Co_{70}Fe_{30}$ and $Ni_{80}Fe_{20}$ at 10 K, as measured previously [21] and given in Fig. 3, the spin polarization of the Heusler-Co_2MnSi layer is determined to be 61% and hence exceeds that of $Ni_{80}Fe_{20}$ by 14% and that of $Co_{70}Fe_{30}$ by 24%. It is observed that the TMR-effect amplitude of Co_2MnSi rapidly declines with increasing temperature and reaches only about 33% at RT, while that of the two reference systems declines more slowly. Preparing the MTJ without the V layer leads to a loss of the (110)-texture resulting in a polycrystalline

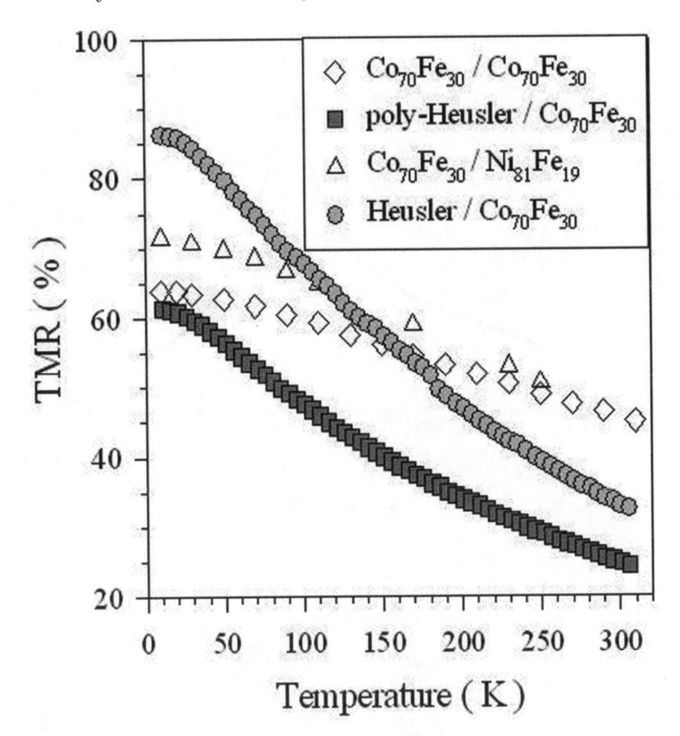

Fig. 7. Temperature dependence of the TMR-effect amplitude comparing four different MTJs measured at a bias voltage of 10 mV with identically made tunnel barriers: (●) V (42 nm) / Co_2MnSi (100 nm) / AlO_x (1.8 nm) / $Co_{70}Fe_{30}$ (5.1 nm) $Mn_{83}Ir_{17}$ (10 nm), (■) Co_2MnSi (100 nm) / AlO_x (1.8 nm) / $Co_{70}Fe_{30}$ (5.1 nm) $Mn_{83}Ir_{17}$ (10 nm) reference system I: (◇) $Mn_{83}Ir_{17}$ (10 nm) $Co_{70}Fe_{30}$ (6 nm) / AlO_x (1.8 nm) / $Co_{70}Fe_{30}$ (6 nm) and reference system II: (△) $Mn_{83}Ir_{17}$ (10 nm) $Co_{70}Fe_{30}$ (6 nm) / AlO_x (1.8 nm) / $Ni_{80}Fe_{20}$ (4 nm)

Co_2MnSi layer as was proven by XRD and in a TMR-effect amplitude vs. temperature behavior which is below that of the two reference systems as is additionally shown in Fig. 7.

Obviously, without the texture formation assisting V-underlayer, Co_2MnSi growths polycrystalline on the amorphous SiO_2 substrate. Although experimentally not verified yet, this result could have huge impact on the potential preparation of a MTJ containing Co_2MnSi as lower as well as upper ferromagnetic electrode. Without the texture formation assisting V-underlayer it can be assumed that Co_2MnSi when used as an upper ferromagnetic electrode will also be polycrystalline when growing on the amorphous or partly amorphous AlO_x barrier and would not have the 61% spin polarization. The maximum TMR-effect amplitude attainable is related to the bias voltage dependence of the TMR-effect as is shown for 20 K in Fig. 8a. At low temperatures it is considerably stronger than that for both reference systems

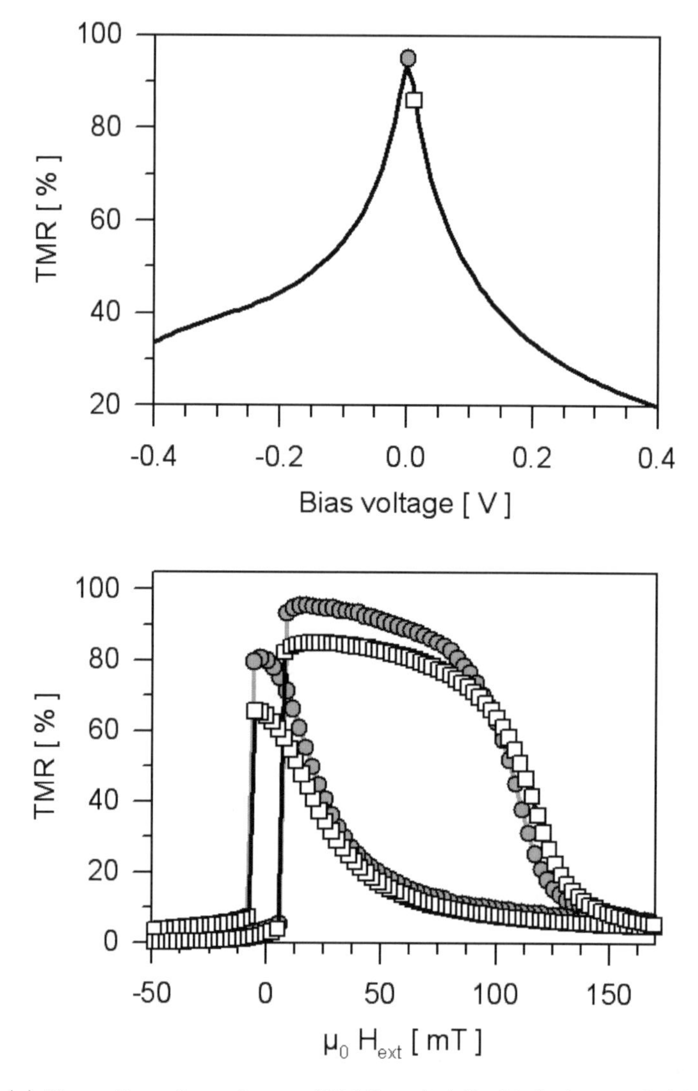

Fig. 8. (**a**) Bias voltage dependence of V (42 nm) / Co_2MnSi (100 nm) / AlO_x (1.8 nm) / $Co_{70}Fe_{30}$ (5.1 nm) $Mn_{83}Ir_{17}$ (10 nm) MTJs at 20 K. (**b**) corresponding TMR characteristics at 20 K at 1 mV (●) and 10 mV (□), respectively. The TMR-effect amplitudes are 95% (●) and 86% (□) resulting in a Co_2MnSi spin polarization of 66% (●) and 61% (□), respectively

and is characterized by a significant asymmetry between positive and negative bias. TMR characteristics measured at 1 mV and 10 mV are presented in Fig. 8b and yield TMR-effect amplitudes of 94.6% and 86% leading to a spin polarization of 65.5% and 61%, respectively. A conclusive explanation of the TMR temperature and bias voltage dependence is rather complex because of the variety of different contributions to the total conductance like direct tunneling including band structure effects, magnons and phonons assisted tunneling, unpolarized conductance via defect states in the barrier or spin scattering on paramagnetic ions. The analysis of the data with respect to all these contributions is under way, thus in the following we focus on the magnetic and microstructural quality of the Co_2MnSi / AlO_x – interface.

4 The Magnetic Switching of the Ferromagnetic Electrodes in MTJs with Integrated Ferromagnetic Co_2MnSi

To ensure that the two ferromagnetic electrodes in the MTJ – Co_2MnSi as lower and $Co_{70}Fe_{30}$ as upper electrode – are well separated their magnetic switching has been analyzed. The TMR-characteristics $TMR(H_{ext})$ [30] as well as the magnetization vs. field curve $M(H_{ext})$ [31] can be calculated via:

$$TMR(H_{ext}) = A_{TMR}(T)\frac{1}{2}\left[1 - \cos(\theta_{Co_2MnSi} - \theta_{Co_{70}Fe_{30}})\right] \tag{3}$$

where θ_{Co_2MnSi} and $\theta_{Co_{70}Fe_{30}}$ are the angles between the magnetization of the Co_2MnSi or $Co_{70}Fe_{30}$ layer and the external magnetic field, respectively and $A_{TMR}(T)$ is the experimental TMR-effect amplitude determined at temperature T and:

$$\begin{aligned} M(H_{ext}) = &\frac{M_{Co_2MnSi}t_{Co_2MnSi}}{t_{Co_2MnSi} + t_{Co_{70}Fe_{30}}}\cos(\theta_{Co_2MnSi}) \\ &+ \frac{M_{Co_{70}Fe_{30}}t_{Co_{70}Fe_{30}}}{t_{Co_2MnSi} + t_{Co_{70}Fe_{30}}}\cos(\theta_{Co_{70}Fe_{30}}) \end{aligned} \tag{4}$$

with t_i and M_i being the individual layer thickness and magnetization of the Co_2MnSi and $Co_{70}Fe_{30}$ layer. The corresponding angles and describing the magnetic switching of both magnetizations relative to the external field orientation can be found by minimization of the free energy density, which can be formulated based on an extension of the Stoner-Wohlfarth model [32], at each external field value H_{ext}:

$$\begin{aligned} E = &\mu_0 H_{ext} M_{Co_2MnSi} t_{Co_2MnSi} + \mu_0 H_{ext} M_{Co_{70}Fe_{30}} t_{Co_{70}Fe_{30}} \\ &- K_{eff,Co_2MnSi} t_{Co_2MnSi} \sin(2\theta_{Co_2MnSi} - \Psi_{1^\circ \ldots 180^\circ})^2 \\ &- K_{eff,Co_{70}Fe_{30}} t_{Co_{70}Fe_{30}} \sin(\theta_{Co_{70}Fe_{30}} - \Psi_{1^\circ \ldots 180^\circ})^2 \\ &- J_{ex}\cos(\theta_{Co_{70}Fe_{30}}) \end{aligned} \tag{5}$$

The first and second part of (5) denote the Zeeman energies of the two ferromagnetic layers. The K_i are the effective anisotropies of the Co_2MnSi and $Co_{70}Fe_{30}$ layer taking into account that Co_2MnSi has a cubic crystalline

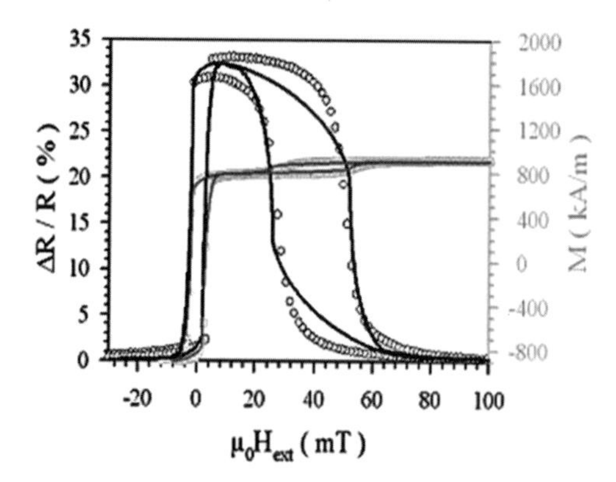

Fig. 9. Measured TMR-characteristics (open circles) and corresponding hysteresis loop (gray squares) at RT of a V (42 nm) / Co_2MnSi (100 nm) / AlO_x (1.8 nm) / $Co_{70}Fe_{30}$ (5.1 nm) $Mn_{83}Ir_{17}$ (10 nm) MTJs. Both curves where calculated employing equations 3, 4 and 5. The results are plotted as black or gray lines, respectively

anisotropy and $Co_{70}Fe_{30}$ which is pinned to the antiferromagnetic $Mn_{83}Ir_{17}$ effectively shows a uniaxial anisotropy. J_{ex} describes the strength of the exchange-bias between the antiferromagnetic $Mn_{83}Ir_{17}$ and $Co_{70}Fe_{30}$. The index of Ψ indicates that the differently orientated grains within the plane of two ferromagnetic layers have been considered by averaging over 180 easy axis, respectively. Taking the sputtered ferromagnetic layer thickness, the measured saturation magnetization $M_{Co_2MnSi} = 881$ kA/m of Co_2MnSi and the literature value of $M_{Co_{70}Fe_{30}} = 1538$ kA/m [33] for $Co_{70}Fe_{30}$ as input parameters for the energy minimization procedure, the best match between measured and calculated TMR-characteristics and corresponding hysteresis loop of the MTJ stack at RT shown in Fig. 9 is achieved with $K_{eff,Co_2MnSi} = 2.89$ kJ/m^3 and $K_{eff,Co_{70}Fe_{30}} = 20.1$ kJ/m^3 together with an exchange-bias energy of $J_{ex} = 0.3$ mJ/m^2. In case of MTJs with polycrystalline Co_2MnSi layers of identical thickness and saturation magnetization the same procedure yields an effective anisotropy for Co_2MnSi of $K_{eff,Co_2MnSi} = 1.5$ kJ/m^3.

Cooling the MTJ down to 10 K results in the TMR-characteristics presented in Fig. 10. The TMR-effect amplitude is increased to 86%. Calculating the TMR-characteristic at 10 K by taking the theoretical magnetization at 0 K for Co_2MnSi and $Co_{70}Fe_{30}$ to be $M_{Co_2MnSi} = 1018$ kA/m [13] and $M_{Co_{70}Fe_{30}} = 1764$ kA/m [33] increase in the corresponding anisotropy and exchange-bias energies can be quantified to $K_{eff,Co_2MnSi} = 7.1$ kJ/m^3, $K_{eff,Co_{70}Fe_{30}} = 77.4$ kJ/m^3 and $J_{ex} = 0.52$ mJ/m^2. From these results it can be concluded that the Co_2MnSi as well as the pinned $Co_{70}Fe_{30}$ layer are magnetically well separated over the whole temperature range and do not cause major limitation of the TMR-effect amplitude. Thus, the magnetic switching of both ferromagnetic electrodes is sufficiently separated.

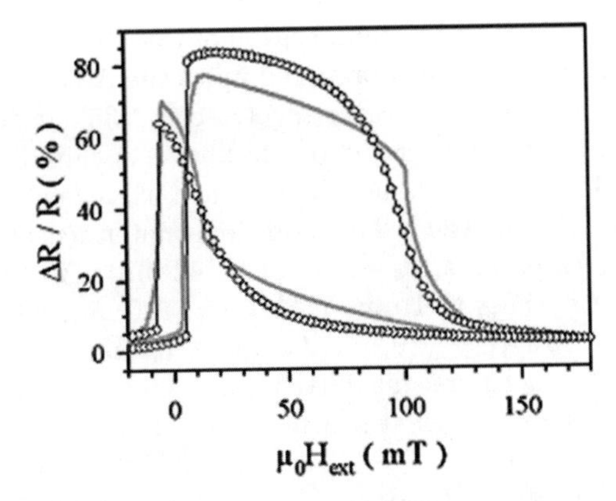

Fig. 10. Measured TMR-characteristics (black circles) at 10 K of a V (42 nm) / Co_2MnSi (100 nm) / AlO_x (1.8 nm) / $Co_{70}Fe_{30}$ (5.1 nm) $Mn_{83}Ir_{17}$ (10 nm). The TMR-characteristics were calculated employing equations 4 and 5. The results are plotted as gray lines, respectively

5 Quality of the Co_2MnSi / AlO_x – Interface

Concerning the oxygen affinity of especially Si and Mn in the Co_2MnSi-Heusler compound the Co_2MnSi / AlO_x – interface quality is an important issue so as to increase the TMR-effect amplitude. Figure 11 summarizes the formation enthalpies [34] of possible oxides at the Co_2MnSi / AlO_x – interface, which can either be formed during the plasma oxidation or later upon annealing. Whereas the formation of the chemically stable Al_2O_3 is favored in comparison to pure Co- and Si-oxides, the formation probability of Mn- and Mn_xSi_y-oxides, especially that of Mn_2SiO_4, is fairly comparable to that

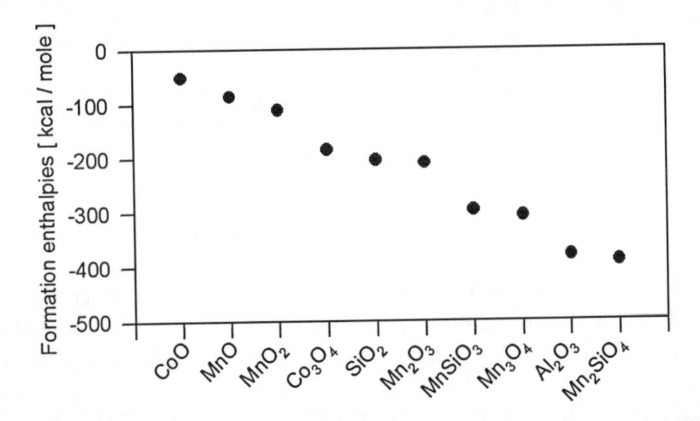

Fig. 11. Comparison of the formation enthalpies of different potential oxide-phases, taken form [34], formed at the Co_2MnSi / AlO_x – interface during plasma oxidation of the AlO_x-barrier or upon annealing

of Al_2O_3. In order to resolve the composition of Co_2MnSi at the AlO_x-barrier half MTJs, which have only been prepared up to the AlO_x-barrier, have been investigated utilizing Auger electron spectroscopy [35]. It turned out, that the plasma oxidation process step (3), see above, required to establish the AlO_x-barrier originates a Mn – and Si – enriched oxide layer of about 0.6 nm in between Co_2MnSi and AlO_x. This additional interfacial oxide layer leads to a 400 times increased area resistance of V (42 nm) / Co_2MnSi (100 nm)/ AlO_x (1.8 nm) / $Co_{70}Fe_{30}$ (5.1 nm) / $Mn_{83}Ir_{17}$ (10 nm) when compared to reference system I: $Mn_{83}Ir_{17}$ (10 nm) $Co_{70}Fe_{30}$ (6 nm) / AlO_x (1.8 nm) / $Co_{70}Fe_{30}$ (6 nm). Furthermore, the barrier parameters, such as barrier height, barrier width and asymmetry factor, which are compared in Table 1 indicate a much thicker barrier in case of V (42 nm) / Co_2MnSi (100 nm) / AlO_x (1.8 nm) / $Co_{70}Fe_{30}$ (5.1 nm) / $Mn_{83}Ir_{17}$ (10 nm). These results support the conclusion that the Co_2MnSi containing MTJs are not characterized by a simple AlO_x tunnel barrier but more likely by a step like one of type $MnSi_xO_y$ / AlO_x as is exemplary sketched in Fig. 12. Hence it can be assumed that the Mn- and Si- atoms displaced to the AlO_x-barrier interface leaving behind a compositionally disturbed Co_2MnSi surface region.

Table 1. Comparison of the tunnel barrier parameters which have been determined [36] by fitting the corresponding differential IV curves to the Brinkman model [37]

MTJs:	barrier parameters		
	barrier height [eV]	barrier width [nm]	asymmetry [eV]
V (42 nm) / Co_2MnSi (100 nm) / AlO_x (1.8 nm) / $Co_{70}Fe_{30}$ (5.1 nm) / $Mn_{83}Ir_{17}$ (10 nm)	2.0	2.8	-0.4
$Mn_{83}Ir_{17}$ (10 nm) / $Co_{70}Fe_{30}$ (6 nm) / AlO_x (1.8 nm) / $Co_{70}Fe_{30}$ (6 nm)	2.9	1.8	1.2

To evaluate the magnetization of this region, which in turn is an indicator for the DOS at E_F and the resulting spin polarization, X-ray magnetic circular dichroism (XMCD) has been employed [35]. Measuring the magnetic moment of a Co_2MnSi-single layer as a function of the layer thickness, compare with Fig. 13b, shows that it remains about constant for layer thickness equal or above 8 nm. Below a layer thickness of 8 nm the magnetic moment rapidly decreases. A similar behavior is found when determining the element specific relative XMCD signal A_{total} for Mn and Co [35] which is a measure of the total magnetic moment given in units of μ_B per atom and number of 3*d*-holes, if the XMCD sum rules were applicable without any restriction. As

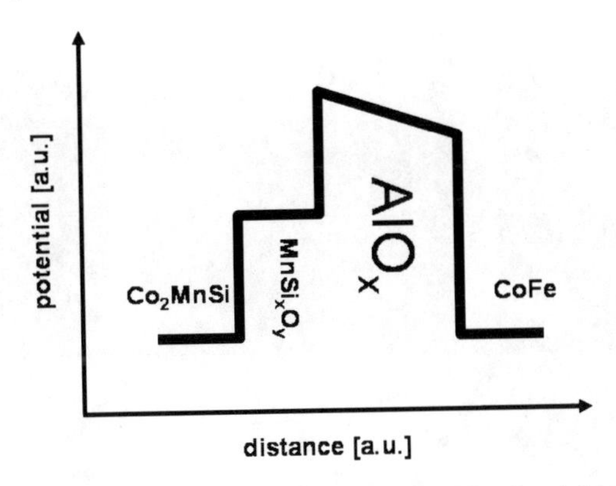

Fig. 12. Model of the potential step like $MnSi_xO_y$ / AlO_x barrier

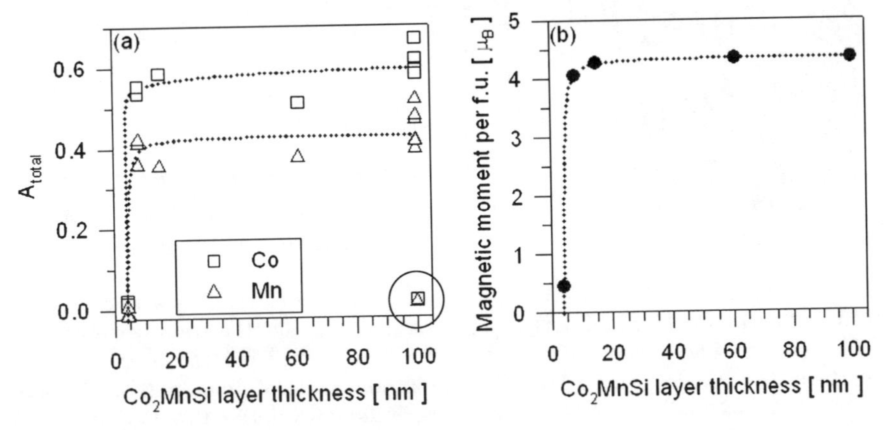

Fig. 13. (**a**) Relative XMCD signal A_{total} of Mn and Co at the Co_2MnSi / AlO_x – interface in a half MTJ-stack consisting of V (42 nm) / Co_2MnSi (X nm) / AlO_x (1.8 nm) as a function of the Co_2MnSi layer thickness. Gray encircled is the relative XMCD signal A_{total} of a half MTJ-stack which has not been annealed and hence is characterized by a completely disordered Heusler layer. (**b**) Corresponding magnetic moment of half MTJ-stack consisting of V (42 nm) / Co_2MnSi (X nm) / AlO_x (1.8 nm) as a function of the Co_2MnSi layer thickness

it is shown in Fig. 13a, the magnetic moment of Mn as well as of Co at the Co_2MnSi – AlO_x interface remains about constant down to a Co_2MnSi layer thickness of 8 nm and rapidly deteriorates below. Consequently, the evolution of the magnetic bulk and interface moment of Co_2MnSi with increasing layer thickness nicely correlate, leading to the conclusion that Co_2MnSi layers above 8 nm are well ordered, resulting in a comparable spin polarization and TMR effect amplitude.

However the corresponding Co_2MnSi layer thickness dependence of the TMR effect amplitude measured at RT and presented in Fig. 14 is very

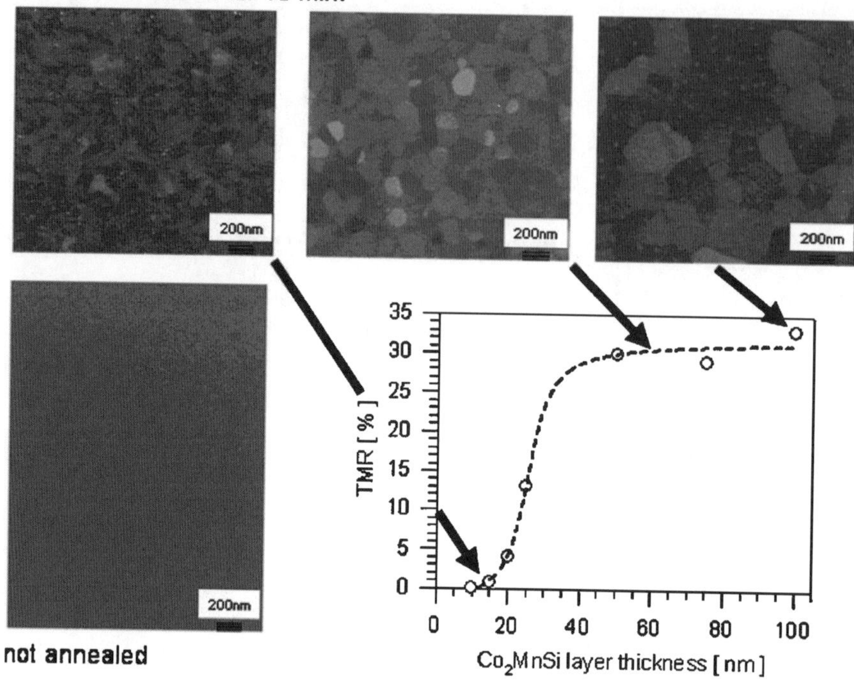

Fig. 14. SEM pictures, taken through the AlO_x layer of half MTJ-stacks consisting of V (42 nm) / Co_2MnSi (X nm) / AlO_x (1.8 nm), showing the grain size evolution of the Co_2MnSi layer upon annealing as a function of the Co_2MnSi layer thickness. In correlation, the resulting TMR-effect amplitudes at RT of similar full MTJ-stacks made of V (42 nm) / Co_2MnSi (100 nm) / AlO_x (1.8 nm) / $Co_{70}Fe_{30}$ (5.1 nm) $Mn_{83}Ir_{17}$ (10 nm) are summarized as a function of the Co_2MnSi layer thickness

different from that of the magnetic bulk and interface moment. This can be attributed to the fact that the TMR-effect is much more sensitive to local ordering while the magnetic moment averages over the magnetic interaction length which in case of the soft magnetic Co_2MnSi is probably over several decades of nm. Scanning electron micrographs shown in Fig. 14 reveal that the magnitude of the TMR-effect amplitude is strongly correlated with the grain size of corresponding Co_2MnSi layers. It can be concluded that the grain size depends on the layer thickness itself which can be taken as a hint of the underlying grain boundary energy stored therein. Thus, upon annealing the stored grain boundary energy is the driving force for grain growth. In turn, the grain growth obviously assists the atomic ordering of the Co_2MnSi on nm-scale as can be concluded from the evolution of the TMR-effect amplitude with increasing grain size up to a Co_2MnSi layer thickness of 50 nm where the TMR-effects starts to saturate. To elaborate the issue of local disorder, one serious defect in Co_2MnSi is the Co antisite formation where Co atoms

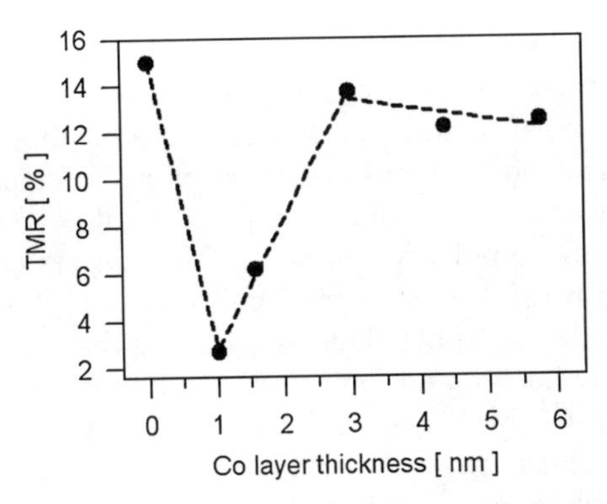

Fig. 15. TMR-effect amplitudes at RT of full MTJ-stacks made of V (42 nm) / Co_2MnSi (100 nm) / Co (X nm) / AlO_x (1.8 nm) / $Co_{70}Fe_{30}$ (12 nm) with intercalated Co layers of different thickness. The $Co_{70}Fe_{30}$ top electrodes were not exchange biased

are occupying Mn sites. According to band calculations [38] this results in the presence of DOS at E_F within the spin down $3d$ band tuning it from insulating to metallic character. Consequently, the 100% spin polarization of Co_2MnSi is lost and so is a significant TMR effect amplitude. The Co antisite formation can be promoted by intercalating thin Co layers in between the Co_2MnSi layer and the AlO_x-barrier allowing the excess Co atoms to occupy Mn atomic sites within a surface region of the Co_2MnSi layer upon annealing. The evolution of the TMR effect amplitude as a function of the Co intercalation layer thickness has been determined and is given in Fig. 15. It should be pointed out, that the upper magnetic $Co_{70}Fe_{30}$ electrode in these MTJs investigated were not exchange biased by an antiferromagnetic $Mn_{83}Ir_{17}$ layer. Hence, the resulting TMR effect amplitude is about half of the RT value obtained for V (42 nm) / Co_2MnSi (100 nm)/ AlO_x (1.8 nm) / $Co_{70}Fe_{30}$ (5.1 nm) / $Mn_{83}Ir_{17}$ (10 nm). Starting from a reference value of 15% for V (42 nm) / Co_2MnSi (100 nm)/ AlO_x (1.8 nm) / $Co_{70}Fe_{30}$ (5.1 nm) the TMR effect amplitude is instantaneously dropping to about 2.5% when a Co layer of 1 nm is intercalated. For Co layers above 3 nm the TMR effect amplitude has recovered to a value of about 13% indicating that the TMR effect is based on the new MTJ Co (3 nm) / AlO_x / $Co_{70}Fe_{30}$ (12 nm). The drastic change in the TMR effect amplitude can be taken as a strong hint that local atomic disorder at the Co_2MnSi / AlO_x barrier interface is one major obstacle towards the experimental realization of 100% spin polarization.

6 Crystallography and Atomic Ordering of Co_2MnSi

So far, it was argued that a large TMR-effect amplitude is inherently connected to a high degree of atomic order. However, nothing is said about the ordering mechanisms upon annealing nor was it possible to determine the degree of atomic order. From crystallographic point of view, the evolution of the long-range order parameter as a function of annealing time and temperature can be determined by monitoring the intensity change of superlattice reflections in X-ray or neutron diffraction patterns of polycrystalline samples.

As was shown above polycrystalline Co_2MnSi layers can be prepared by sputtering it without the V seed layer. Due to the complexity of the unit cell of Co_2MnSi the crystallographic planes originating the superlattice reflections are not immediately known and have firstly to be identified.

The unit cell of Co_2MnSi with a lattice constant of a = 0.5670 nm [39] contains 16 atoms which are located on four interpenetrating fcc sublattices – A, B, C and D – where each of the sublattices is occupied by atoms of one element – A by Co, B by Mn, C by Co and D by Si – when fully ordered. This arrangement corresponds to the $L2_1$ structure type and is drawn in Fig. 16. This $L2_1$ structure gives rise to non-zero Bragg reflections only when the Miller indices of the corresponding scattering planes are either all even or all odd. The planes with all even Miller indices can further be divided in those for which $(h+k+l)/2$ is odd and those for which $(h+k+l)/2$ is even. The intensities of these allowed reflections can be calculated by the squares of the relevant structure factors as given by [40]:

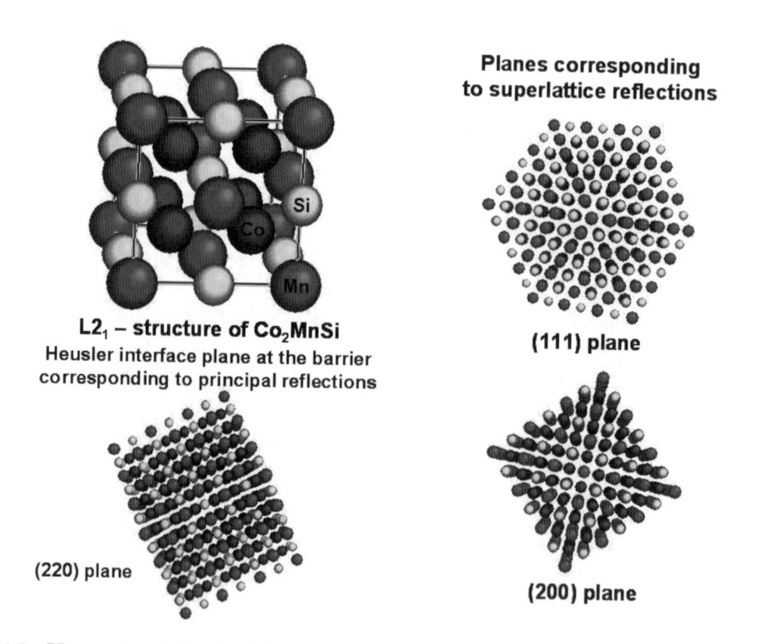

Fig. 16. Unit cell of Co_2MnSi together with the atomic arrangement of the two superlattice planes (111) and (200) and that of the principal plane (220)

$$h,k,l \text{ all odd, for example} : F^2(111) = 16[(f_A - f_C)^2 + (f_B - f_D)^2]$$
$$\frac{h+k+l}{2} = 2n+1, \text{ for example} : F^2(200) = 16(f_A - f_B + f_C - f_D)^2$$
$$\frac{h+k+l}{2} = 2n, \text{ for example} : F^2(220) = 16(f_A + f_B + f_C + f_D)^2 \quad (6)$$

where the f_i are the averaged atomic scattering factors for the four sublattices sites. The squared structure factor $F^2(220)$ is characterized by the sum of all averaged atomic scattering factors for the four sublattices sites. Hence, even in the presence of complete disorder the resulting reflection intensity will remain unchanged and identifies this class of scattering planes as the principal reflection. In contrast, the intensities of the two other classes of scattering planes $F^2(111)$ and $F^2(200)$ are very sensitive to any disorder process whereby one atom of one sublattice is interchanged by one atom of another sublattice and identify the superlattice reflections. Figure 16 additionally shows the atomic arrangements of principal and superlattice planes of which the (220) – plane is the interface of Co_2MnSi next to the AlO_x-barrier.

Since Co_2MnSi is a ternary alloy it is not possible to describe the state of order by one single order parameter as is usually done for binary alloyed phases. On the contrary, the possible ways of disorder have to be associated with a certain disorder parameter α which defines the fraction of either x or y atoms not on the correct sublattice. The four sublattices present in Co_2MnSi allow six most probable types of disorder as is summarized in Table 2.

Inserting these f_i summarized in Table 2 into equation 6 enables to calculate the intensity change of the superlattice reflections (111) and (200) following all six probable routes of disorder. Figure 17 shows the resulting evolution of the intensity change of each way of disorder as a function of the disorder parameter. Comparing all six ways of disorder, their corresponding intensity changes of the superlattice reflections (111) and (200) strongly differ except that of way 2 and 4, which are about equal. Thus, measuring the X-ray diffraction intensity of the superlattice reflections (111) and (200) of polycrystalline Co_2MnSi yields a fingerprint of the underlying atomic ordering which can then be correlated to corresponding saturation magnetization and/or TMR data. Such a correlation is made in Fig. 18 where the normalized X-ray diffraction intensities of the (111) and (200) superlattice reflection are plotted versus the saturation magnetization of 100 nm polycrystalline Co_2MnSi single layers at RT. While the normalized intensity of the (111) superlattice reflection is decreasing with decreasing saturation magnetization that of the (200) superlattice reflection is about unchanged. Interpreting a lower saturation magnetization value with a higher degree of disorder this situation can only be compared with the third route of disorder, where Mn-atoms of sublattice B are interchanged with Si-atoms of sublattice D, assuming that the disorder parameter is well below 0.5, which is justified by the fact that the saturation magnetization is already above 70% of the theoretical value.

Table 2. Averaged atomic scattering factors of the four sublattices A, B, C and D after [40]. f_{Co}, f_{Mn}, and f_{Si} are the corresponding atomic scattering factors

Way of Disorder:	Scattering factor of f_A sublattice A	Scattering Factor of f_B sublattice B	Scattering Factor of f_B sublattice C	Scattering Factor of f_D sublattice D
(1) Co(A)-Mn(B) interchange	$(1-\alpha)f_{Co}$ $+\alpha f_{Mn}$	$(1-\alpha)f_{Mn}$ $+\alpha f_{Co}$	f_{Co}	f_{Si}
(2) Co(A)-Si(D) interchange	$(1-\alpha)f_{Co}$ $+\alpha f_{Si}$	f_{Mn}	f_{Co}	$(1-\alpha)f_{Si}$ $+\alpha f_{Co}$
(3) Mn(B)-Si(D) interchange	f_{Co}	$(1-\alpha)f_{Mn}$ $+\alpha f_{Si}$	f_{Co}	$(1-\alpha)f_{Si}$ $+\alpha f_{Mn}$
(4) Co(A)-Mn(B)-Co(C) interchange	$(1-\frac{\alpha}{2})f_{Co}$ $+\frac{\alpha}{2}f_{Mn}$	$(1-\alpha)f_{Mn}$ $+\alpha f_{Co}$	$(1-\frac{\alpha}{2})f_{Co}$ $+\frac{\alpha}{2}f_{Mn}$	f_{Si}
(5) Co(A)-Si(D)-Co(C) interchange	$(1-\frac{\alpha}{2})f_{Co}$ $+\frac{\alpha}{2}f_{Si}$	f_{Mn}	$(1-\frac{\alpha}{2})f_{Co}$ $+\frac{\alpha}{2}f_{Si}$	$(1-\alpha)f_{Si}$ $+\alpha f_{Co}$
(6) random disorder in all sublattices	$(1-\frac{2}{3}\alpha)f_{Co}$ $+\frac{\alpha}{3}(f_{Mn}+f_{Si})$	$(1-\alpha)f_{Mn}$ $+\frac{\alpha}{3}(2f_{Co}+f_{Si})$	$(1-\frac{2}{3}\alpha)f_{Co}$ $+\frac{\alpha}{3}(f_{Mn}+f_{Si})$	$(1-\alpha)f_{Si}$ $+\frac{\alpha}{3}(2f_{Co}+f_{Mn})$

Correlating the saturation magnetization with the corresponding spin polarizations at 10 K in Fig. 19 it turns out that the spin polarization scales with the resulting saturation magnetization and shows only high values when the saturation is already close to the theoretical value. Since it has already been mentioned that the saturation magnetization is less sensitive to local ordering the question arises whether the TMR-effect amplitude can be taken as a local probe to explore the degree of order. Although the TMR-effect amplitude is rather sensitive to the Co_2MnSi interface region adjacent to the AlO_x tunnel barrier when compared to the volume sensitive XRD measurements the intensities of the (200) and (111) reflections are correlated to the TMR-effect amplitudes by the same approach as was used for the saturation magnetization. From Fig. 20 it can be seen that the intensity of the (111) superlattice reflection slightly increased with decreasing TMR-effect amplitude whereas that of the (200) superlattice reflection is decreasing. This situation matches second and fourth way of disorder, the Co-Mn antisite formation.

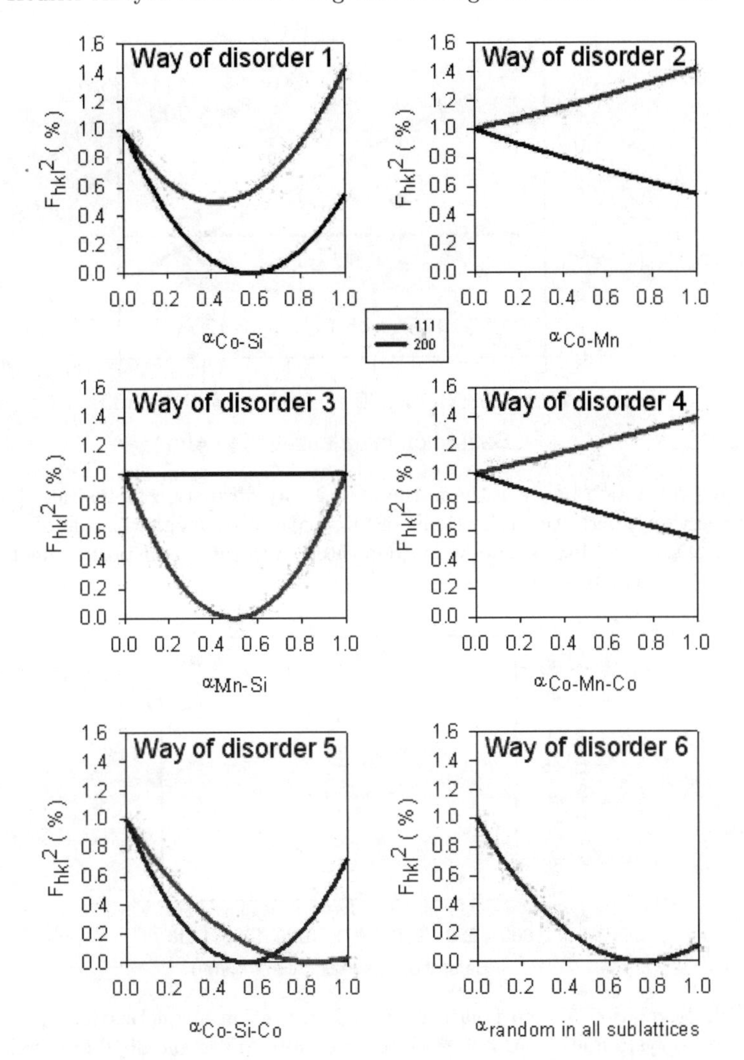

Fig. 17. Summary of the evolution of the intensity change of the superlattice reflections (111) and (200) along the six most probable ways to disorder the $L2_1$ structure of Co_2MnSi. The intensities are drawn as a function of the respective disorder parameter i and normalized to the intensity value of the state of full atomic order where $i = 0$ (see Table 2 for the definitions)

Finally, it can be concluded that two ordering mechanisms found are responsible for large resulting saturation magnetizations and large TMR-effect amplitudes. On a large nanometer scale the Co_2MnSi layer is ordered with respect to the correct occupation of the Mn- and Si- sublattice which results in large saturation magnetization. This process is followed or completed by

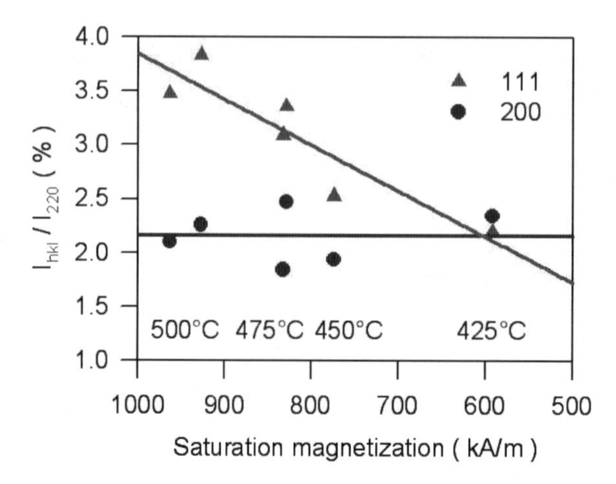

Fig. 18. Correlation between the normalized X-ray diffraction intensities I_{111} and I_{200} of the two superlattice reflections and the saturation magnetization of 100 nm polycrystalline Co_2MnSi single layers deposited onto SiO_2 and post annealed for 70 min at different temperatures

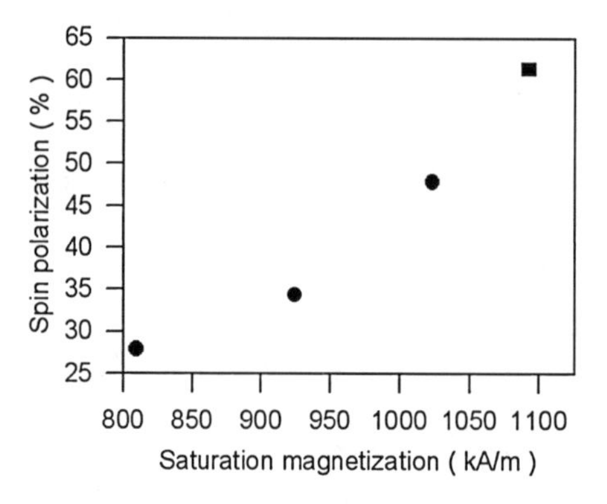

Fig. 19. Correlation of spin polarization and saturation magnetization at 10 K of 100 nm polycrystalline Co_2MnSi (●) integrated into MTJs and of 100 nm textured Co_2MnSi (■) integrated into MTJs

local ordering so as to minimize present Co-Mn antisite defects and results in large TMR-effect amplitudes.

In conclusion, it has been demonstrated that the full-Heusler compound Co_2MnSi integrated as one of the magnetic electrodes in technological relevant MTJs is characterized by a spin polarization which clearly exceeds that of $Ni_{80}Fe_{20}$ and $Co_{70}Fe_{30}$ at 10 K. That about 66% is currently the maximum spin polarization attainable at low temperatures for an AlO_x barrier thickness of 1.8 nm has mainly be associated with two experimental results. Firstly, the formation of a step like barrier which is already created during

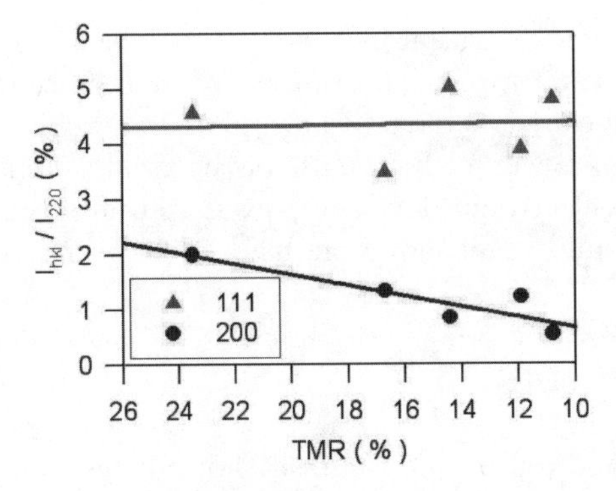

Fig. 20. Correlation between the normalized X-ray diffraction intensities I111 and I200 of the two superlattice reflections and the RT- TMR-effect amplitude of full MTJ-stacks made of Co_2MnSi (100 nm) / AlO_x (1.8 nm) / $Co_{70}Fe_{30}$ (5.1 nm) $Mn_{83}Ir_{17}$ (10 nm) with polycrystalline Co_2MnSi layers

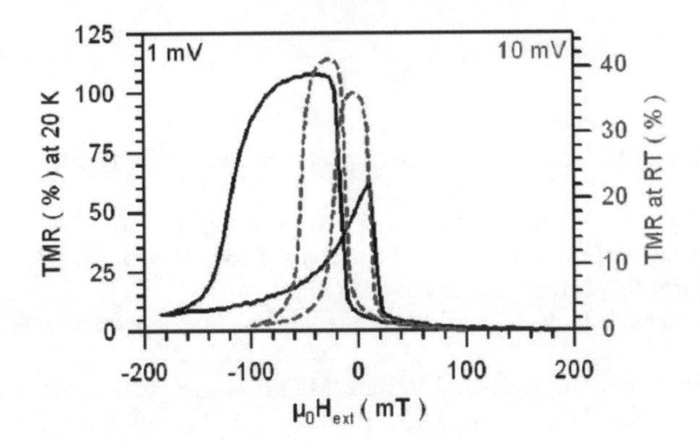

Fig. 21. Record TMR characteristic at RT (41%) and 20K (108%) of a full MTJ-stack made of V (42 nm) / Co_2MnSi (100 nm) / Al (initially about 2 nm) / $Co_{70}Fe_{30}$ (5.1 nm) $Mn_{83}Ir_{17}$ (10 nm) with an initially thicker Al layer. The 108% at 20 K yields a Co_2MnSi spin polarization of 72%. The bias voltage used is given as inset

the plasma oxidation while preparing the AlO_x tunnel barrier and is a direct consequence of the oxygen affinity of the Heusler elements Mn and Si. Consequently, when preparing a MTJ with an initially thicker Al layer the plasma oxidation assisted formation of such a step like barrier is less pronounced resulting in our current record TMR value of 108% at 20K associated with a spin polarization of 72% as is presented in Fig. 21.

Secondly, it is difficult to avoid Co antisite formation at the Co_2MnSi / AlO_x – barrier interface, which has been simulated by the intercalation of thin Co layers in between Co_2MnSi and AlO_x.

Consequently, new oxidation methods have to be tested so as to improve the barrier microstructure. In addition, new Heusler compounds with less oxidation sensitive elements have to be explored as well.

However, the gate way for future use of the great potential of the Heusler alloys for magnetoelectronic devices or even spin-injection contacts to elemental and compound semiconductors has widely been opened.

Acknowledgements

This work was supported by the Deutsche Forschungsgemeinschaft. The Advanced Light Source is supported by the Director, Office of Science, Office of Basic Energy Sciences, Materials Sciences Division, of the U.S. Department of Energy under Contract No. DE-AC03-76SF00098 at Lawrence Berkeley National Laboratory.

References

1. P. Grünberg, R. Schreiber, Y. Pang, M. B. Brodsky, and H. Sowers: Phys. Rev. Lett. **57**, 2442 (1986)
2. M.N. Baibich, J.M. Broto, A. Fert, F. Nguyen Van Dau, F. Petroff, P. Eitenne, G. Creuzet, A. Friederich, and J. Chazelas: Phys. Rev. Lett. **61**, 2472 (1988)
3. M. Julliere: Phys. Lett. **54**A, 225 (1975)
4. J.S. Moodera, L.R. Kinder, T.M. Wong, and R. Meservey: Phys. Rev. Lett. **74**, 3273 (1995)
5. T. Miyazaki and N. Tezuka: J. Magn. Magn. Mater. **139**, L231 (1995)
6. .A. Wolf, D.D. Awschalom, R.A. Buhrman, J.M. Daughton, S. von Molnár, M.L. Roukes, A.Y. Chtchelkanova, and D.M. Treger: Science **294**, 1488 (2001)
7. J.M.D. Coey and C.L. Chien: MRS Bulletin **28**, 720 (2003)
8. C. Palmstrom: MRS Bulletin **28**, 725 (2003)
9. R.A. de Groot. F.M. Mueller, P.G. van Engen, and K.H.J. Buschow: Phys. Rev. Lett. **50**, 2024 (1983)
10. J. Tobola, J. Pierre, S. Kaprzyk, R.V. Skolozdra, and M.A. Kouacou: J. Phys. Condens. Matter **10**, 1013 (1998)
11. J. Tobola and J. Pierre: J. Alloys Comp. **296**, 243 (2000)
12. I. Galanakis, P. H. Dederichs, and N. Papanikolaou: Phys. Rev. B **66**, 134428 (2002)
13. S. Ishida, T. Masaki, S. Fujii, and S. Asano: Physica B **245**, 1 (1998)
14. A. Ayuela, J. Enkovaara, K. Ullakko, and R.M. Nieminen: J. Phys. Condens. Matter **11**, 2017 (1999)
15. A. Deb and Y. Sakurai: J. Phys. Condens. Matter **12**, 2997 (2000)
16. I. Galanakis, P. H. Dederichs, and N. Papanikolaou: Phys. Rev. B **66**, 174429 (2002)
17. J. Nogues and I.K. Schuller: J. Magn. Magn. Mater. **192**, 203 (1999)
18. A.E. Berkowitz and K. Takano: J. Magn. Magn. Mater. **200**, 552 (1999)
19. C.T. Tanaka, J. Nowak, and J.S. Moodera: J. Appl. Phys. **86**, 6239 (1999)

20. K. Inomata, S. Okamura, R. Goto, and N. Tezuka: Jpn. J. Appl. Phys. **42**, L419 (2003)
21. A. Thomas, Ph.D. thesis, Department of Physics, University of Bielefeld (2004)
22. P.J. Brown, K.U. Neumann, P.J. Webster, and K.R.A. Ziebeck: J. Phys. Condens. Matter **12**, 1827 (2000)
23. R. Meservey, D. Pereskevopoulus, and P.M. Tedrow: Phys. Rev. Lett. **37**, 858 (1976)
24. R.C. O'Handley. In: *Modern Magnetic Materials*, (John Wiley & Sons, New York 2000) p. 100
25. L.J. Singh, Z.H. Barber, Y. Miyoshi, Y. Bugoslavsky, W.R. Branford, and L.F. Cohen: unpublished, preprint arXiv:cond-mat/031116
26. S. Kämmerer, A. Thomas, A. Hütten, and G. Reiss: Appl. Phys. Lett. **85**, 79 (2004)
27. A. Hütten, S. Kämmerer, J. Schmalhorst, A. Thomas, and G. Reiss: Phys. Stat. Sol. (a),) **201**, 3271 (2004)
28. S. Kämmerer, S. Heitmann, D. Meyners, D. Sudfeld, A. Thomas, A. Hütten, and G. Reiss: J. Appl. Phys. **93**, 7945 (2003)
29. A. Thomas, H. Brückl, M.D. Sacher, J. Schmalhorst, and G. Reiss: J. Vac. Sci. Technol. B **21**, 2120 (2003)
30. J.C. Slonczewski: Phys. Rev. B **39**, 6995 (1989)
31. A. Hütten, T. Hempel, S. Heitmann, and G. Reiss: Phys. Stat. Sol. (a) **189**, 327 (2002)
32. E.C. Stoner and E.P. Wohlfarth: Phil. Trans. R. Soc. Lond. A **240**, 599 (1948)
33. R.M. Bozorth. In: *Ferromagnetism* (IEEE Press, New York 1978) p. 194
34. *Lange's handbook of chemistry*, ed by J. Dean, (McGraw-Hill Book Company, New York 1973)
35. J. Schmalhorst, S. Kämmerer, M. Sacher, G. Reiss, A. Hütten, and A. Scholl: Phys. Rev. B **70**, 024426 (2004)
36. S. Kämmerer: Ph.D. thesis, Department of Physics, University of Bielefeld (2004)
37. W.F. Brinkman, R.C. Dynes, and J.M. Rowell: J. Appl. Phys. **41**, 1915 (1970)
38. S. Picozzi, A. Continenza, and A.J. Freeman: Phys. Rev. B **69**, 094423 (2004)
39. Reference database, International Centre for Diffraction Data (1999)
40. P.J. Webster: Contemp. Phys. **10**, 559 (1969)

Half-Metallic Ferromagnetism and Stability of Transition Metal Pnictides and Chalcogenides

Bang-Gui Liu

Institute of Physics, Chinese Academy of Sciences, P.O. Box 603, Beijing 100080, China; Beijing National Laboratory for Condensed Matter Physics, Beijing 100080, China
bgliu@aphy.iphy.ac.cn
Department of Physics, University of California at Berkeley, Berkeley, CA 94720, USA
bgliu@berkeley.edu

Abstract. It is highly desirable to explore robust half-metallic ferromagnetic materials compatible with important semiconductors for spintronic applications. A state-of-the-art full potential augmented plane wave method within the density-functional theory is reliable enough for this purpose. In this chapter we review theoretical research on half-metallic ferromagnetism and structural stability of transition metal pnictides and chalcogenides. We show that some zincblende transition metal pnictides are half-metallic and the half-metallic gap can be fairly wide, which is consistent with experiment. Systematic calculations reveal that zincblende phases of CrTe, CrSe, and VTe are excellent half-metallic ferromagnets. These three materials have wide half-metallic gaps, are low in total energy with respect to the corresponding ground-state phases, and, importantly, are structurally stable. Half-metallic ferromagnetism is also found in wurtzite transition metal pnictides and chalcogenides and in transition-metal doped semiconductors as well as deformed structures. Some of these half-metallic materials could be grown epitaxially in the form of ultrathin films or layers suitable for real spintronic applications.

1 Introduction

1.1 Background

It is believed that spintronics, or spin-based electronics, will result in new multi-functional electronic devices and even would lead to next generation of high-performance computers [1, 2]. Since basic units of current computer chips already reach submicrometer scales, future practical devices may be tens of nanometers or smaller in size. For these purposes one needs satisfactory spin materials that must be robust enough not only structurally but also functionally at these scales. High spin polarization seems to be necessary in real nanoscale materials for this goal. Fortunately, half-metallic ferromagnets can potentially satisfy these conditions [3–9]. Their electronic bands for one spin are metallic but those for the other spin are non-metallic if spin-orbit coupling is neglected [3], as demonstrated in Fig. 1 of the introductory chapter. The spin-orbit coupling is quite small for many materials in this category and, as a result, high spin polarization can be achieved at quite high temperatures. The half-metallic ferromagnetism is found experimentally in half-heusler alloys [3, 4], CrO_2 [5], Fe_3O_4 [6], some Manganese oxides [7], and other materials [8]. Very high (nearly 100%) spin polarization already is obtained experimentally in some of them at quite high temperatures [4, 5].

On the other hand, since current electronics and computers are based on semiconductor technology, it is highly desirable to explore half-metallic materials which are compatible with current important semiconductors. Of course, one wants to achieve good compatibility both in crystal structure and in electronic structure. This compatibility should be helpful to fabricating stable composite structures and improving interfacial destructive effect of the spin polarization in them. In addition, it should be easier to design devices inheriting the achievement of current semiconductor technology. One can obtain these half-metallic materials by replacing respectively the group-II or group-III atoms in II-VI or III-V semiconductors with some appropriate transition metal atoms. One also hopes that the half-metallic materials obtained in this way can inherit the crystalline structures of the original semiconductors, such as the zincblende structure that most important semiconductors such as GaAs share. This structure consists of two face-center cubic lattices, with one being displaced from the other by a quarter of the lattice constant in the body diagonal direction. The two face-center cubic sublattices are occupied respectively by group-III (II) atoms or group-V (VI) atoms in the original semiconductors. For the transition metal pnictides and chalcogenides with zincblende structure, one face-center cubic sublattice is occupied by the group-V or -VI atoms but the other sublattice is occupied by the transition metal atoms. The transition metal atoms in these materials play the role of the group-III or group-II atoms in the III-V or II-VI semiconductors. Therefore, there is almost no direct bonding between transition metal atoms. An atom in one sublattice is bonding with the four nearest atoms belonging to the other sublattice. This binary structure can also be considered to be a reduced form of the half-heusler structure, being obtained by keeping atoms at the x and y sites but removing atoms at the z sites, as demonstrated in Fig. 2 in the introductory chapter. Another interesting crystal structure realized by some semiconductors is the wurtzite structure. It is also a binary structure. Its local bonding environment is the same as that of the zincblende structure, but its crystal symmetry is hexagonal.

1.2 Historical Review

Zincblende phase of MnAs is the first example under intensive investigation [10]. The majority-spin band structure of the zincblende MnAs is metallic and the minority-spin bottom of conduction bands is crossed by the Fermi level, and therefore it is only a nearly half metal, not a true half metal at its equilibrium lattice constant [10, 11]. Zincblende manganese pnictides, from MnS to MnBi, are studied systematically [10–12]. The MnSb and MnBi are shown to be half-metallic [12]. One of the most important advances in this direction is the discovery of half-metallic ferromagnetism in zincblende thin films and multilayers of CrAs [9, 13]. Its half-metallic ferromagnetism is established both theoretically and experimentally, and the Curie temperature can be higher than 400 K [9, 13]. Since the thickness of the samples range up to 5 unit cells, it seems to be not far from real applications. In addition, ultrathin zincblende layers of CrSb and disk-like zincblende nanodots of MnAs are also fabricated experimentally [14, 15]. Their Curie temperatures can reach 300 and 280 K, respectively. Theoretically, the surface effect of the CrAs is studied systematically [16]; the zincblende CrSb is shown to be a robust half-metallic ferromagnet [17]. Furthermore, various explorations are performed in order to study their physical properties and to seek new half metals [18–21]. The mechanism of the half-metallic ferromagnetism is clarified in term of the electronic structures of these compounds [12, 22]. High Curie transition temperatures are predicted for some of these half-metallic materials using first-principle calculations [23]. Effects of spin-orbit coupling are studied theoretically for these materials [24]. Some heterostructures, such as (Ga,Mn)As, are predicted theoretically to be half-metallic [25]. Some multilayers toward real applications, such as CrAs/GaAs/CrAs trilayers, are explored theoretically, too [26]. Many other transition metal compounds are found theoretically to be half-metallic [27–29].

It is interesting that only the three zincblende transition metal compounds have been successfully fabricated and the best thin film samples, achieved in the case of the CrAs, are limited to only 5 unit cells in thickness [9, 13]. Zincblende phases of other transition metal pnictides cannot be fabricated experimentally although great effort has been made [30]. One wonders why the thickness of these zincblende thin films are so small and why zincblende phases of the other transition metal pnictides cannot be realized experimentally. These questions can be answered partly by examining the structural stability of the three zincblende phases [31]. First of all, these zincblende phases are substantially higher in total energy than the corresponding ground state phases with the lowest total energies. The metastable energy, defined as the total energy difference per formula unit from the corresponding ground state, is 0.9 eV or higher for each of all these zincblende compound phases. This metastable energy is too high to achieve thin films or layered samples with larger thickness [12, 22, 31]. On the other hand, it is shown that some of the zincblende phases, such as the MnSb and MnBi, lose energy when their crystalline structures are deformed so as to be not stable at all [32]. The MnAs, CrAs, and CrSb are shown to be stable against crystalline deformations, but are very soft [22, 31]. This implies that there exists essential difficulty for improving on materials of these compounds. This is consistent with recent experimental observations [33]. One has to turn to other compounds for better materials.

Zincblende CrSe, CrTe, and VTe are much better in these aspects than the zincblende pnictides [22]. Their widest half-metallic gaps (to be defined in the following) can reach 0.88 eV. Their metastable energies are 0.3 $\sim$ 0.5 eV only, substantially lower than the zincblende pnictides. Their stability against crystalline deformations is improved substantially with respect to the zincblende pnictides. Most important is that their stability is better than that of the zincblende CrAs, the best experimentally realized half metal with the zincblende structure. This enhances the hope that better zincblende half-metallic samples would be fabricated on appropriate substrates. Another theoretical approach is also used to study these half-metallic materials, showing that some of these materials in the form of tetragonal-deformed zincblende ultrathin films, maybe of a few unit cells in thickness, may be likely realized epitaxially on appropriate substrates [34]. The low metastable energies and good stability against crystal deformations both should be necessary if one wants to grow thin films or layered samples, with a thickness of 5 unit cells or more, suitable for real applications [22].

Half metals with other crystalline structures are interesting since they could enrich flexibility in device design. We have shown using accurate first-principle calculations that nine wurtzite phases of the binary transition metal pnictides and chalcogenides are good half metals [39]. Furthermore, magnetic thin films with thickness reaching 80 nm have been successfully fabricated by doping Cr into ZnTe semiconductor with the zincblende structure [35,36]. Accurate first-principle calculations show that half-metallic ferromagnetism can be formed in the ZnTe and CdTe semiconductors with appropriate doping of chromium, manganese, and vanadium [37, 38]. In some cases, deformed zincblende structures are low in total energy and half-metallicity persists in the distorted structures realized in real environments of hetero-epitaxial ultrathin films or multilayers of the zincblende materials [31].

1.3 Technical Details

The theoretical work reviewed above was done with various methods within the density-functional theory [40]. Results shown in this chapter are calculated using the Vienna package WIEN2k [41]. This is a full-potential (linear) augmented plane wave plus local orbital method within the density functional theory [40]. We use the generalized gradient approximation (GGA) [42] for exchange-correlation potential in most of our calculations. The local density approximation (LDA) is also used for comparison purpose [43]. Relativistic effects are taken into account within the scalar approximation. The spin-orbit coupling is neglected because it has little effects on our results. 3000 k points in the Brillouin zone are used for the zincblende structure and 2000 k points for the nickel-arsenide structure. For calculating the shear modulus constants of the zincblende phases, we use 6000 k points or more. The parameter $R_{mt} * K_{max}$ is set to 8.0 and the maximal angular quantum number l_{max} for the expansion of potentials and wave functions in the muffin tins is set to 10. The self-consistent calculations are considered to be converged only when the integrated charge difference per formula unit, $\int |\rho_n - \rho_{n-1}| dr$, between input charge density ($\rho_{n-1}(r)$) and output ($\rho_n(r)$) is less than 0.0001.

We assume that electronic structure of majority spin be metallic in the following definition. When majority-spin bands are metallic, a minority-spin gap at the Fermi energy level is necessary to a half-metallic gap but it alone is not enough to produce a half-metallic gap. The Fermi level must be in the minority-spin gap to make a half metal. We define E_c (or $-E_v$) as the energy of conduction band bottom (or valence band top) of minority spin with respect to the Fermi level E_F. The half-metallic gap E_g of a half metal is defined as the smaller one of E_c and E_v. If E_g becomes negative, we should set it to zero because this means actually that the half-metallic gap is closed [12, 22]. This definition works also for those half metals in which majority-spin bands are nonmetallic and minority-spin bands metallic. The metastable energy of a metastable phase is defined to be its total energy difference per formula unit from the corresponding ground state phases, where the two total energies must be calculated with corresponding equilibrium lattice constants, respectively.

The remaining part of this chapter is organized as follows. In next section we shall discuss half-metallic transition-metal pnictides with zincblende structure and analyze their structural stability. In Sect. 3 we shall discuss half-metallic ferromagnetism and stability of zincblende transition-metal chalcogenides. In Sect. 4 we shall discuss other materials with half-metallic ferromagnetism. We shall pay attention mainly to wurtzite transition-metal pnictides and chalcogenides, transition-metal-doped ZnTe and CdTe, and some deformed zincblende phases. Finally we shall present a conclusion in Sect. 5.

2 Zincblende Transition-Metal Pnictides

2.1 Electronic Structures and Half-Metallic Ferromagnetism

Zincblende phase of MnAs is first studied to seek spintronic materials compatible with current semiconductor technology [10] since GaAs is one of the most important binary semiconductors. It has been established independently by several groups around the world that the zincblende MnAs at its theoretical equilibrium lattice constant is not a true half metal, but can be considered to be a nearly half metal because its Fermi energy level crosses its minority-spin bottom of conduction bands [10, 11]. Its electronic structure in an energy window of 9.5 eV around the Fermi energy level is shown in Fig. 1. The majority-spin part is indeed typically metallic, but the minority-spin part is also metallic rather than nonmetallic as a half metal requires. It is clear

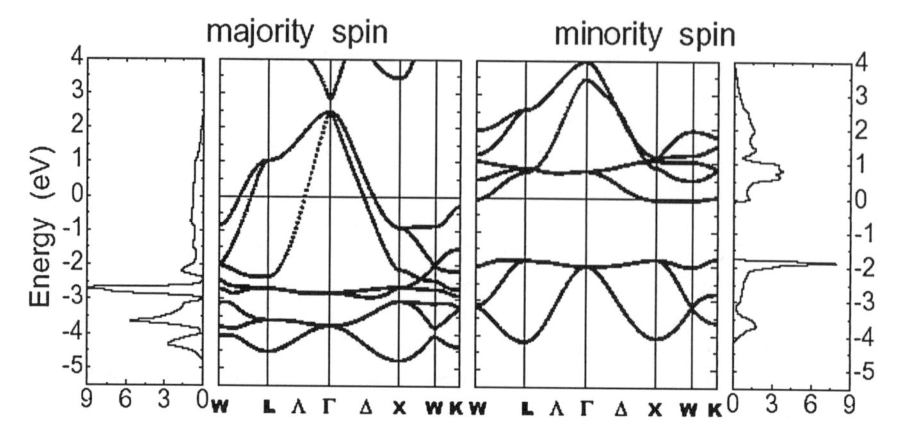

Fig. 1. Electronic structure, between −5.5 and 4 eV around the Fermi level, of zincblende half-metallic MnAs at its equilibrium lattice constant. Spin-dependent density of states (units of states/eV per formula) are shown in column 1 (*majority spin*) and 4 (*minority spin*). Spin-dependent bands are shown in column 2 (*majority spin*) and 3 (*minority spin*). The letters at the bottom indicate the high symmetry points in the Brillouin zone. The Fermi level crosses the bottom of spin-down conduction bands

that the zincblende MnAs at its equilibrium lattice constant is not a true half metal. Because the Fermi level crosses the bottom of the conduction bands, half-metallic ferromagnetism can be obtained for zincblende MnAs by enlarging the lattice constant. After the zincblende MnAs, zincblende MnP, MnSb, and MnBi have been studied systematically [11,12]. Their band structures are similar to each other except for their Fermi energy positions, but the bands around the Fermi energies, from the MnP to MnBi, systematically become narrower. This trend can be explained by considering the increasing trend of ionic radii from P to Bi, because larger ionic radii lead to larger lattice constants. As a result of this band trend, the zincblende MnP is a normally metal, but the zincblende MnSb and MnBi at their equilibrium lattice constants are half metallic. On the experimental side, interesting disk-like zincblende nanodots of MnAs have been fabricated successfully [15], but other zincblende manganese pnictides have not yet been fabricated [21,30]. These zincblende nanodots are quite large in size [15]. Their height can reach 50 Å, their diameters can reach 250 Å, and the Curie temperature can be 280 K. On the other hand, interesting multilayers including zincblende MnAs layers have been proposed [25]. Although the zincblende MnAs is not truly half-metallic, this experimental achievement can be considered to be a good starting point, encouraging for seeking better materials.

The successful fabrication of zincblende thin films of CrAs and the discovery of half-metallic ferromagnetism in them is particularly encouraging [9]. The Curie transition temperature has been proved to reach and exceed 400 K.

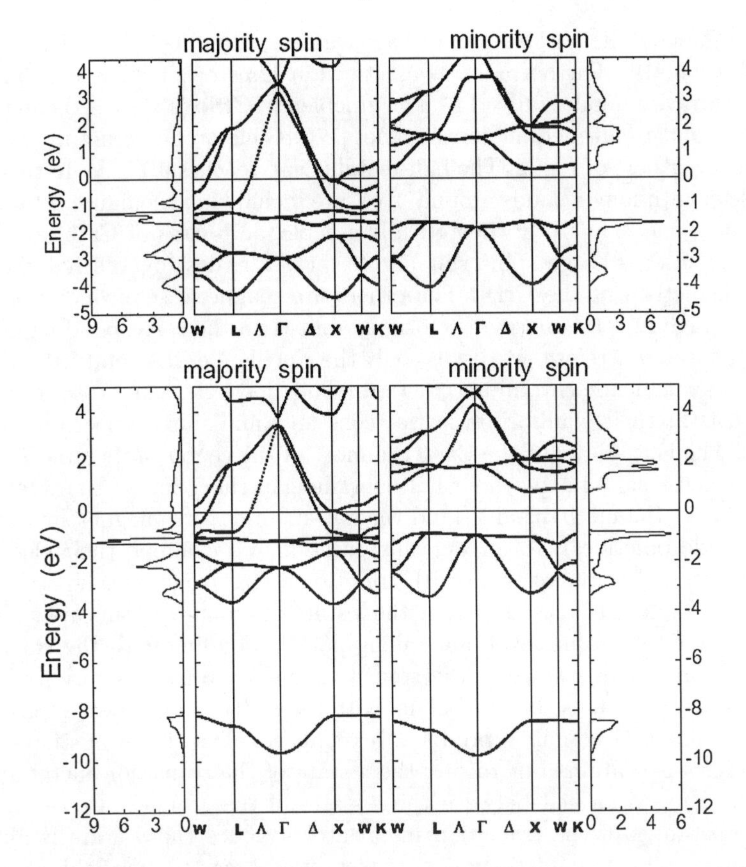

Fig. 2. Electronic structure of zincblende half-metallic CrAs (upper part, between −5 and 4.5 eV around the Fermi level) and CrSb (lower part, between −12 and 5 eV) at their equilibrium lattice constants. Spin-dependent density of states (units of state/eV per formula) are shown in column 1 (*majority spin*) and 4 (*minority spin*). Spin-dependent bands are shown in column 2 (*majority spin*) and 3 (*minority spin*). The letters at the bottom indicating the points of high symmetry in the Brillouin zones

A thickness of 5 unit cells has been achieved. It is believed reasonably that the middle part of these thin films with this or larger thickness should be the same as what one expects in the corresponding bulk crystal material of the same compounds. This implies that one can study the internal part of these thin films or layers by using the electronic structure of the corresponding crystals. Most theoretical work in this field has followed this approach. The electronic structure of the zincblende CrAs at its equilibrium lattice constant is shown in the upper part of Fig. 2. The majority-spin band structure is similar to that of the zincblende MnAs, but the Fermi level is in the minority-spin gap, which indicates that the zincblende CrAs is a true half metal with a half-metallic gap of 0.46 eV. Its surface properties have been examined for real spintronic application [16]. Furthermore, magnetic thin films of zincblende CrSb have been fabricated epitaxially [14]. The zincblende CrSb is shown theoretically to be a robust half-metallic ferromagnet [17]. Its electronic structure is shown in the lower part of Fig. 2. The half-metallic gap reaches 0.77 eV. In this case, the deep antimony *s* bands around -9 eV are included in the figure. Extending this idea, one may study theoretically zincblende phases of CrBi and vanadium pnictides. Large spin-orbit effects, however, destroy the half-metallic ferromagnetism of the CrBi. The others are magnetic semiconductors and are too high in total energy per formula unit than the corresponding ground state phases. Therefore, we discuss only the zincblende CrAs and CrSb for the series of zincblende chromium pnictides. They have the same total magnetic moments at their equilibrium lattice constants and similar electronic structures. The Fermi level of the CrSb is almost at the center of its minority-spin gap, which results in the wider half-metallic gap than that of the CrAs.

What is the mechanism behind the half-metallic ferromagnetism of these zincblende pnictides? To answer this question, we examine their electronic structures and compare them with traditional ferromagnets such as bcc Fe, hcp Co, and fcc Ni. It is clear that the *s* state of group-V atoms are substantially below the Fermi level, not taking part in bonding with the transition metal atoms. The *s* state of the transition metal atoms is hybridized with the *d* states, and must be orthogonal with the *s* state of the nearest group-V atoms. This orthogonality produces a repulsive effect, lowering the *s* state of the group-V atoms but raising the *s* state of the transition metal atoms. The bonding is formed between the *p* states of the group-V atoms and the *d* plus *s* states of the transition metal atoms. Since the *d* state is already split into the e_g doublet and t_{2g} triplet by the crystalline field, it is the t_{2g} triplet that mainly contributes to the bonding with the *p* electrons of the group-V atoms at the nearest sites. As a result, the bonding is formed mainly between the *p* electrons of the group-V atoms and the t_{2g} *d* electrons of the transition metal atoms at the nearest sites. The basic ferromagnetism is driven by the exchange splitting of the *d* electrons, which is the same as the ferromagnetism in Fe, Co, and Ni. The *s* bands of the traditional ferromagnets are around the Fermi levels, as the *d* bands do, so that there exists no band gap for both spins. The clear difference between the half-metallic ferromagnet and the traditional ferromagnet is that the transition metal *s* state of the former is raised substantially above the Fermi level, giving place to the minority-spin gap around the Fermi level of the minority-spin bands. This makes a key point for the half-metallic ferromagnetism. The other key factor is the favorable distribution of the density of state of the majority-spin bands. It pins the Fermi energy right in the minority-spin gap to make a clear half-metallic gap. The three factors together make the half-metallic ferromagnetism of zincblende pnictides.

2.2 Structural Stability and Its Relation to Experimental Status

All the electronic structures presented are calculated using corresponding equilibrium lattice constants. Now we show how to determine the equilibrium lattice constant, and thereby study metastable energies and structural stability against crystal deformations. For cubic crystals there are three independent crystal deformations: (1) volume change with the crystal symmetry fixed; (2) tetragonal deformation with the volume fixed; and (3) body diagonal deformation with the volume fixed [22, 44]. All three independent elastic moduli, bulk modulus B and two shear moduli C' and C_{44}, of a cubic crystal can be obtained in term of its total energy as functions of these crystal deformations. For these materials the second deformation clearly needs less energy than the other two. As a result, the elastic constants C' is usually much smaller than the other two and becomes the key quantity for determining structural stability of these materials. The ground state phase of these materials assumes the nickel arsenide structure with ferromagnetic or antiferromagnetic order, and the zincblende phase is substantially higher in energy. We study only the bulk crystal deformation for the nickel arsenide and the wurtzite structures. A total energy curve against the volume per formula unit can be calculated for each phase. The local minimal energy defines the equilibrium energy and the equilibrium volume of the phase. There may be several such curves for one compound. Each corresponds to one of the phases. The global minimal energy is the ground state energy of the compound. The metastable energy of a zincblende phase is defined with respect to the ground state energy. We take CrSb as an example [17]. Total energies of its zincblende phase and its ground state phase as functions of volumes per formula are shown in Fig. 3. It is clear that the equilibrium zincblende phase

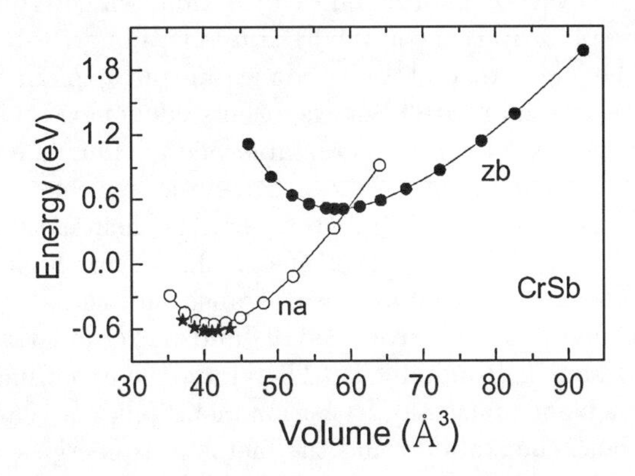

Fig. 3. Relative total energies per formula of zincblende ferromagnetic (*solid circle*), nickel-arsenide ferromagnetic (*hollow circle*), and nickel-arsenide antiferromagnetic (*solid star*) phases of CrSb as functions of cell volumes (Å^3) per formula [17]. The minimal volume for a phase is defined as its equilibrium volume

Table 1. The theoretical equilibrium lattice constants a, moments m, half-metallic gaps E_g, meta-stable energies E_t, and shear moduli C' of main zincblende transition-metal pnictides

Name	a (Å)	m (μ_B)	E_g (eV)	E_t (eV)	C' (GPa)
MnBi	6.40	4	0.46	0.9	– [32]
MnSb	6.166	4	0.15	0.9	– [32]
MnAs	5.717	3.91	–	0.9	~3 [31]
CrAs	5.659	3	0.46	0.9	5 [22]
CrSb	6.138	3	0.77	1.1	~1 [31]

is about 1 eV higher in total energy per formula unit than the ground state energy [17,45]. Using the equilibrium lattice constant, one can determine the magnetic moment uniquely. For each of the compounds we can obtain their equilibrium lattice constants, metastable energies, and magnetic moments in the same way. They are summarized in Table 1. Note that a half metal has an integer magnetic moment in units of Bohr magneton μ_B.

Also shown in Table 1 are the shear moduli C' of the zincblende transition metal pnictides. A positive shear modulus C' implies that the crystal material is stable against the crystalline deformations, because the other two moduli are much larger and therefore become less relevant to the stability. For the same reason, they are not included in the following tables and figures. The complete total energy curves of the zincblende MnAs, CrAs and CrSb against the tetragonal crystalline deformation are shown in Fig. 4. They are stable, being in contrast to the cases of the unstable MnSb and MnBi [32]. The best stability is obtained in the case of the zincblende phase of CrAs. The zincblende MnAs is also convincingly stable. The energy curve of the zincblende CrSb is very flat for small deformation, which indicates that the zincblende phase of CrSb is theoretically stable within error bars but should be very soft. The theoretical stability alone can not explain why the three zincblende phases are fabricated but the others cannot be obtained however much effort one has made. The metastable energy must be considered as well. The zincblende vanadium pnictides, magnetic semiconductors with tiny minority-spin gaps, cannot be fabricated because their metastable energies are larger than 1.5 eV. The spin-orbit effects will destroy high spin polarization in any of intermetallic compounds of transition metals and bismuth. In addition, zincblende phases of iron, cobalt, and nickel pnictides are normal ferromagnets without half-metallicity. Therefore, the experimental situation in fabricating zincblende phases of transition metal pnictides can be explained by considering both the stability and the metastable energies of them.

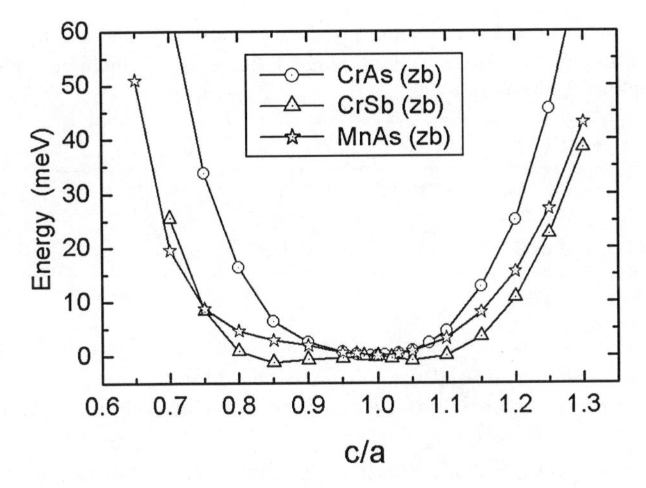

Fig. 4. Total energies per formula as functions of c/a for tetragonal-deformed zincblende structures of CrAs (*circle*), CrSb (*triangle*), and MnAs (*star*), where c is lattice constant in the z direction and a that in the x or y direction [31]. The total energies are set to zero for $c/a = 1$ in all the three cases

3 Zincblende Transition-Metal Chalcogenides

3.1 Energetics and Magnetic Moments

As it is shown in last section, half-metallic ferromagnetism is predicted theoretically in many zincblende chromium and manganese pnictides, but only three of them have been fabricated experimentally. This can be explained by comparing both their shear moduli and their metastable energies. Even for the zincblende MnAs, CrAs, and CrSb, their metastable energies are larger than 0.9 eV and their theoretical shear moduli are not larger than 5 GPa. These should imply it is very difficult, if not impossible, to experimentally fabricate thicker films or layered samples for real applications. It is highly desirable to seek better half-metallic ferromagnets. For this purpose, it is natural to explore all zincblende phases of 3d-transition-metal chalcogenides. As we have shown [22], there exist only three zincblende half-metallic ferromagnets, namely CrSe, CrTe and VTe phases, among all these zincblende transition-metal chalcogenide compounds.

The ground-state phase of CrTe is a metallic ferromagnet with the hexagonal nickel-arsenide structure. Its experimental lattice constants are $a = 3.998$ Å and $c = 6.254$ Å [46, 47] and the experimental Curie temperature is $T_c = 340$ K [46]. The ground-state phase of CrSe is an antiferromagnet with the nickel-arsenide structure. Its experimental lattice constants are $a = 3.674$ Å and $c = 6.001$ Å [46, 48] and the Neel temperature is located at 320 K by specific heat measurement [49]. There have been no experimental reports on VTe, but it is shown by comparing the total energies of various phases that the

Table 2. The theoretical equilibrium lattice constants a, moments m, half-metallic gaps E_g, meta-stable energies E_t, and shear moduli C' of zincblende half-metallic transition-metal chalcogenides [22]

Name	a (Å)	m (μ_B)	E_g (eV)	E_t (eV)	C' (GPa)
CrSe	5.833	4	0.61	0.31	5.6
CrTe	6.292	4	0.88	0.36	5.5
VTe	6.271	3	0.31	0.53	9.9

nickel-arsenide ferromagnetic phase, with equilibrium lattice constants 4.13 and 6.07 Å, is the ground-state phase in this case. Systematic calculations show that the equilibrium lattice constants of the zincblende half-metallic ferromagnetic phases of CrSe, CrTe, and VTe are 5.833, 6.292, and 6.271 Å, respectively. The equilibrium lattice constants, magnetic moments per formula unit, half-metallic gaps, metastable energies, and shear moduli C' of the zincblende half-metallic phases are summarized in Table 2. The magnetic moments in units of Bohr magneton are integer, being 4 for the CrSe and CrTe, and 3 for VTe. This reflects the fact that chromium or vanadium atom has six or five electrons respectively for the $4s$ plus $3d$ states. Two of them are used to bond with the tellurium or selenium atom, respectively, and the remaining electrons (4 for chromium, 3 for vanadium) are polarized completely. Another issue is to check if the ferromagnetic orders of the zincblende phases achieve the lowest total energy when compared with other magnetic orders. Total energies of all antiferromagnetic orders with upto 8 atoms in primitive cells have been calculated for all the three cases of zincblende CrSe, CrTe, and VTe. It is found that the ferromagnetic phases are the lowest ones for all the three zincblende phases of CrSe, CrTe, and VTe, respectively. All the total energies of the zincblende and nickel arsenide phases of these compounds as functions of their cell volumes per formula unit are presented in Fig. 5. The half-metallic ferromagnetic phases presented in this figure are indeed lower than any corresponding phase of any antiferromagnetic order. It is clear that the zincblende phases of CrTe and CrSe are 0.26 and 0.36 eV higher in total energy per formula unit than the corresponding nickel-arsenide ground-state phases. These metastable energies are substantially smaller than those of the zincblende transition metal pnictides discussed above.

3.2 Electronic Structure and Half-Metallic Ferromagnetism

The total density of states and the band structures, between -6 eV (or -5.5 eV) and $+4$ eV with respect to the Fermi levels, of the zincblende CrSe, CrTe, and VTe at their equilibrium lattice constants are shown in Fig. 6. Their energy levels are scaled in a unified way in order to compare the bands and DOS conveniently. The electronic structures are similar to each other, and are similar to those of the zincblende half-metallic transition-metal pnictides.

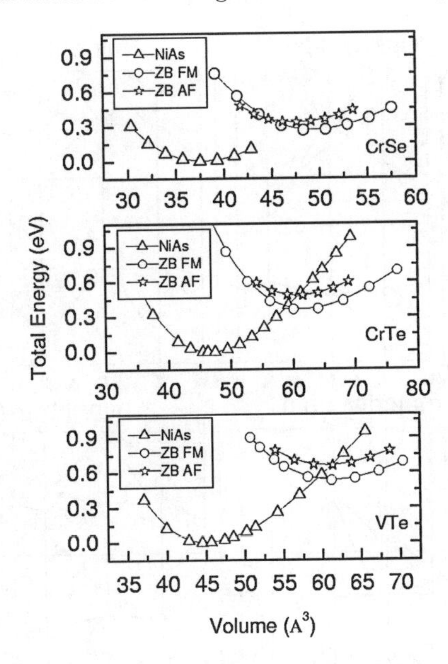

Fig. 5. Relative total energies per formula unit of ferromagnetic zincblende (*circle*), ferromagnetic zincblende (*star*), and nickel-arsenide ground-state (*triangle*) phases of CrSe (*upper panel*), CrTe (*middle panel*), and VTe (*lower panel*) as functions of cell volumes (Å^3) per formula [22]. The minimal volume for a phase is defined as its equilibrium volume

The spin-dependent density of states projected in atomic muffin tins and in interstitial regions are shown respectively in Fig. 7. The two figures together are very useful to clarify the bonding property and the atomic state characters of the bands of the zincblende half-metallic materials. The transition metal d states are split into a triplet t_{2g} and a doublet e_g, separate from each other. The filled bands below -2 eV, both majority-spin and minority-spin, are originated from the polarized covalent bonding between the p electrons of tellurium (or selenium) and the d electrons of chromium (or vanadium) with the t_{2g} character. They, especially the minority-spin ones, are dominated by tellurium (or vanadium) p states, which implies that the bonding polarization is very strong. The majority-spin e_g bands remain very narrow and are filled. For both spins, the doublet e_g bands are above the triplet p-dominated bands. The transition metal s state is moved up, almost above the e_g bands. The majority-spin triplet bands around the Fermi levels are mainly originated from the transition metal t_{2g} states. These are essential for forming the half-metallic ferromagnetism in these three good half metals, which is the same as those of the transition metal pnictides. It is safe to conclude that the mechanism of the half-metallic ferromagnetism in the three chalcogenides is the same as that in the pnictides discussed above.

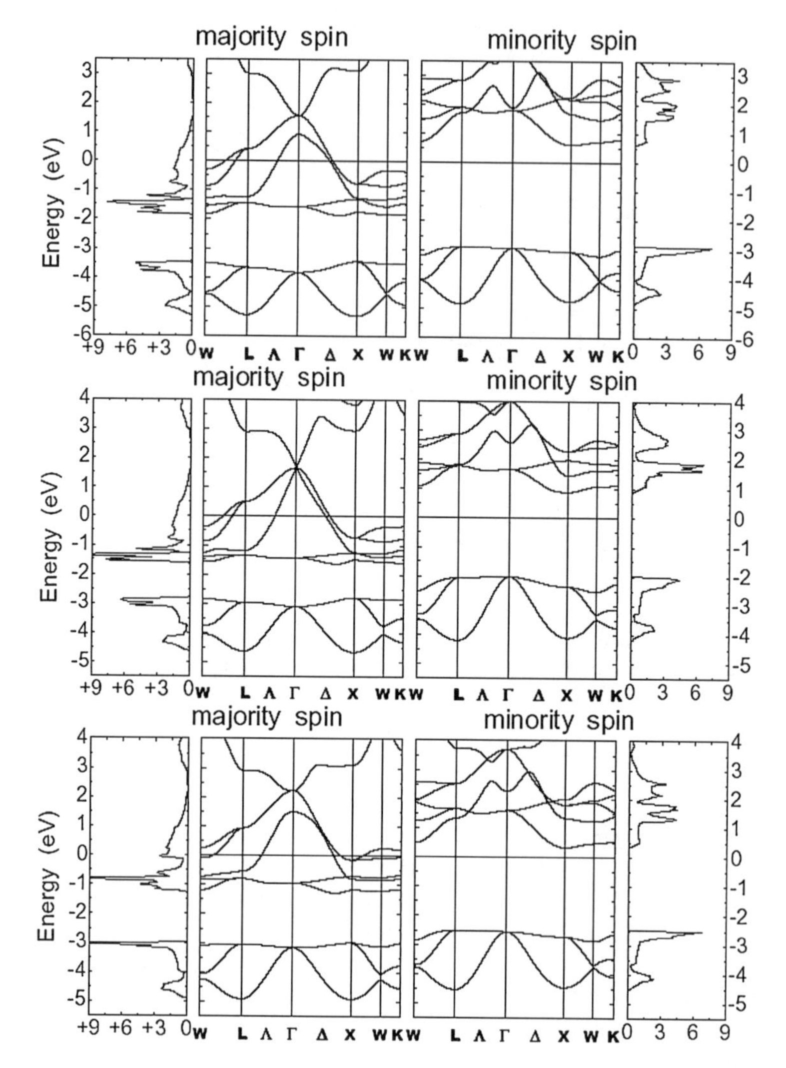

Fig. 6. Electronic structures, around the Fermi level, of zincblende half-metallic CrSe (*upper row*), CrTe (*middle row*), and VTe (*lower row*) at their equilibrium lattice constants. Spin-dependent density of states (units of state/eV per formula unit) are shown in column 1 (*majority spin*) and 4 (*minority spin*). Spin-dependent bands are shown in column 2 (*majority spin*) and 3 (*minority spin*), with the letters at the bottom indicating the points of high symmetry in the Brillouin zone

3.3 Mechanical Stability and Potential to be Realized

As we have discussed in the first subsection, the three zincblende half-metallic chalcogenides have substantially smaller metastable energies than the zincblende pnictides, rendering them much more favorable to experimental

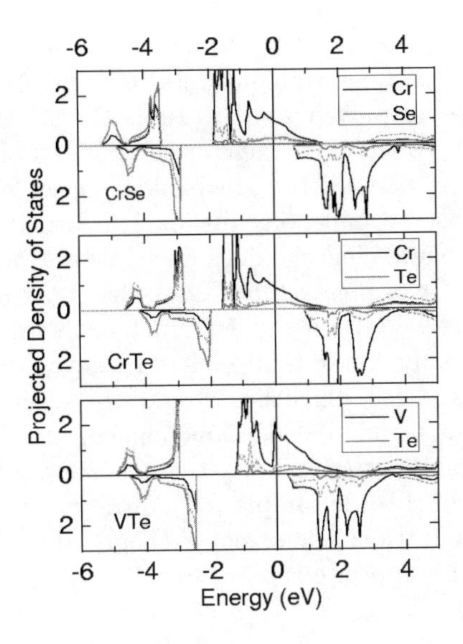

Fig. 7. Spin-dependent density of states (DOS, in units of states/eV per formula) of zincblende CrSe (*upper panel*), CrTe (*middle panel*), and VTe (*lower panel*) projected in transition metal muffin tins (*black*), group-VI muffin tins (*grey*), and interstitial regions (*grey dash*). The upper part in each panel is for majority spin and the lower part for minority spin

realization. We now address such a question: whether or not are these three zincblende phases mechanically stable enough so as to have better potential to be fabricated experimentally [22, 44]? Shown in Fig. 8 are the total energies of the three zincblende half-metallic phases against the tetragonal strains,

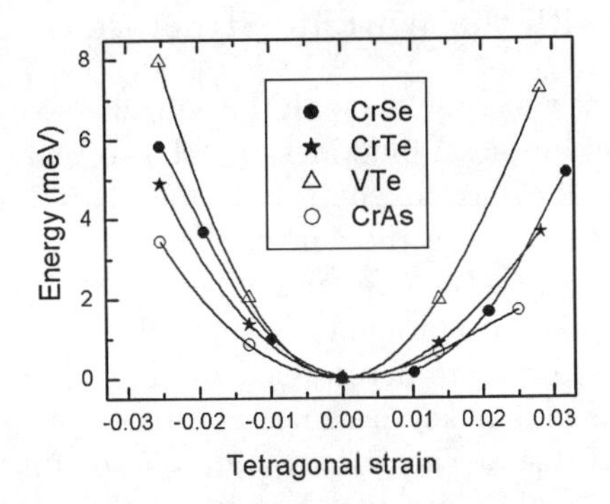

Fig. 8. Total energies of the zincblende phases of CrSe (*dot line plus solid circles*), CrTe (*solid line plus stars*), VTe (*long dash plus triangles*) and CrAs (*dash line plus open circles*) as functions of tetragonal strains with cell volumes fixed [22]. The total energies are set to zero for zero strain (or standard equilibrium zincblende structures)

with the zincblende CrAs added for comparison. In fact, all the tetragonal and body diagonal shear elastic constants, C' and C_{44}, have been calculated. The C' values are given also in Table 2, but the C_{44}, being quite large, is less relevant to examine their mechanical stability and therefore is not shown. It is clear that all the zincblende phases are convincingly stable against the tetragonal and body diagonal deformations. Quantitatively, the three transition-metal chalcogenides have larger tetragonal shear moduli than the zincblende CrAs phase. Since GaAs has bulk and tetragonal shear modulus of 61.3 and 29.8 GPa, the three zincblende transition-metal chalcogenides are substantially softer than GaAs. However, they are harder than the zincblende CrAs phase which has already been successfully fabricated. Because the three zincblende half-metallic chalcogenides have much smaller metastable energies and better stability than the zincblende CrAs and CrSb, we believe that the zincblende samples of half-metallic CrSe, CrTe and VTe suitable for real applications, or some of them, could be realized through epitaxial growth on appropriate substrates.

4 Half-Metallic Ferromagnetism in Other Structures

In the above sections we have discussed binary half metals with the zincblende structure. Half-metallic ferromagnetism may also be achieved in other crystalline structures or doped materials. In the following we shall discuss binary half metals with the wurtzite structure, ternary half metals obtained by doping transition metal atoms into semiconductors ZnTe and CdTe, and half-metals in deformed zincblende structures as realized in epitaxial ultrathin films.

4.1 Half Metals with the Wurtzite Structure

In order to explore new half metals with the wurtzite structure, we have studied all the 3d-transition-metal pnictides and chalcogenides with the wurtzite structure. Half-metallic ferromagnetism is found in nine transition-metal pnictides and chalcogenides in the wurtzite structure, namely MnSb, CrAs, CrSb, VAs, VSb, CrSe, CrTe, VSe, and VTe. For each of them, the volume and lattice constant ratio c/a are both optimized. This implies that the equilibrium constants a and c are determined by minimizing the total energies of the wurtzite phases. The key parameters of the nine wurtzite half metals are summarized in Table 3. The ground-state phase of most transition-metal pnictides and chalcogenides assumes hexagonal NiAs structure [50–55], the other two being MnP structure [56,57]. The metastable energy (E_t) is rather small for some of them. For example, the metastable energy E_t reaches 0.26 eV in the case of the CrSe. Spin-dependent band structure of the wurtzite

Table 3. The equilibrium lattice constants, bulk moduli (B), magnetic moments per formula unit (m), half-metallic gaps (E_g), and metastable energies per formula unit (E_t) of the nine wurtzite phases of transition-metal pnictides and chalcogenides [39]

Name	Lattice a/c (Å)	B (Gpa)	m (μ_B)	E_g (eV)	E_t (eV)
MnSb	4.379 / 7.138	44.9	4	0.23	0.95
CrAs	4.002 / 6.531	68.4	3	0.52	0.9
CrSb	4.343 / 7.088	48.2	3	0.96	0.99
VAs	4.035 / 6.583	70.6	2	0.53	1.6
VSb	4.393 / 7.170	49.9	2	0.92	1.2
CrSe	4.122 / 6.785	61.9	4	0.63	0.31
CrTe	4.458 / 7.234	42.3	4	0.97	0.34
VSe	4.097 / 6.690	57.9	3	0.25	0.47
VTe	4.423 / 7.285	49.1	3	0.60	0.56

CrTe with its equilibrium lattice constants, as the presentative of the nine wurtzite half metals, is shown in Fig. 9. The spin-dependent density of states projected in chromium and tellurium muffin tins and in interstitial region are shown in Fig. 10. The number of the bands are doubled with respect to that of

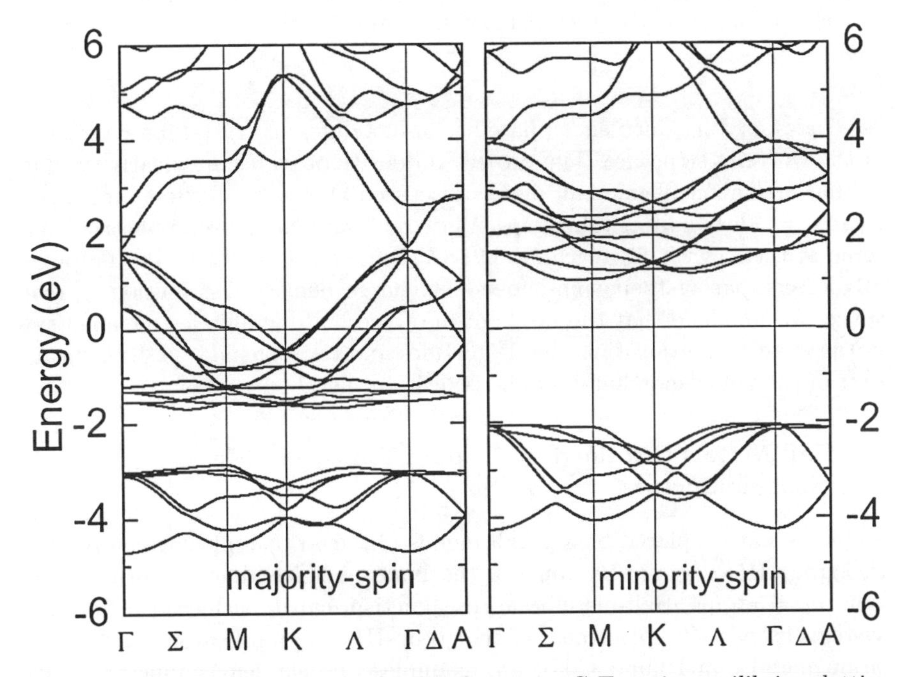

Fig. 9. Spin-dependent band structure of wurtzite CrTe at its equilibrium lattice constants. The left panel is for majority spin and the right panel for minority spin. The letters at the bottom are used to indicate the points of high symmetry in the Brillouin zone

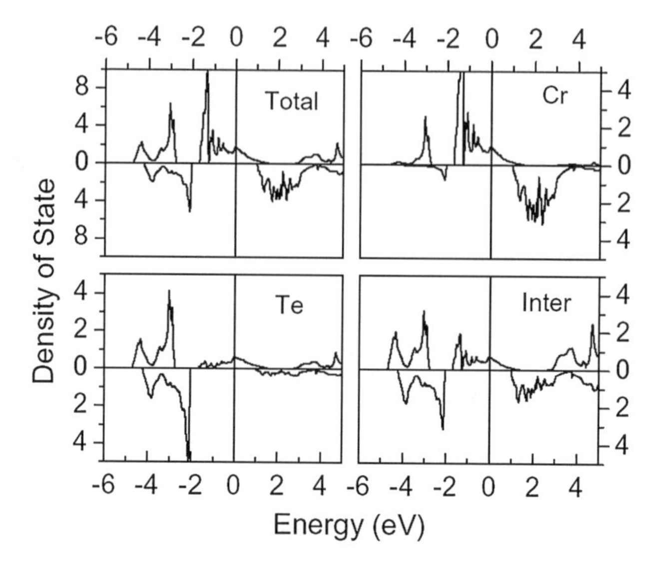

Fig. 10. Spin-dependent densities of states (DOS) of wurtzite CrTe at its equilibrium lattice constants. The left-top panel is the total DOS, the right-top panel the DOS projected in chromium muffin tin, the left-bottom panel the DOS projected in tellurium muffin tin, and the right-bottom panel the DOS projected in interstitial region. The upper part in each panel is for majority spin and the lower part for minority spin

the corresponding zincblende phase because of different crystalline symmetry of the wurtzite structure. The projected density of states are clearly similar to that of the zincblende half metals discussed above. In all these nine cases there exist clear half-metallic gaps. Wide half-metallic gaps are observed for some of them, with E_g reaching 0.97 eV in the case of the CrTe. Through further comparing their spin-dependent charge density distribution in real space, we conclude that the mechanism of the half-metallic ferromagnetism in these wurtzite transition metal pnictides and chalcogenides is the same as that of the zincblende half-metallic transition metal ones.

4.2 Half Metals Obtained by Doping Transition Metals into Semiconductors

So far we have explored binary half metals obtained by replacing completely the group-III or group-II atoms of the binary semiconductors with transition metal atoms. Half-metallic ferromagnetism can be achieved in ternary systems by partial replacement of the group-III or group-II atoms with transition metals. ZnTe and CdTe are examples of such semiconductors since upto 33% Cr and more Mn have been doped into ZnTe or CdTe [35, 36]. The spin-dependent DOS and bands of $Zn_{1-x}Cr_xTe$ and $Cd_{1-x}Cr_xTe$ with $x = 0.75$ and 0.25 are presented in Fig. 11. All these zincblende-based ternary

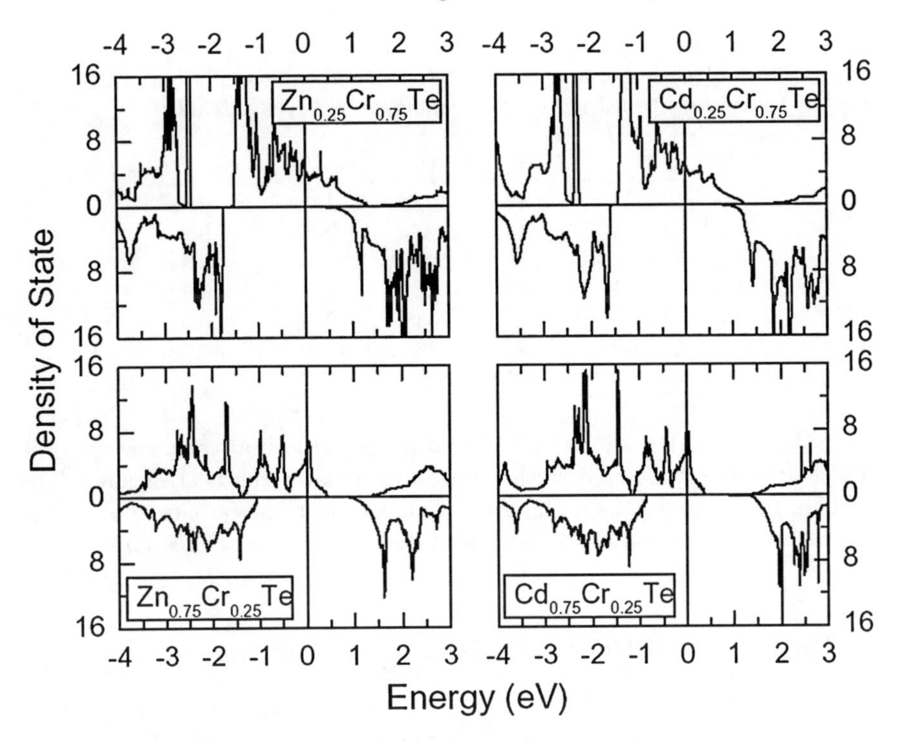

Fig. 11. Spin-dependent densities of states (DOS, in units of states/eV per formula) of the ZnTe (*left column*) and CdTe (*right column*) with Cr doping levels 75% (*upper row*) and 25% (*lower row*). The upper part in each panel is for majority spin and the lower part for minority spin

compounds are half-metallic ferromagnets with clear half-metallic gaps, with the Fermi levels being within the gaps of their minority-spin bands. The moments per chromium atom, the equilibrium lattice constants a, the gaps of minority-spin bands G_{MIS}, and the half-metallic gaps are summarized in Table 4. Assume that cadmium or zinc atoms occupy the corner and face-center sites of the unit cell of CdTe or ZnTe. The $x = 0.25$ ($x = 0.75$) is realized by doping at the corner (face-center) sites of the unit cell. The volumes are optimized but the positions of the transition metal atoms are not. Also shown in Table 4 are those half metals obtained with chromium, manganese, and vanadium doping into ZnTe in the same way. One manganese (chromium or vanadium) atom contributes $5\mu_B$ ($4\mu_B$ or $3\mu_B$) to the total moment when half-metallic ferromagnetism is obtained, and contributes less when half-metallic ferromagnetism is broken. Optimizing the positions of the transition metal atoms does not change the electronic structures significantly, but changes half-metallicity in some critical cases. Manganese doped CdTe can be half-metallic with full structural optimization, but vanadium doped CdTe loses half-metallicity when full structural optimization is made. Other

Table 4. Magnetic moments per magnetic atom (m in units of μ_B), equilibrium lattice constants (a), minority-spin gaps (G_{MIS}), and half-metallic gaps (E_g) of the six zincblende-based transition-metal compounds of ZnTe and CdTe

Compound	m (μ_B)	a (Å)	G_{MIS} (eV)	E_g (eV)
$Zn_{0.75}Cr_{0.25}Te$	4	6.213	1.92	0.86
$Zn_{0.25}Cr_{0.75}Te$	4	6.244	2.10	0.36
$Zn_{0.75}Mn_{0.25}Te$	5	6.221	1.57	0.54
$Zn_{0.75}V_{0.25}Te$	3	6.223	1.95	0.33
$Cd_{0.75}Cr_{0.25}Te$	4	6.548	2.36	0.30
$Cd_{0.25}Cr_{0.75}Te$	4	6.363	2.28	0.72

doping levels can be realized with appropriate structural construction. The magnetic atoms bond only with tellurium atoms at the nearest sites. The mechanism of the half-metallic ferromagnetism in these ternary compounds should be similar to that of the binary half-metallic pnictides and chalcogenides.

4.3 Half Metals in Deformed Zincblende Structures

The zincblende half-metallic pnictides and chalcogenides are often grown as ultrathin films epitaxially on semiconductors or as ultrathin layers between some semiconductors [9, 13–15, 27, 28]. It is unavoidable that they are subject to some crystal deformations. Therefore, it is of real interest to determine whether the half-metallic ferromagnetism is robust enough to survive such crystal deformations. The magnetic moment of the nearly half-metallic zincblende MnAs is enhanced a little when it is subject to volume-conserving tetragonal deformations, but the moments of the true zincblende half metals, such as zincblende CrAs and CrSb, persist to be integers in units of Bohr magneton (μ_B) when their lattice constant ratio c/a changes substantially. The spin-dependent DOS of the zincblende CrAs, CrSb, and MnAs with c/a ratios of 0.85 and 1.15 are presented in Fig. 12. It is clear that the spin-dependent DOS around the Fermi level, and thus the half-metallic gaps, in the cases of CrAs and CrSb change very little even under such large deformations. Half-metallic ferromagnetism persists even when the crystalline symmetry is further reduced to orthogonal symmetry. Their half-metallic gaps, however, get smaller and are even closed with the orthogonal deformation increasing. The bonding property changes slightly under deformations. Therefore, the half-metallic ferromagnetism in the zincblende transition metal pnictides and chalcogenides is robust enough to survive even large crystalline deformations in real environments.

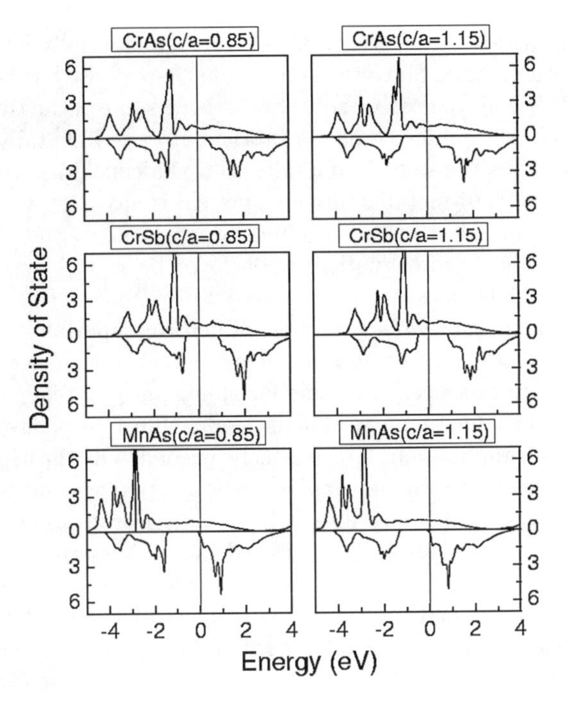

Fig. 12. Spin-dependent densities of states (DOS, in units of states/eV per formula unit) of tetragonal-deformed zincblende CrAs (*upper row*), CrSb (*middle row*), and MnAs (*lower row*) from their equilibrium lattice constants. The left panels are made for $c/a = 0.85$ and the right panels are made for $c/a = 1.15$. The upper part in each panel is for majority spin and the lower part for minority spin

5 Conclusion

We have briefly introduced the background of half metals compatible with important semiconductors and outlined the main findings and advance that have been achieved in the last five years. We have reviewed theoretical results on half-metallic ferromagnetic zincblende transition-metal pnictides and chalcogenides. Their structural stability is examined with connection to the potentials and possibility of experimental fabrication of them. The research in this field begins with zincblende MnAs because of its close relation with GaAs. It is a nearly half metal, not a true half metal. Zincblende CrAs and CrSb are true half-metallic ferromagnets with good stability. All these zincblende pnictides are at least 0.9 eV higher in total energy per formula unit than their ground-state phases. Nevertheless, half-metallic zincblende films and layers of CrAs with thickness of upto five unit cells have been fabricated. The good stability and low metastable energy together can explain why only zincblende MnAs, CrAs, and CrSb have been realized experimentally but the others have not. The half-metallic zincblende CrSe, CrTe, and VTe are shown to be outstanding for their better stability and substantially lower metastable energies than the others. Since the zincblende CrAs phase has been fabricated successfully, it is believed reasonably that at least some of the three zincblende half-metallic chalcogenides could be fabricated experimentally. In addition, nine wurtzite transition-metal pnictides and chalcogenides are found to be good half metals. Half-metallic ferromagnetism is also achieved while doping some transition metal atoms into semiconductors ZnTe and CdTe. In some cases, half-metallic ferromagnetism persists to be robust enough when the zincblende materials are subject to large deformations.

The advance in the search for semiconductor-compatible half metals over last years has been very encouraging. However, much more work is needed in order to obtain practical materials for spintronics and for novel computer chips in the future. On the experimental side, thin film or layered materials suitable for real applications are urgently needed. On the theoretical side, it is highly desirable to further explore new compounds and new structures for better electronic structure and half-metallic ferromagnetism. Since half metals have to be grown on appropriate semiconductors, the interfacial effects are very important and must be further studied. Other aspects of these half metals and their composite structures need be studied too. Being metastable phases, the half-metallic zincblende or wurtzite materials reviewed above can only be realized in the form of ultrathin films or layered samples, which should be sufficient for future nanoscale applications.

Acknowledgments

The author is grateful to Professor David G. Pettifor, Dr. Ya-Qiong Xu, Dr. Wen-Hui Xie, Ms Li-Jie Shi, and Mr. Yong Liu for their collaborations in researches concerned; and to Professor Steven G. Louie, Dr. Pei-Hong Zhang, Dr. Jeffrey B. Neaton, and Dr. Wei-Dong Luo for their many kind helps during the author's stay in University of Californian at Berkeley. This work is supported in parts by Chinese Department of Science and Technology under the National Key Projects of Basic Research (No. G1999064509), by Nature Science Foundation of China (Nos. 90406010 and 60021403), and by Berkeley Scholars Program, University of California at Berkeley, Berkeley, CA 94720, USA.

References

1. S.A. Wolf, D.D. Awschalom, R.A. Buhrman, J.M. Daughton, S. von Molnar, M.L. Roukes, A.Y. Chtchelkanova, and D.M. Treger: Science **294**, 1488 (2001); I.S. Osborne, Science **294**, 1483 (2001)
2. D.D. Awschalom and J.M. Kikkawa: Physics Today **52**, No. 6, 33 (1999); W.E. Pickett and J.S. Moodera: Physics Today **54**, No. 5, 39 (2001).

3. R.A. de Groot, F.M. Mueller, P.G. van Engen, and K.H.J. Buschow: Phys. Rev. Lett. **50**, 2024 (1983)
4. J.W. Dong, L.C. Chen, C.J. Palmstrom, R.D. James, and S. McKernan: Appl. Phys. Lett. **75**, 1443 (1999)
5. S.M. Watts, S. Wirth, S. von Molnar, A. Barry, and J.M.D. Coey: Phys. Rev. B **61**, 9621 (2000)
6. A. Yanase and H. Siarori: J. Phys. Soc. Jpn. **53**, 312 (1984); F.J. Jedema, A.T. Filip, and B. van Wees: Nature **410**, 345 (2001); S. Soeya, J. Hayakawa, H. Takahashi, K. Ito, C. Yamamoto, A. Kida, H. Asano, and M. Matsui: Appl. Phys. Lett. **80**, 823 (2002)
7. J.M.D. Coey, M. Viret, and S. von Molnar: Adv. Phys. **48**, 167 (1999)
8. K.-I. Kobayashi, T. Kimura, H. Sawada, K. Terakura, and Y. Tokura, Nature **395**, 677 (1998)
9. H. Akinaga, T. Manago, and M. Shirai: Jpn. J. Appl. Phys. **39**, L1118 (2000); M. Shirai: Physica E **10**, 143 (2000)
10. S. Sanvito and N.A. Hill: Phys. Rev. B **62**, 15553 (2000)
11. A. Continenza, S. Picozzi, W.T. Geng, and A.J. Freeman: Phys. Rev. B **64**, 085204 (2001); Y.J. Zhao, W.T. Geng, A.J. Freeman, and B. Delley: Phys. Rev. B **65**, 113202 (2002)
12. Y.-Q. Xu, B.-G. Liu, and D.G. Pettifor: Phys. Rev. B. **66**, 184435 (2002)
13. M. Mizuguchi, H. Akinaga, T. Manago, K. Ono, M. Oshima, M. Shirai, M. Yuri, H.J. Lin, H.H. Hsieh, and C.T. Chen: J. Appl. Phys. **91**, 7917 (2002)
14. J.H. Zhao, F. Matsukura, K. Takamura, E. Abe, D. Chiba, and H. Ohno: Appl. Phys. Lett. **79**, 2776 (2001)
15. K. Ono, J. Okabayashi, M. Mizuguchi, M. Oshima, A. Fujimori, and H. Akinaga: J. Appl. Phys. **91**, 8088 (2002)
16. I. Galanakis: Phys. Rev. B **66**, 012406 (2002)
17. B.-G. Liu: Phys. Rev. B **67**, 172411 (2003)
18. T. Akimoto, Y. Moritomo, A. Nakamura, and N. Furukawa: Phys. Rev. Lett. **85**, 3914 (2000); J.M.D. Coey and M. Venkatesan: J. Appl. Phys. **91**, 8345 (2002); I. Galanakis, P.H. Dederichs, and N. Papanikolaou: Phys. Rev. B **66**, 134428 (2002)
19. C.M. Fang, G.A. de Wijs, and R.A. de Groot: J. Appl. Phys. **91**, 8340 (2002)
20. P. Ravindran, A. Delin, P. James, B. Johansson, J.M. Wills, R. Ahuja, and O. Eriksson: Phys. Rev. B **59**, 15680 (1999)
21. T. Plake, M. Ramsteiner, V.M. Kaganer, B. Jenichen, M. Kästner, L. Däweritz, and K.H. Ploog: Appl. Phys. Lett. **80**, 2523 (2002); S. Sugahara and M. Tanaka, Appl. Phys. Lett. **80**, 1969 (2002)
22. W.-H. Xie, Y.-Q. Xu, B.-G. Liu, and D.G. Pettifor: Phys. Rev. Lett. **91**, 037204 (2003)
23. J. Kubler: Phys. Rev. B **67**, R220403 (2003)
24. Ph. Mavropoulos, K. Sato, R. Zeller, P.H. Dederichs, V. Popescu, and H. Ebert: Phys. Rev. B **69**, 054424 (2004)
25. S. Sanvito: Phys. Rev. B **68**, 054425 (2003); J. Stephens, J. Berezovsky, J.P. McGuire, L.J. Sham, A.C. Gossard, and D.D. Awschalom: Phys. Rev. Lett. **93**, 097602 (2004)
26. C.Y. Fong, M.C. Qian, J.E. Pask, L.H. Yang, and S. Dag: Appl. Phys. Lett. **84**, 239 (2004); O. Bengone, O. Eriksson, J. Fransson, I. Turek, J. Kudrnovsky, and V. Drchal: Phys. Rev. B **70**, 035302 (2004)

27. J.E. Pask, L.H. Yang, C.Y. Fong, W.E. Pickett, and S. Dag: Phys. Rev. B **67**, 224420 (2003)
28. S. Picozzi, T. Shishidou, A.J. Freeman, and B. Delley: Phys. Rev. B **67**, 165203 (2003)
29. H. Shoren, F. Ikemoto, K. Yoshida, N. Tanaka and K. Motizuki: Physica E **10**, 242 (2001); I. Galanakis and P. Mavropoulos: Phys. Rev. B **67**, 104417 (2003); B. Sanyal, L. Bergqvist, and O. Eriksson: Phys. Rev. B **68**, 054417 (2003)
30. for examples: H. Akinaga, M. Mizuguchi, K. Ono, and M. Oshima: Appl. Phys. Lett. **76**, 357 (2000); M. Mizuguchi, H. Akinaga, K. Ono, and M. Oshima: Appl. Phys. Lett. **76**, 1743 (2000)
31. L.-J. Shi and B.-G. Liu: J. Phys.: Condens. Matter **17**, 1209 (2005)
32. J.-C. Zheng and J.W. Davenport: Phys. Rev. B **69**, 144415 (2004)
33. V.H. Etgens, P.C. de Camargo, M. Eddrief, R. Mattana, J.M. George, and Y. Garreau: Phys. Rev. Lett. **92**, 167205 (2004)
34. Y.-J. Zhao and A. Zunger: Phys. Rev. B **71**, 132403 (2005)
35. H. Saito,V. Zayets, S. Yamagata, Y. Suzuki, and K. Ando: J. Appl. Phys. **91**, 8085 (2002); H. Saito, V. Zayets, S. Yamagata, and K. Ando: Phys. Rev. B **66**, R081201 (2002)
36. H. Saito, V. Zayets, S. Yamagata, and K. Ando: Phys. Rev. Lett. **90**, 207202 (2003)
37. W.-H. Xie and B.-G. Liu: J. Appl. Phys. **96**, 3559 (2004)
38. Y. Liu and B.-G. Liu, unpublished
39. W.-H. Xie, B.-G. Liu, and D.G. Pettifor: Phys. Rev. B **68**, 134407 (2003)
40. P. Hohenberg and W. Kohn: Phys. Rev. **136**, B864 (1964); W. Kohn and L.J. Sham: Phys. Rev. **140**, A1133 (1965)
41. P. Blaha, K. Schwarz, P. Sorantin, and S.B. Trickey: Comp. Phys. Comm. **59**, 399 (1990)
42. J.P. Perdew, K. Burke, and M. Ernzerhof: Phys. Rev. Lett. **77**, 3865 (1996)
43. J.P. Perdew and Y. Wang: Phys. Rev. B **45**, 13244 (1992)
44. P.J. Craievich, M. Weinert, J.M. Sanchez, and R.E. Watson: Phys. Rev. Lett. **72**, 3076 (1994)
45. P. Radhakrishna and J.W. Cable: Phys. Rev. B **54**, 11940 (1996)
46. G.I. Makovetski: Sov. Phys. Solid State **28**, 447 (1986)
47. J. Dijkstra, H.H. Weitering, C.F. van Bruggen, C. Haas, and R.A. de Groot: J. Phys.: Condens. Matter **1**, 9141 (1989); J. Dijkstra, C.F. van Bruggen, C. Haas, and R. A. de Groot, J. Phys.: Condens. Matter **1**, 9163 (1989)
48. F.K. Lotgering and E.W. Gorter: J. Phys. Chem. Solids **3**, 238 (1957); L.M. Corliss, N. Elliott, J.M. Hastings, and R.L. Sass: Phys. Rev. **122**, 1402 (1961)
49. I. Tsubokawa, J. Phys. Soc. Jpn. 15, 2243 (1960).
50. W.J. Takei, D.E. Cox, and G. Shirane: Phys. Rev. **129**, 2008 (1963); A.I. Snow: Phys. Rev. **85**, 365 (1952); P. Radhakrishna and J.W. Cable: Phys. Rev. B **54**, 11940 (1996)
51. F.K. Lotgering and E.W. Gorter: J. Phys. Chem. Solids **3**, 238 (1957); L.M. Corliss, N. Elliott, J.M. Hastings, and R.L. Sass: Phys. Rev. **122**, 1402 (1961)
52. J. Dijkstra, H.H. Weitering, C.F. van Bruggen, C. Haas, and R.A. de Groot: J. Phys.:Condens. Matter **1**, 9141 (1989); J. Dijkstra, C.F. van Bruggen, C. Haas, and R.A. de Groot: J. Phys.:Condens. Matter **1**, 9163 (1989); S. Ohta, T. Kanomata, T. Kaneko, and H. Yoshida: J. Phys.:Condens. Matter **5**, 2759 (1993); H. Shoren, F. Ikemoto, K. Yoshida, N. Tanaka, and K. Motizuki: Physica E **10**, 242 (2001)

53. G.I. Makovetski: Sov. Phys. Solid State **28**, 447 (1986)
54. Hoschel and Klemm: Z. Anorg. Chem. **242**, 49 (1939)
55. Grison and Beck: Acta Crystallogr. **15**, 807 (1962)
56. Nowotny and Arstad: Z. Phys. Chem. (Leipzig) **38**, 461 (1937)
57. Bachmayer and Nowotny: Monatsch. Chem. **86**, 741 (1955)

Materials Design and Molecular-Beam Epitaxy of Half-Metallic Zinc-Blende CrAs and the Heterostructures

Hiro Akinaga[1], Masaki Mizuguchi[1], Kazutaka Nagao[2], Yoshio Miura[2], and Masafumi Shirai[2]

[1] Nanotechnology Research Institute, National Institute of Advanced Industrial Science and Technology, 1-1-1 Umezono, Tsukuba, Ibaraki 305-8568, Japan
akinaga.hiro@aist.go.jp

[2] Research Institute of Electrical Communication, Tohoku University, 2-1-1 Katahira, Aoba-ku, Sendai 980-8577, Japan
shirai@riec.tohoku.ac.jp

Abstract. Zinc-blende half-metallic ferromagnets are promising materials in order to open up a new world of semiconductor spintronics. We design a new class of half-metallic ferromagnets, the zinc-blende transition-metal mono-pnictides, using *ab-initio* calculations based on the density-functional theory. The calculations show that the total-energy difference between the ferromagnetic and the antiferromagnetic states for the zinc-blende CrAs is the largest among all the studied compounds and the highly spin-polarized electronic band structure is almost unaffected from the spin-orbit interaction. We further study the properties of zinc-blende CrAs/GaAs multilayers theoretically and show that they keep a high spin polarization. Experimental realization of the previously nonexistent zinc-blende CrAs thin film has been achieved by molecular-beam epitaxy. The crystallographic analysis is presented together with the magnetic properties of the epitaxial film. The morphological structure and the magnetic properties change sensitively depending on the substrate temperature during the growth. The room-temperature saturation magnetization of the film grown under optimum conditions reaches about 3 μ_B per formula unit, which agrees with the theoretically predicted value and reflects the half-metallic behavior. The epitaxial growth of multilayers consisting of 2–4 monolayers of CrAs and 2–4 monolayers GaAs with atomically flat surfaces and interfacess is demonstrated.

1 Introduction

Control of the charge degree of freedom of electrons in semiconductors brought about the current success of semiconductor electronics. On the other hand, the spin degree of freedom, normally used in the magnetization control of magnetic storage devices, has been neglected in semiconductors, because the energies of two kinds of spin states are almost degenerated and the difference is smeared out at room temperature. Recent progress in semiconductor science and technology, however, enables us to reduce the size of the semiconductor device to the nano-scale dimension, where the spin-dependent properties of electrons can be controlled and exploited. The demonstration of the carrier-induced ferromagnetism in a semiconductor promoted the emergence of the new research field in the semiconductor electronics, known as spin-electronics or spintronics [1]. The spatiotemporal spin coherence can be widely manipulated at will in a semiconductor, which is one of the great advantages in semiconductor-based spin-electronics, but the manipulation can not be widely performed in a metallic spin-electronic material.

The recent advances in an epitaxial growth technology of semiconductors offers various magnetic semiconductors. Among them, a series of ferromagnetic semiconductors, such as (Ga,Mn)As, has been intensively studied, because of the good compatibility with III-V semiconductor technologies [2]. One of the issues to be solved in this research field is to increase the ferromagnetic transition temperature, the Curie temperature T_C, up to room temperature. Ferromagnetic-metal/semiconductor hybrid structures have been also received considerable attention [3]. The advantage of the hybrid structure is that one can expect the novel spin-dependent functionality at room temperature, because the Curie temperature is higher than room temperature. Spin-polarized electron injection from the ferromagnetic metal and the manipulation in the semiconductor is one of the building blocks in the field of semiconductor-based spin-electronics [1]. Various attempts have been reported, although the spin-injection efficiency was very low at room temperature. The recipe to obtain the hybrid structure with the atomically-flat interface and no by-product has been reported in the combination of various ferromagnetic metals and GaAs, since GaAs provided the epitaxial growth of the ferromagnetic metal. The electronic properties at the interface, however, have not yet been well controlled. The spin-dependent electronic transparency through the interface depends, without doubt, sensitively not only on the crystallographic morphology but also the crystallographic symmetry and the chemical bonding etc. at the heterointerface.

What kind of ferromagnetic metal is the most suitable material for the spin-polarized electron injection into GaAs? First, the growth condition of the ferromagnetic metal should be compatible to that of GaAs. Second, to select a certain k-vector of injected electrons, the material which can be grown epitaxially on GaAs is preferable. The combination of Fe and GaAs is a well-known ferromagnetic metal/semiconductor hybrid structure to meet these demands [4,5]. Nevertheless, this combination is not the best solution for the spin injection. The chemical bonding of Fe-As and Fe-Ga may degrade the spin polarization of electrons, when the electron passes through the interface, because these combinations provide para- and antiferro-magnetic materials. To avoid the by-product at the interface and to realize the atomically flat interface, the material should be composed of the same element of Ga or As. Besides these demands, the same crystal symmetry of the ferromagnetic metal, namely the zinc-blende crystal symmetry, will bring us a great advantage, since the electronic band structure does not receive any perturbation arising from the change of the symmetry at the interface. The high spin polarization at the Fermi level of the ferromagnetic metal is surely required, although none of previously reported half-metallic ferromagnets measure up to the demands mentioned above. Actually, various half-metallic ferromagnets have been reported, such as Heusler alloy [6–12], CrO_2 [13–16], perovskite manganites [17, 18], while the experimentally observed spin polarization is degraded substantially except for CrO_2 [19–21]. The reduction of the spin polarization has been attributed to the atomic disorder, the by-product and the breaking crystallographic symmetry at the interface with a non-magnetic material. None of these half-metallic ferromagnets can play a role as the spin-polarized electron source for GaAs [22].

Thus, zinc-blende half-metallic ferromagnets are strongly required in order to open up a new world of semiconductor spin-electronics. We design the new class of half-metallic ferromagnets, zinc-blende transition-metal mono-pnictides, by ab initio calculations based on the density-functional theory [23, 24]. The calculations show that the total-energy difference between the ferromagnetic and the antiferromagnetic states of zinc-blende CrAs is largest among them and the highly spin-polarized electronic band structure is almost unaffected even though the spin-orbit interaction is taken into account. Experimental realization of the previously nonexistent zinc-blende CrAs thin film was achieved by molecular-beam epitaxy (MBE). The crystallographic analysis is presented together with the magnetic properties of the epitaxial film in this article. The morphological structure and the magnetic properties change sensitively depending on the substrate temperature during the growth. The room-temperature saturation magnetization of the film grown under the optimum condition reaches about 3 μ_B per formula unit, which agrees well with the theoretically predicted value reflecting the half-metallic behavior. These experimental results are presented in the following section. We further study the properties of zinc-blende CrAs/GaAs multilayers theoretically and show that the high spin polarization does not degrade by the formation of the multilayer [25]. The epitaxial growth of multilayers consisting of 2–4 monolayers of CrAs and 2–4 monolayers GaAs with the atomically flat surface and interfaces is presented together with these results of theoretical calculations in Sect. 3.

2 Zinc-Blende Transition-Metal Mono-Pnictides

2.1 Theoretical Design of Zinc-Blende Transition-Metal Mono-Pnictides

First-principles electronic band-structure calculations are based on the density-functional theory within the local spin-density approximation (LSDA) for the exchange-correlation energy and potential. The calculation were carried out by using the full-potential linearized augmented-plane-wave

(FLAPW) method. In the FLAPW method, the potential, the charge and the spin densities in the crystal were treated with no shape approximation. The basis-functions were expanded in terms of the spherical harmonics up to $\ell = 7$ inside each muffin-tin sphere and they were expressed as plane-waves in the interstitial region. About 300 basis-functions per formula unit were taken into account. Relativistic effects were included through the so-called scalar-relativistic treatment in the calculation of radial wave functions in each muffin-tin sphere. Then, the spin-orbit interaction is included using the second variational treatment to the scalar-relativistic calculation.

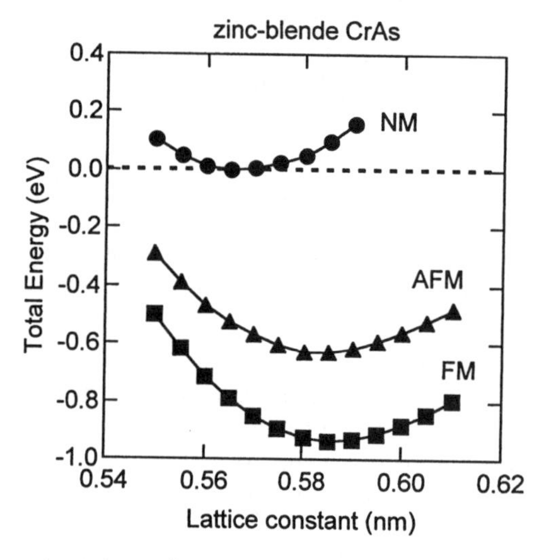

Fig. 1. Lattice constant dependence of total energies of zinc-blende CrAs in non-magnetic (NM), antiferromagnetic (AFM) and ferromagnetic (FM) states are shown by circles, triangles and squares, respectively

First, the total energies of the zinc-blende CrAs in the non-magnetic (NM), antiferromagnetic (AFM) and ferromagnetic (FM) states are shown in Fig. 1. These energies are plotted as a function of the lattice constant. As clearly shown in the figure, the ferromagnetic state becomes stable for a wide range of the lattice constant. The calculations predict theoretically that the thermal-equilibrium lattice constant will be about 0.58 nm. This result facilitates the thin-film growth of zinc-blende CrAs, because the lattice mismatch of about 3% comparing to that of GaAs is small enough to realize the epitaxial growth on a GaAs substrate. In the case of transition-metal mono-arsenides, no magnetic ordered state could be obtained for TiAs, CoAs and NiAs. Figure 2 shows that the ferromagnetic state becomes stable only in the case of zinc-blende VAs, MnAs and CrAs. Furthermore, the total energy difference between the ferromagnetic and the antiferromagnetic states, ΔE, which is a measure of the magnetic coupling between the nearest neighboring

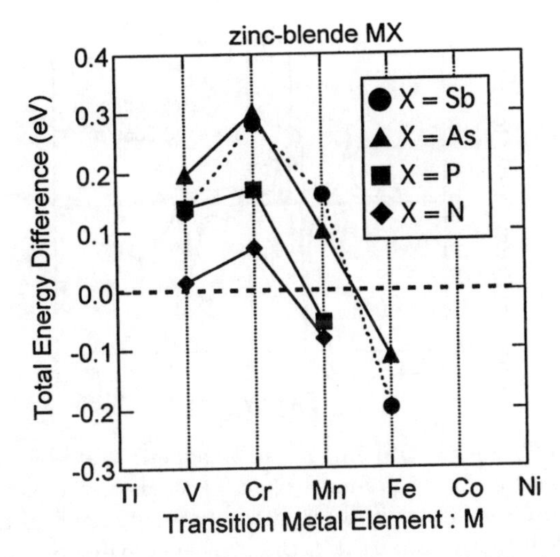

Fig. 2. Total energy difference, defined as E (AFM) – E (FM) where E (AFM) and E (FM) mean the total energies in the antiferromagnetic and ferromagnetic states, respectively, calculated in zinc-blende MX. M and X represent transition-metal and V-group elements, respectively

transition-metal spins, is evaluated at about 0.3 eV for the zinc-blende CrAs. This is the largest value among the transition-metal mono-pnictides. The strong ferromagnetic coupling between the neighboring Cr spins is attributed to the large hybridization between Cr $3d$ and As $4p$ orbitals. The Curie temperature of the zinc-blende CrAs is estimated at about 1800 K within the mean-field approximation of the Heisenberg model, where the Curie temperature is calculated using the exchange coupling constant between the nearest neighboring Cr spins, J, evaluated from $E\Delta E$.

Sakuma investigated theoretically the stability of ferromagnetism in zinc-blende CrAs [26]. His first-principle calculations showed a large effective exchange constant of 150 meV for lattice constants above the experimental one of GaAs. Such an exchange constant corresponds to a very high Curie temperature of 1200 K within the framework of the molecular field approximation. The stability of the ferromagnetism as a function of the lattice constant was reported by Sanyal et al. for values between the experimental constants of GaAs and InAs [27]. The Curie temperature was estimated adopting mean field theory at 1320 K and 1100 K for the zinc-blende CrAs and for the lattice constants of GaAs and InAs, respectively. The Monte Carlo simulation gave relatively lower Curie temperatures of 980 K and 790 K for the same lattice constants. These values are notably higher than the room temperature, indicating the suitability of the material for spin-electronic device applications. This fact contrasts with the case of the zinc-blende MnAs, which shows

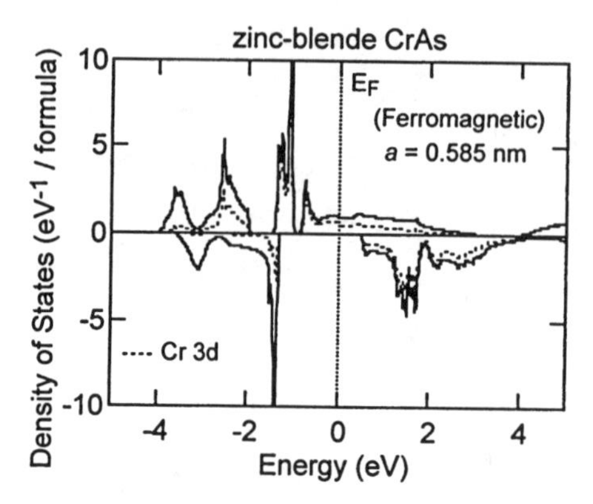

Fig. 3. Density of states (DOS) in the ferromagnetic state of zinc-blende CrAs with the lattice constant of 0.585 nm. Broken lines denote the partial DOS for the Cr 3*d* orbitals. Note that the spin-orbit interaction is ignored in this figure, since the effect is so small that it is buried in the present drawing of the figure

the half-metallic property only in the case that the lattice constant exceeds 0.58 nm [28].

As shown in Fig. 3, zinc-blende CrAs possesses a half-metallic band structure. The total magnetic moment was calculated at 3 μ_B per unit formula of CrAs, reflecting the half-metallic behavior. In the minority spin state, the Fermi level is located in the energy gap of about 2 eV, which is larger than for other existent half-metallic ferromagnets. The wavefunction near the top of the valence-band is constructed mainly from the As 4*p* orbitals which hybridize with the Cr 3*d* orbitals. The wavefunction at the conduction-band bottom is constructed from the Cr 3*d* non-bonding orbitals. In the majority spin state, the well-hybridized Cr 3*d* and As 4*p* orbitals , which have the anti-bonding character, cross the Fermi level, showing the metallic behavior. The conduction electrons in the hybridized bands play an important role to stabilize the ferromagnetic state through the double-exchange mechanism. Once the spin-orbit interaction is included in band structure calculation, mixing of both spin states takes place. The spin polarization is reduced from 100%. For zinc-blende CrAs, however, the spin polarization is preserved as high as 99.8% [29]. The fact that the spin polarization does not degrade seriously by the spin-orbit interaction is attributed to the large minority-spin gap of about 2 eV, which is about one order of magnitude larger than the spin-orbit coupling constant of Cr 3*d* and As 4*p* electrons. A similar result (99.6%) was also obtained by a fully relativistic calculation based on the screened KKR method [30]. The high spin polarization may enable us to detect the spin polarization even in the case that the spin injection is performed into a semiconductor through a diffusive transport regime [31].

The other Cr mono-pnictides, zinc-blende CrSb, CrP and CrN, have a possibility to become half-metallic, as shown in Fig. 2. The total-energy difference, ΔE, becomes 0.28 eV for zinc-blende CrSb which is predicted to possess the half-metallic band structure. It is noted that the thin-film growth of zinc-blende CrSb was achieved by MBE, although the zinc-blende crystal structure remained only up to a few monolayers [32]. For zinc-blende CrP, ΔE is only 0.17 eV. The spin polarization at the Fermi level of the zinc-blende CrP with the theoretical thermal-equilibrium lattice constant of 0.545 nm is quite small. This small spin polarization is considered to be due to the broad Cr 3d bandwidth originating from the relatively short interatomic distance between the nearest neighboring Cr atoms. Only in the case that the lattice constant is expanded above 0.56 nm, the band structure becomes half-metallic. The epitaxial growth technique may help to control the lattice constant of zinc-blende CrP, by choosing a proper substrate to expand the lattice.

2.2 Epitaxial Growth of Zinc-Blende CrAs Thin Film

Epitaxial thin film of zinc-blende CrAs was successfully grown on a GaAs (001) substrate by molecular-beam epitaxy (MBE) [23]. The substrate was set on a molybdenum holder with indium solder. After being degassed at around 400°C in a loading chamber, the substrate was transferred into the MBE chamber without breaking vacuum. A GaAs buffer layer with the thickness of 20 nm was grown on the substrate at around 600°C in the MBE chamber. After cooling the substrate temperature down to 200°C, the zinc-blende CrAs thin film was grown on the substrate by opening shutters of Knudsen cells of Cr and As simultaneously. The growth rate was kept at about 0.02 nm/s by controlling the Cr beam pressure and the beam equivalent pressure ratio of As/Cr was set at 100 to 1000. The film for magnetization measurements was capped by a 4 nm Au layer to prevent the oxidization. Figure 4 shows the reflection high-energy electron diffraction (RHEED) patterns from the CrAs surface along [110] and [100] directions of the GaAs substrate. Before the growth of the CrAs film, the RHEED pattern from the GaAs surface

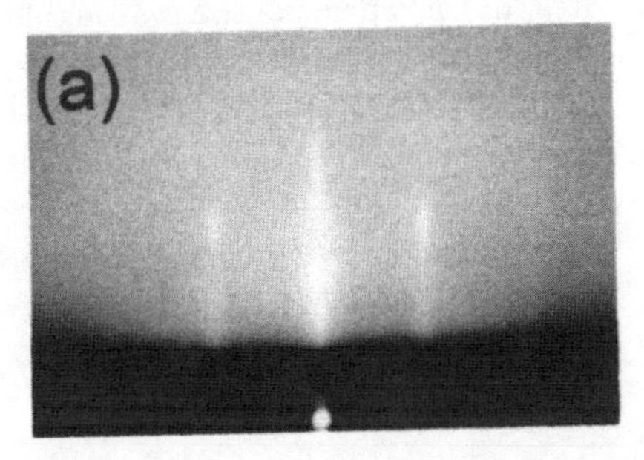

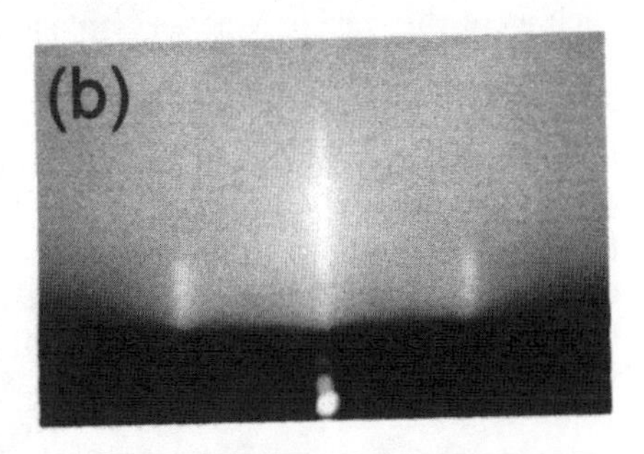

Fig. 4. RHEED patterns from the zinc-blende CrAs surface. The electron-beam incidence is along (**a**) [110] and (**b**) [100]

showed a c(4 × 4) pattern. When the growth started, the reconstruction pattern immediately disappeared and the length of the line became shorted, but the main streaky lines certainly remained as shown in the figure. According to the symmetry of the intensity fluctuation, one can evaluate that the zinc-blende crystal structure remains even after the CrAs deposition. These facts showed that the zinc-blende CrAs thin film forms on the GaAs (001) substrate epitaxially. The change of the interval of RHEED lines was not evidenced between surfaces of the CrAs film and the GaAs substrate. Temporal change of the pattern indicated the existence of the critical thickness. Above the critical thickness, about 3 nm under the present growth conditions, the additional spots appeared in the RHEED pattern along the [110] GaAs direction although the pattern kept the similarity along the [100] GaAs direction. These results showed that an unknown phase and/or the twin structure of zinc-blende CrAs emerged when the nominal film thickness reached 3 nm. The high-resolution transmission electron microscopy (TEM) revealed the same real-space atomic image of the CrAs thin film as the GaAs substrate, when the thickness was thinner than the critical thickness. The growth mode changed according to the substrate temperature [33]. By increasing the substrate temperature up to 300°C, the critical thickness decreased less than 2 nm and the island growth mode was observed.

The local structural properties were characterized by fluorescence extended X-ray absorption fine structure (EXAFS) measurements. The measurements were performed at the beam line BL12C at the Photon Factory in Tsukuba with a Si (111) double crystal monochromator and a bent cylindrical mirror using synchrotron radiation from the 2.5 GeV storage ring. The EXAFS spectra were taken in the fluorescence detection mode at 70 K. The detailed measurement and the procedure of the analyses was described in reference [34]. The best-fit of crystallographic parameters according to the curve fitting of the EXAFS spectra indicated that the central Cr atom in the present CrAs thin film with the nominal thickness of 2 nm was merely coordinated by As atoms, and the coordination number was 4. The curve fitting based on the model assuming the coexisting Cr-As and Cr-Cr bonds, such as in the case of MnP-crystal structures, did not provide the reasonable conclusion, as shown in Fig. 5. Thus, it is considered that the zinc-blende crystal structure is realized in the present CrAs thin film. The numerical analysis also showed that the length of the Cr-As bond was 0.249 nm, which agrees well with the predicted bond length of 0.253 nm calculated from the theoretical thermal-equilibrium lattice constant of zinc-blende CrAs, 0.585 nm. The EXAFS studies proved that the theoretically predicted zinc-blende CrAs is synthesized on the GaAs (001) substrate.

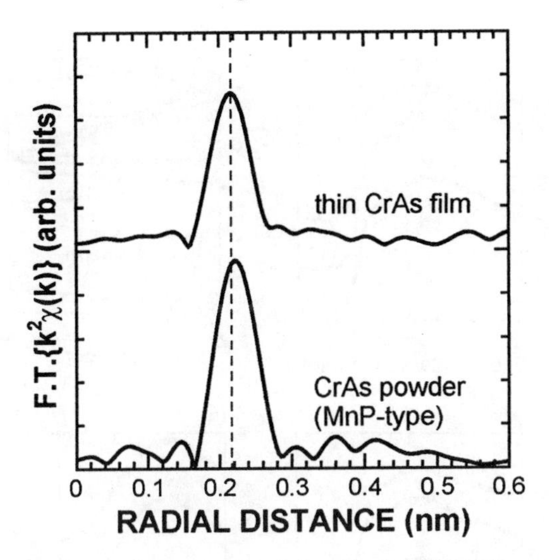

Fig. 5. Fourier transform of experimental Cr K-edge EXAFS spectra of the zinc-blende CrAs thin film and the CrAs powder with the MnP-type crystal structure. The shift of the main peak is observed, indicated by the broken line in the figure

2.3 Electronic and Magnetic Properties of Zinc-Blende CrAs Thin Film

Angle-resolved valence-band photoemission spectra of the zinc-blende CrAs thin film are shown in Fig. 6(a). The experimental procedure is described in reference [35]. The Fermi edge is clearly observed in all the spectra, indicating metallic characteristics. The weakly dispersive band, which is assigned to the majority-spin Cr $3d$ state, is also confirmed at the energy level of 2 eV below the Fermi level. These behaviors agree well with the theoretical calculation, shown in Fig. 6(b). As mentioned in Sect. 2.2, the critical thickness of the epitaxial growth was about 3 nm at the substrate temperature of 200°C. Above the critical thickness, the additional spots indicating the formation of the other phase of CrAs appeared in the RHEED pattern. According to the change of the RHEED pattern, the energy-band dispersion disappears from the spectra, shown in Fig. 6(c). Amorphous-like CrAs film was obtained, when the substrate temperature decreased down to 150°C. From the amorphous-like surface, as shown in Fig. 6(d), the photoemission spectra with no Fermi edge are observed. This result indicates that the film loses the metallic property. The substrate temperature dependence of the photoemission spectra shows that the growth window to realize zinc-blende CrAs is narrow in terms of the substrate temperature variation, reasonably corresponding to the results of crystallographic analyses mentioned in Sect. 2.2, and supports strongly the existence of zinc-blende CrAs.

The magnetic properties of the zinc-blende CrAs film were investigated by a superconducting quantum interference devices (SQUID) magnetometer.

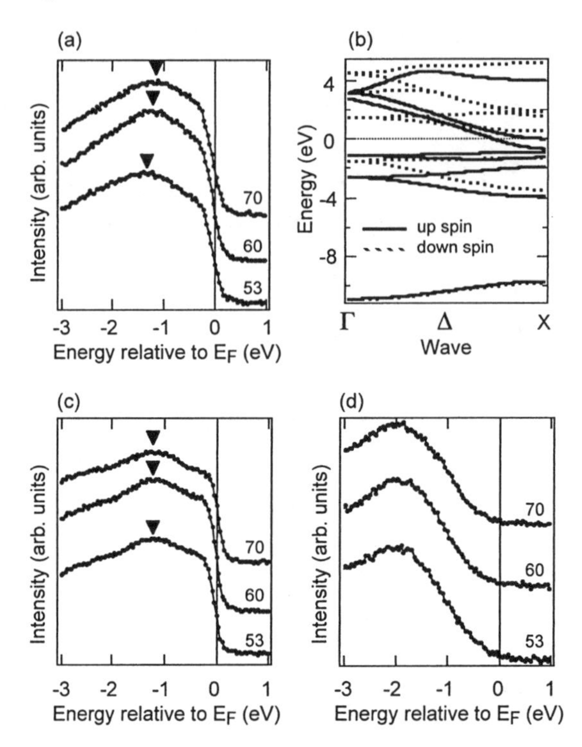

Fig. 6. (**a**) Valence-band photoemission spectra of the zinc-blende CrAs thin film with the nominal thickness of 2 nm. The photon energies were set at 53, 60, and 70 eV, and denoted in the figure. The weak dispersion is indicated by solid triangles. (**b**) The band structure of zinc-blende CrAs calculated by FLAPW method. (**c**) Valence-band photoemission spectra of the CrAs film with the thickness of 30 nm. The energy-band dispersion disappears as presented by solid triangles. (**d**) Valence-band photoemission spectra of the CrAs film with the nominal thickness of 2 nm, but grown at 150°C

The magnetization hysteresis measurements were made at room temperature. The magnetic field was applied parallel to the film plane. The magnetization curve is shown in Fig. 7. The contribution of the GaAs substrate has been subtracted from the data. As shown in the figure, the CrAs thin film shows a ferromagnetic behavior at room temperature. The magnetic field to saturate the magnetization is relatively large (~5 kOe), but the smooth curvature implies that the single phase of CrAs contributes the magnetization. And the small coercivity (~100 Oe) is observed. The saturation magnetization is determined to be 560 kA/m which corresponds to about 3 μ_B per formula unit of CrAs. The saturation magnetization agrees well with the theoretical calculation and can be an evidence showing the half-metallic characteristics of the present zinc-blende CrAs thin film. The Curie temperature has not yet been determined because of the operating-temperature limit of the

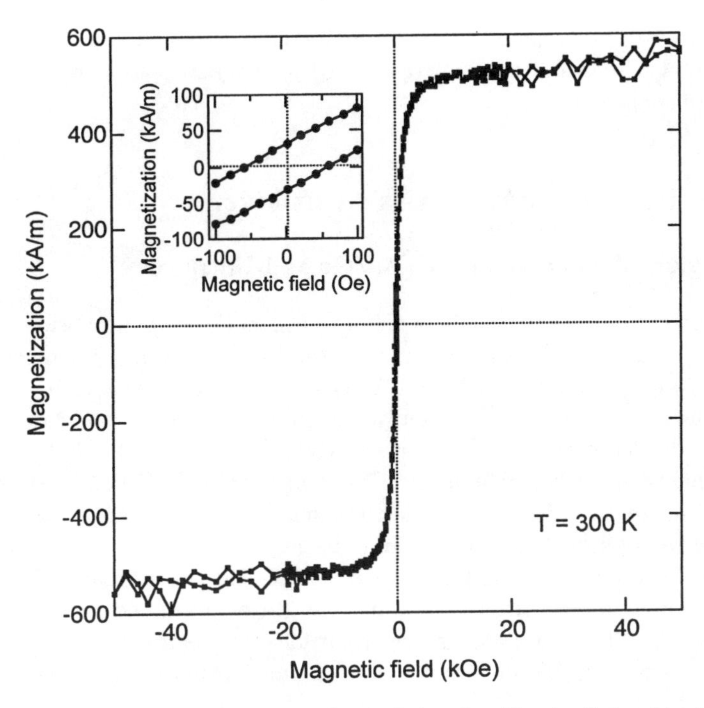

Fig. 7. The magnetization hysteresis loop of the zinc-blende CrAs thin film with the thickness of 2 nm, measured at room temperature (300 K). The magnetic field was applied parallel to the film plane. The inset shows the loop at around zero magnetic field

SQUID magnetometer. The remnant magnetization remained up to 400 K. The thermal equilibrium Cr-As compounds possesses MnP- and Cu_2Sb-type crystal structures. MnP-type CrAs has been known to show a helimagnetic-paramagnetic transition at 265 K [36, 37]. Cr_2As is an antiferromagnet with a Neel temperature of 393 K [38]. None of these thermal-equilibrium compounds shows a ferromagnetic behavior at room temperature [39]. The literature survey supports that the ferromagnetic behavior appeared in the magnetization observation shown in Fig. 7 is due to the existence of the newly-synthesized zinc-blende CrAs.

The morphology of the zinc-blende CrAs thin film was investigated by ex-situ atomic force microscopy (AFM) [33]. Considering the short streak lines of the RHEED patterns shown in Fig. 4, it is easily expected that the growth of the zinc-blende CrAs is not two dimensional but three dimensional mode. Actually, submicron-size terrace structures were observed in the AFM image. Namely, zinc-blende CrAs does not form a continuous film in the present growth conditions. Consequently, the in-plane electric resistance of the film became too large to be measured. Magnetotransport properties, such as the magnetoresistance effect and the Hall effect, have not yet been well investigated so far. Magnetooptical signal coming from the zinc-blende CrAs film was also so small that no result of the magnetooptical properties has been reported.

3 Zinc-Blende CrAs/GaAs Multilayers

3.1 Design of Zinc-Blende CrAs/GaAs Multilayers

We investigated the robustness of the ferromagnetic coupling between the Cr spins [40]. The *ab-initio* simulation code developed by Akai implementing the coherent potential approximation (CPA) within the Korringa-Kohn-Rostoker (KKR) method was used [41]. Only the stoichiometric zinc-blende CrAs is considered. The number of the antisite Cr atoms is equal to that of the antisite As atoms in the calculation. The total energy was calculated for three different magnetic states, namely ferromagnetic, ferrimagnetic and spin-glass. In the ferromagnetic state, all Cr spins are aligned in parallel, while the antisite Cr spins are anti-parallel to those of Cr spins at the ordinary (cation) site in the ferrimagnetic state. In the spin-glass state considered here, half of the randomly distributed Cr atoms have their local magnetic moments along one direction, while the other Cr moments are along the opposite direction, so that the total magnetic moment vanishes. The calculations showed that the ferrimagnetic state is energetically favorable compared to the other magnetic states. This fact indicates that the antisite Cr spins are coupled antiferromagnetically with the nearest neighboring Cr spins at the ordinary site; in other words, the ferromagnetic coupling between the Cr spins at the ordinary sites is robust against the formation of the antisite defects. In the minority-spin energy gap of the ferrimagnetic state of zinc-blende CrAs, impurity-bands composed mainly of antisite Cr $3d$ orbitals appear near the Fermi level, while the spin polarization is not noticeably reduced by the impurity-band formation at least for an antisite Cr concentration less than 5%. This robustness led us to proceed to the material design of zinc-blende CrAs/GaAs multilayers, since the antisite defects may appear at the realistic heterointerface with high probability. The calculations were performed by the Vienna ab initio simulation package [42–44], where planewave basis sets and the projector augmented wave method are used [45, 46]. The KKR-CPA simulation code was also used [41]. Because of the very small effect of the spin-orbit coupling, it was neglected in the calculation described below.

First we investigated the interface between zinc-blende CrAs (001) and GaAs (001), which consist of nine and ten atomic layers, respectively. The interface geometry was chosen so that an As atomic layer is placed between Ga and Cr atomic layers. This interface structure can be experimentally realized by MBE as described in the following section. The assumption of the interface geometry is considered to be reliable, since the multilayer is grown under As exposure. The in-plane lattice parameter was set at 0.565 nm, which is the lattice constant of the GaAs substrate. On the other hand, the lattice parameter in the [001] direction was set at 3.5 nm, which is large enough to accommodate about 1.0 nm of vacuum in addition to the two slabs. The atomic coordinates were fully relaxed except for the two atomic layers at the GaAs surface. This treatment aimed to reduce the effects stemming from the finite thickness of the GaAs slab. The calculation showed that the spin polarization remained very high at the interface and gradually decreased in the GaAs layer [25]. This behavior originates from the same crystallographic symmetry and the similar chemical composition, and indicates that the combination of zinc-blende CrAs and GaAs is suitable to achieve the spin-polarized electron injection into a semiconductor. It is interesting to note that the high spin polarization was observed theoretically only in the case that the zinc-blende CrAs surface was terminated with Cr atoms, not with As atoms, according to the calculations reported by Galanakis [47]. The hetero-junction of GaAs on the As-terminated surface of zinc-blende CrAs prevents the degradation of the spin polarization in the present heterostructure, which is probably due to the structural reconstruction of the As-terminated surface. Next, the electronic band structure of the multilayer consisting of 2 mono-layers (ML) of zinc-blende CrAs and 2 ML of GaAs stacking alternatively in the [001] direction was investigated. In the calculation, the in-plane lattice parameter was set at 0.565 nm. On the other hand, the lattice parameter normal to the plane was fully optimized as well as the atomic coordinates. The resulting lattice parameter was found to be 1.148 nm, which was slightly larger than 1.13 nm, i.e., the unit of the coupled CrAs/GaAs cell elongates along the normal direction. As easily anticipated from the calculation performed in the abovementioned heterostructure, the calculated spin-resolved DOS showed the half-metallic characteristics. The spin polarization remained to some extent even inside the GaAs layer.

In the present calculations for the heterointerface and the multilayer, the atomically flat interface between zinc-blende CrAs and GaAs layers was assumed. However, the interchange of Cr and Ga atoms may occur at the interface with high possibility at the realistic interface. To evaluate the effect of the interchange, the electronic properties of bulk zinc-blende $(\mathrm{Ga}_{1-x}\mathrm{Cr}_x)\mathrm{As}$ was calculated using the KKA-CPA method. The calculation showed that the high spin polarization remains even in the material with $x = 0.5$, corresponding to the completely diffused multilayer with the same number of Cr and Ga (for example, 2 ML CrAs/2 ML GaAs). In real multilayers, the atomic diffusion (the replacement of Cr by Ga) will take place and antisite defects may appear at the interface. Nevertheless, on the basis of the calculations mentioned above, the high spin polarization is rather robust against these imperfections. One can expect that the zinc-blende CrAs/GaAs multilayer will be a hopeful candidate for the spin-polarized electron sources.

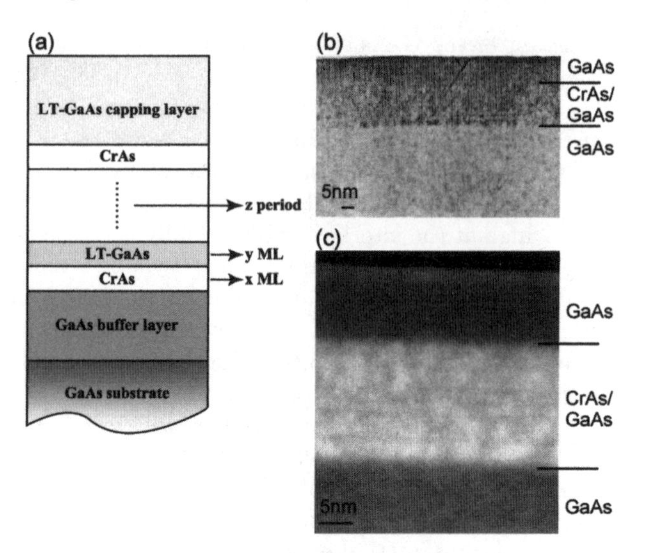

Fig. 8. (**a**) A cross-sectional schematic drawing of the CrAs and GaAs multilayer. The notation LT means "low-temperature grown". (**b**) Cross-sectional TEM image of the sample with the structure of [CrAs (4 ML)/GaAs (4 ML)] × 10 layers, grown on the 20 nm GaAs buffer layer and capped by the 10 nm GaAs layer. (**c**) Dark-field cross-sectional TEM image of the sample with the structure of [CrAs (4 ML)/GaAs (4 ML)] × 10 layers, grown on the 20 nm GaAs buffer layer and capped by the 10 nm GaAs layer

3.2 Epitaxial Growth and the Physical Properties of Zinc-Blende CrAs/GaAs Multilayers

The schematic cross-sectional structure of zinc-blende CrAs/GaAs multilayers is shown in Fig. 8(a). The procedure of the thermal cleaning and the buffer layer growth was the same as those for the thin film described in the Sect. 2.2. The multilayer growth was performed by opening shutters of Knudsen cells of Cr and Ga alternately under the As exposure. Figure 9 shows the substrate temperature dependence of the RHEED-pattern development during the growth. In the case of the substrate temperature of 200°C which was the optimum growth temperature for the single thin film, the RHEED pattern becomes spotty according to the growth [48]. To the strong contrast, in the case of the substrate temperature of 300°C, the pattern remains streaky, rather the intensity increases with increasing the repetition of the couple of CrAs and GaAs layers [49]. It should be noted that the RHEED pattern became completely spotty above the film thickness of about 2 nm in the case that the thin film growth was done at 300°C. The TEM observation revealed that CrAs spheres appeared on the surface in the case of the substrate temperature of 300°C and the film thickness above 2 nm. It is considered that the higher substrate temperature, 300°C, promotes the growth of the flat thin

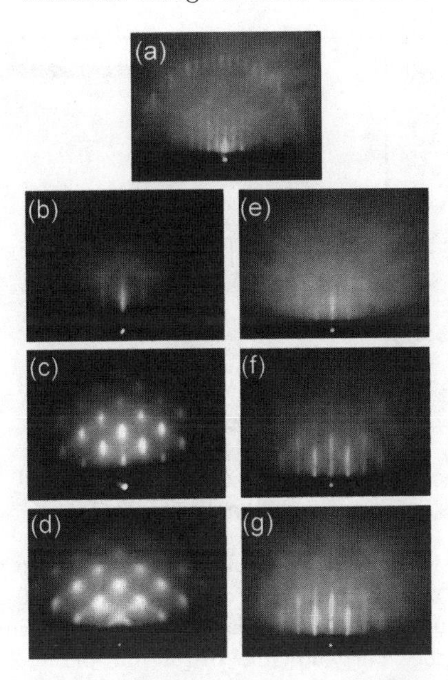

Fig. 9. Substrate temperature dependence of the RHEED-pattern development during the MBE growth of zinc-blende CrAs. (**a**) The RHEED pattern from the GaAs surface prior to the zinc-blende CrAs growth. (**b**)–(**d**) The RHEED patterns from the surface of the zinc-blende CrAs film grown at 200°C with z = 1, 10, and 100 periods, respectively. (**e**)–(**f**) The RHEED patterns from the surface of the zinc-blende CrAs film grown at 300°C with z = 1, 10, and 100 periods, respectively

film but reduces the critical thickness. Fig. 8(b) shows the cross-sectional TEM image of the multilayer with the 10 periods of 4 monolayers (4 ML) CrAs and 4 ML GaAs. Although a crystallographic defect is observed in the low-magnification images, the clear zinc-blende atomic image is proved in the high-resolution TEM image. The atomic image goes through the multilayer region to the GaAs caplayer. In the dark field image of the multilayer (Fig. 8(c)), the periodic contrast showing the 10 times repetition of CrAs and GaAs layers was confirmed. No evident Cr diffusion into the GaAs cap and buffer layers were observed by secondary ion mass spectroscopy (SIMS). The SIMS special resolution in the depth profile was not fine enough to identify each CrAs layer in the multilayer part, while the total thickness of the layer including the Cr atom agreed with the designed thickness.

The magnetic properties of the multilayer with the 100 periods of 2 ML CrAs and 2 ML GaAs were investigated by SQUID [49]. Figure 10 shows the magnetization curve measured at room temperature. The remnant magnetization is as small as that observed in the single film of zinc-blende CrAs shown in Fig. 7. The saturation magnetization is determined to be 400 kA/m, corresponding to about 2 μ_B per formula unit of CrAs. This value is smaller

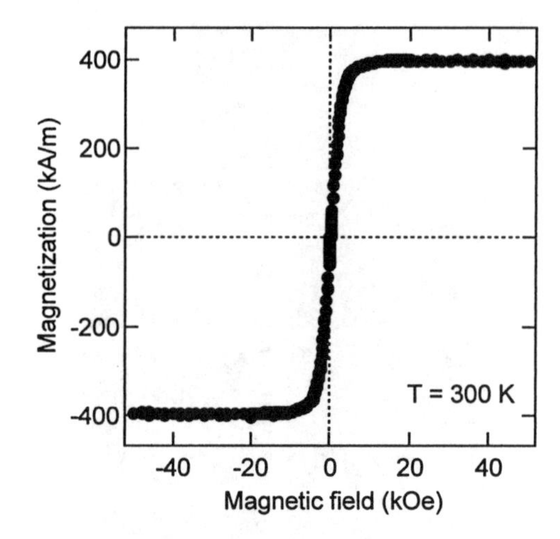

Fig. 10. The magnetization hysteresis loop of the sample with the structure of [CrAs (2 ML)/GaAs (2 ML)] × 100 layers, grown on the 20 nm GaAs buffer layer and capped by the 10 nm GaAs layer, measured at room temperature (300 K). The magnetic field was applied perpendicular to the film plane. The magnetization per volume was calculated by the use of total thickness of all CrAs layers in the sample

than 3 μ_B predicted by theoretical calculation [25]. The smaller magnetization is though to be due to a perturbation of the interface structure which is not taken into account in the present framework of the theoretical prediction [50]. Nevertheless, the multilayer enables us to investigate magneto-transport and magneto-optical properties of zinc-blende CrAs. These measurements are now in progress. In term of the magneto-optical properties, theoretical studies have been already performed for zinc-blende CrAs in the bulk and CrAs/GaAs (100) multilayers on the basis of their electronic band-structures obtained by the density-functional calculation [51]. Huge magneto-optical Kerr effect (MOKE) is obtained in the ultraviolet region (4–8 eV). The optical transitions from occupied Cr 3*d* states to unoccupied Cr 4*p* states mainly contribute to the MOKE in relevant photon energies. The enhancement of the MOKE is attributed to the characteristic photon-energy dependence of diagonal matrix elements of dielectric function.

4 Prospect – How to Make CrAs Work –

In this review article, we have given the successful example of material design and the sequent experimental realization of the designed material. This new type of half-metallic ferromagnets are the zinc-blende transition-metal mono-pnictides. For the zinc-blende CrAs, *ab-initio* calculations predict a very high spin polarization at the Fermi level, almost 100%, even though the spin-orbit interaction is taken into account. Furthermore, the spin polarization is robust against the formation of antisite defects, the hetero-epitaxial junction formation with GaAs and the atomic diffusion at the interface. These results show that zinc-blende CrAs is a suitable material for the spin-polarized electron injection into GaAs-based semiconductors.

In contrast to the dream-like theoretical prediction, the thin film growth of zinc-blende CrAs was achieved by treading a thorny path. The very thin critical thickness of the epitaxial growth and the three dimensional growth prevented us from investigating the physical properties, except for the total magnetic moment showing indirectly the half-metallic behavior of the obtained thin film. Here, the epitaxial growth technology of a semiconductor helped us fabricate the short-period zinc-blende CrAs/GaAs multilayer structure, where *ab-initio* calculations showed the high spin polarization remained throughout the short-period multilayer. Physical properties, such as the spin polarization at the Fermi level, will be revealed in the near future. We believe that the close relationship between a theoretician and an experimental researcher will incubate zinc-blende half-metallic ferromagnets effectively and rapidly, and these materials will play a major role in the field of spin-electronics.

Acknowledgements

The authors would like to thank T. Manago, K. Ono, J. Okabayashi, M. Oshima for their valuable discussion. The EXAFS analyses shown in Fig. 5 were performed by Dr. H. Ofuchi (Nagoya Univ.).

References

1. H. Akinaga and H. Ohno: IEEE Trans. Nanotechnology **1**, 19 (2002)
2. H. Ohno: Science **281**, 951 (1998)
3. G.A. Prinz: Science **250**, 1092 (1990)
4. M. Zolfl, M. Brickmann, M. Kohler, S. Kreuzer, T. Schweinbock, S. Miethaner, F. Bensch, and G. Bayreuther: J. Magn. Magn. Mater. **175**, 16 (1997)
5. O. Wunnicke, P. Mavropoulos, and P.H. Dederichs: J. Supercond.: Inc. Nov. Magn. **16**, 171 (2003)
6. R.A, de Groot, F.M. Mueller, P.G. Van Engen, and K.H.J. Buschow: Phys. Rev. Lett. **50**, 2024 (1983)
7. C.N. Borca, T. Komesu, H. Jeong, P.A. Dowben, D. Ristoiu, C. Hordequin, J. Pierre, and J.P. Nozieres: Appl. Phys. Lett. **77**, 88 (2000)
8. T. Ambrose, J.J. Krebs, and G.A. Prinz: Appl. Phys. Lett. **76**, 3280 (2000)
9. W. Van Roy, J. De Boeck, B. Brijs, and G. Borghs: Appl. Phys. Lett. **77**, 4190 (2000)
10. K. Inomata, S. Okamura, R. Goto, N. Tezuka: Jpn. J. Appl. Phys. **42**, L419 (2003)
11. Y. Miura, K. Nagao, and M. Shirai: Phys. Rev. B **69**, 144413 (2004)

12. H. Kubota, J. Nakata, M. Oogane, Y. Ando, A. Sakuma, and T. Miyazaki: Jpn. J. Appl. Phys. **43**, L894 (2004)
13. K. Schwarz: J. Phys. F: Met. Phys. **16**, L211 (1986)
14. K.P. Kamper, W. Schmitt, G. Guntherodt, R.J. Gambino, and R. Ruf: Phys. Rev. Lett. **59**, 2788 (1987)
15. W.J. DeSisto, P.R. Broussard, T.F. Ambrose, B.E. Nadgorny, and M.S. Osofsky: Appl. Phys. Lett. **76**, 3789 (2000)
16. Y. Ji, G.J. Strijkers, F.Y. Yang, C.L. Chien, J.M. Byers, A. Anguelouch, G. Xiao, and A. Gupta: Phys. Rev. Lett. **86**, 5585 (2001)
17. J.-H. Park, E. Vescovo, H.-J. Kim, C. Kwon, R. Ramesh, and T. Venkatesan: Nature **392**, 794 (1998)
18. K.-I. Kobayashi, T. Kimura, H. Sawada, K. Terakura, and Y. Tokura: Nature **395**, 677 (1998)
19. R.J. Soulen Jr., J.M. Byers, M.S. Osofsky, B. Nadgorny, T. Ambrose, S.F. Cheng, P.R. Broussard, C.T. Tanaka, J. Nowak, J.S. Moodera, A. Barry, and J.M.D. Coey: Science **282**, 85 (1998)
20. Y.S. Dedkov, M. Fonine, C. Konig, U. Rudiger, G. Guntherodt, S. Senz, and D. Hesse: Appl. Phys. Lett. **80**, 4181 (2002)
21. D.J. Huang, L.H. Tjeng, J. Chen, C.F. Chang, W.P. Wu, S.C. Chung, A. Tanaka, G.Y. Guo, H.-J. Lin, S.G. Shyu, C.C. Wu, and C.T. Chen: Phys. Rev. B **67**, 214419 (2003)
22. Theoretical studies predicted that III-V semiconductor-based ferromagnetic semiconductors show a half-metallic property, and the possibility of room-temperature ferromagnetism. Although the ferromagnetic transition temperature is below room temperature, the high spin-polarization at the Fermi level has been proved in a magnetic tunnel junction composed of (Ga,Mn)As. See, T. Dietl, H. Ohno, F. Matsukura, J. Cibert, and D. Ferrand: Science **287**, 1019 (2000); K. Sato and H. Katayama-Yoshida: Jpn. J. Appl. Phys. **40**, L485 (2001); M. Tanaka and Y. Higo: Phys. Rev. Lett. **87**, 26602 (2001)
23. H. Akinaga, T. Manago, and M. Shirai: Jpn. J. Appl. Phys. **39**, L1118 (2000)
24. M. Shirai: Physica E **10**, 143 (2001); J. Appl. Phys. **93**, 6844 (2003)
25. K. Nagao, M. Shirai, and Y. Miura: J. Appl. Phys. **95**, 6518 (2004)
26. A. Sakuma: J. Phys. Soc. Jpn. **71**, 2534 (2002)
27. B. Sanyal, L. Bergqvist, and O. Eriksson: Phys. Rev. B **68**, 054417 (2003)
28. Y.J. Zhao, W.T. Geng, A.J. Freeman, and B. Delley: Phys. Rev. B **65**, 113202 (2002)
29. M. Shirai, K. Ikeuchi, H. Taguchi, and H. Akinaga: J. Supercond. **16**, 27 (2003)
30. Ph. Mavropoulos, K. Sato, R. Zeller, P.H. Dederichs, V. Popescu, and H. Ebert: Phys. Rev. B **69**, 054424 (2004)
31. G. Schmidt, D. Ferrand, L.W. Molenkamp, A.T. Filip, and B.J. van Wees: Phys. Rev. B **62**, R4790 (2000)
32. J. H. Zhao, F. Matsukura, K. Takamura, E. Abe, D. Chiba, and H. Ohno: Appl. Phys. Lett. **79**, 2776 (2001)
33. M. Mizuguchi, H. Akinaga, T. Manago, K. Ono, M. Oshima, and M. Shirai: J. Magn. Magn. Mater. **239**, 269 (2002)
34. H. Ofuchi, M. Mizuguchi, K. Ono, M. Oshima, H. Akinaga, and T. Manago: Nuclear Instruments and Methods in Physics Research B **199**, 227 (2003)
35. M. Mizuguchi, K. Ono, M. Oshima, J. Okabayashi, H. Akinaga, T. Manago, and M. Shirai: Surface Review and Letters, **9**, 331 (2002)

36. H. Bollar and A. Kallel: Solid State Commun. **9**, 1699 (1971)
37. T. Suzuki and H. Ido: J. Appl. Phys. **73**, 5686 (1993)
38. H. Watanabe, Y. Nakagawa, and K. Sato: J. Phys. Soc. Jpn. **20**, 2244 (1965)
39. M. Yuzuri: J. Phys. Soc. Jpn. **15**, 2007 (1960)
40. M. Shirai, M. Seike, K. Sato, and H. Katayama-Yoshida: J. Magn. Magn. Mater. **272-276**, 344 (2004)
41. H. Akai and P.H. Dederichs, Phys. Rev. B **47**, 8739 (1993)
42. G. Kresse and J. Hafner: Phys. Rev. B **47**, R558 (1993); *ibid* **49**, 14251 (1994)
43. G. Kresse and J. Furthmuller: Comput. Mater. Sci. **6**, 15 (1996)
44. G. Kresse and J. Furthmuller: Phys. Rev. B **54**, 11169 (1996)
45. P.E. Blochl: Phys. Rev. B **50**, 17953 (1994)
46. G. Kresse and D. Joubert: Phys Rev. B **59**, 1758 (1999)
47. I. Galanakis, Phys. Rev. B **66**, 012406 (2002)
48. M. Mizuguchi, H. Akinaga, T. Manago, K. Ono, M. Oshima, M. Shirai, M. Yuri, H.J. Lin, H.H. Hsieh, and C.T. Chen: J. Appl. Phys. **91**, 7917 (2002)
49. H. Akinaga and M. Mizuguchi: J. Phys.: Condens. Matter, in press
50. O. Bengone, O. Eriksson, J. Fransson, I. Turek, J. Kudrnovsky, and V. Drchal: Phys. Rev. B **70**, 035302 (2004)
51. M. Shirai, Y. Miura, K. Nagao, and E. Kulatov: unpublished

Lecture Notes in Physics

For information about earlier volumes please contact your bookseller or Springer LNP Online archive: springerlink.com

Vol.631: D. J. W. Giulini, C. Kiefer, C. Lämmerzahl (Eds.), Quantum Gravity, From Theory to Experimental Search

Vol.632: A. M. Greco (Ed.), Direct and Inverse Methods in Nonlinear Evolution Equations

Vol.633: H.-T. Elze (Ed.), Decoherence and Entropy in Complex Systems, Based on Selected Lectures from DICE 2002

Vol.634: R. Haberlandt, D. Michel, A. Pöppl, R. Stannarius (Eds.), Molecules in Interaction with Surfaces and Interfaces

Vol.635: D. Alloin, W. Gieren (Eds.), Stellar Candles for the Extragalactic Distance Scale

Vol.636: R. Livi, A. Vulpiani (Eds.), The Kolmogorov Legacy in Physics, A Century of Turbulence and Complexity

Vol.637: I. Müller, P. Strehlow, Rubber and Rubber Balloons, Paradigms of Thermodynamics

Vol.638: Y. Kosmann-Schwarzbach, B. Grammaticos, K. M. Tamizhmani (Eds.), Integrability of Nonlinear Systems

Vol.639: G. Ripka, Dual Superconductor Models of Color Confinement

Vol.640: M. Karttunen, I. Vattulainen, A. Lukkarinen (Eds.), Novel Methods in Soft Matter Simulations

Vol.641: A. Lalazissis, P. Ring, D. Vretenar (Eds.), Extended Density Functionals in Nuclear Structure Physics

Vol.642: W. Hergert, A. Ernst, M. Däne (Eds.), Computational Materials Science

Vol.643: F. Strocchi, Symmetry Breaking

Vol.644: B. Grammaticos, Y. Kosmann-Schwarzbach, T. Tamizhmani (Eds.) Discrete Integrable Systems

Vol.645: U. Schollwöck, J. Richter, D. J. J. Farnell, R. F. Bishop (Eds.), Quantum Magnetism

Vol.646: N. Bretón, J. L. Cervantes-Cota, M. Salgado (Eds.), The Early Universe and Observational Cosmology

Vol.647: D. Blaschke, M. A. Ivanov, T. Mannel (Eds.), Heavy Quark Physics

Vol.648: S. G. Karshenboim, E. Peik (Eds.), Astrophysics, Clocks and Fundamental Constants

Vol.649: M. Paris, J. Rehacek (Eds.), Quantum State Estimation

Vol.650: E. Ben-Naim, H. Frauenfelder, Z. Toroczkai (Eds.), Complex Networks

Vol.651: J. S. Al-Khalili, E. Roeckl (Eds.), The Euroschool Lectures of Physics with Exotic Beams, Vol.I

Vol.652: J. Arias, M. Lozano (Eds.), Exotic Nuclear Physics

Vol.653: E. Papantonoupoulos (Ed.), The Physics of the Early Universe

Vol.654: G. Cassinelli, A. Levrero, E. de Vito, P. J. Lahti (Eds.), Theory and Appplication to the Galileo Group

Vol.655: M. Shillor, M. Sofonea, J. J. Telega, Models and Analysis of Quasistatic Contact

Vol.656: K. Scherer, H. Fichtner, B. Heber, U. Mall (Eds.), Space Weather

Vol.657: J. Gemmer, M. Michel, G. Mahler (Eds.), Quantum Thermodynamics

Vol.658: K. Busch, A. Powell, C. Röthig, G. Schön, J. Weissmüller (Eds.), Functional Nanostructures

Vol.659: E. Bick, F. D. Steffen (Eds.), Topology and Geometry in Physics

Vol.660: A. N. Gorban, I. V. Karlin, Invariant Manifolds for Physical and Chemical Kinetics

Vol.661: N. Akhmediev, A. Ankiewicz (Eds.) Dissipative Solitons

Vol.662: U. Carow-Watamura, Y. Maeda, S. Watamura (Eds.), Quantum Field Theory and Noncommutative Geometry

Vol.663: A. Kalloniatis, D. Leinweber, A. Williams (Eds.), Lattice Hadron Physics

Vol.664: R. Wielebinski, R. Beck (Eds.), Cosmic Magnetic Fields

Vol.665: V. Martinez (Ed.), Data Analysis in Cosmology

Vol.666: D. Britz, Digital Simulation in Electrochemistry

Vol.667: W. D. Heiss (Ed.), Quantum Dots: a Doorway to Nanoscale Physics

Vol.668: H. Ocampo, S. Paycha, A. Vargas (Eds.), Geometric and Topological Methods for Quantum Field Theory

Vol.669: G. Amelino-Camelia, J. Kowalski-Glikman (Eds.), Planck Scale Effects in Astrophysics and Cosmology

Vol.670: A. Dinklage, G. Marx, T. Klinger, L. Schweikhard (Eds.), Plasma Physics

Vol.671: J.-R. Chazottes, B. Fernandez (Eds.), Dynamics of Coupled Map Lattices and of Related Spatially Extended Systems

Vol.672: R. Kh. Zeytounian, Topics in Hyposonic Flow Theory

Vol.673: C. Bona, C. Palenzula-Luque, Elements of Numerical Relativity

Vol.674: A. G. Hunt, Percolation Theory for Flow in Porous Media

Vol.675: M. Kröger, Models for Polymeric and Anisotropic Liquids

Vol.676: I. Galanakis, P. H. Dederichs (Eds.), Halfmetallic Alloys